DIGITAL VIDEO HACKS™

Other resources from O'Reilly

Related titles

- Digital Photography Hacks
- iPod & iTunes Hacks
- DVD Studio Pro 3: In the Studio
- Digital Video Pocket Guide
- Digital Audio Essentials
- Adobe Encore DVD: In the Studio

Hacks Series Home

hacks.oreilly.com is a community site for developers and power users of all stripes. Readers learn from each other as they share their favorite tips and tools for Mac OS X, Linux, Google, Windows XP, and more.

oreilly.com

oreilly.com is more than a complete catalog of O'Reilly books. You'll also find links to news, events, articles, weblogs, sample chapters, and code examples.

oreillynet.com is the essential portal for developers interested in open and emerging technologies, including new platforms, programming languages, and operating systems.

Conferences

O'Reilly brings diverse innovators together to nurture the ideas that spark revolutionary industries. We specialize in documenting the latest tools and systems, translating the innovator's knowledge into useful skills for those in the trenches. Visit *conferences.oreilly.com* for our upcoming events.

Safari Bookshelf (*safari.oreilly.com*) is the premier online reference library for programmers and IT professionals. Conduct searches across more than 1,000 books. Subscribers can zero in on answers to time-critical questions in a matter of seconds. Read the books on your Bookshelf from cover to cover or simply flip to the page you need. Try it today with a free trial.

DIGITAL VIDEO HACKS™

Joshua Paul

O'REILLY®

Beijing • Cambridge • Farnham • Köln • Paris • Sebastopol • Taipei • Tokyo

Digital Video Hacks™
by Joshua Paul

Published by O'Reilly Media, Inc., 1005 Gravenstein Highway North, Sebastopol, CA 95472.

O'Reilly books may be purchased for educational, business, or sales promotional use. Online editions are also available for most titles (*safari.oreilly.com*). For more information, contact our corporate/institutional sales department: (800) 998-9938 or *corporate@oreilly.com*.

Editor:	Brian Sawyer	**Production Editor:**	Philip Dangler
Series Editor:	Rael Dornfest	**Cover Designer:**	Mike Kohnke
Executive Editor:	Dale Dougherty	**Interior Designer:**	David Futato

Printing History:

June 2005: First Edition.

This book uses RepKover™, a durable and flexible lay-flat binding.

ISBN: 0-596-00946-1
[C]

Contents

Credits

About the Author

Joshua Paul has more than 10 years of experience delivering programming for both cable and network television. He has produced prime-time specials for Fox Television, worked as a Producer for Sony Pictures Entertainment, and produced a variety of projects for companies throughout Los Angeles. He specializes in post-production processes and organizing projects that deal with large amounts of raw footage. During the past decade, he has witnessed, participated in, and pushed the growth and adoption of digital video within the entertainment industry.

Joshua is also cofounder of Overhyped Technologies, LLC (*http://www.overhyped.com*), a company that provides software and service solutions to entertainment production companies. Overhyped's primary service enables producers to log, search, and manage thousands of hours of digital video in real time, from anywhere. The company's software and services have helped produce the television shows *Extreme Makeover: Home Edition*, *Growing Up Gotti*, and *Nanny 911*, among many others. As cofounder of the company, he has acted as a consultant for all of the company's clients, often helping to guide them through the tough post-production process.

Joshua is an active member of the Producer's Guild of America, a published author, and a speaker. He has two patents pending concerning methods of digital video distribution. When not consumed with work, he lives, loves, and enjoys life with his wife and son.

Contributors

The following people contributed their hacks, writing, and inspiration to this book:

- Richard Baguley is a freelance journalist who writes about computers and technology. Prior to freelancing, he worked as a Senior Associate

Editor at *PC World* magazine in San Francisco. He has also worked for various technology magazines in the UK, including *Amiga Format*, *Amiga Shopper*, *Internet Today*, and *Internet Magazine*. While it is true that all of the aforementioned publications closed after he left, he contends that this had nothing to do with their inability to function without him. In addition to writing for magazines and web sites such as PC World and JIWire.com, Richard runs the video-editing web site Videotastic (*http://www.videotastic.com*). Richard lives in the San Francisco Bay area with his wife, Kath, his French Bulldog, Fester, and an ever-changing variety of cats (some of which are permanent, some of which are fosters looking for homes). Two of those cats graciously agreed to model for illustrations in this book in exchange for extra wet food and catnip.

- Michael Dean is editing a digital filmmaking book for O'Reilly and writing for *Make* magazine. He created the $30 School book series, including writing *$30 Film School*, wrote the novel *Starving in the Company of Beautiful Women*, directed the film *D.I.Y. or Die: How to Survive as an Independent Artist*, and took it on a tour of the U.S. and Europe. He also directed the film *Hubert Selby, Jr.: It/ll be better tomorrow* and produced the DVD *Living Through Steve Diet Goedde*. While working in the music industry, he toured the world with the rock band Bomb and put out 13 records on several labels, ranging from D.I.Y. to indie to Warner Brothers. He has been interviewed on NPR, BBC, and NBC, and his books have been reviewed favorably on VH1 and in *Maximum Rock 'n' Roll* magazine. Michael tours and lectures on D.I.Y. art, filmmaking, and production.
- Arthur J. Dustman IV has been doing home automation for about three years. He has expanded from the X10 standard package to a 5×5 room to hold all his equipment. He uses HAL2000, Adicon Ocelot with C-max, nine Secu-16s, one Secu16-IR, 94 Relays, one scm-810 mixer, nine pzm-10 microphones, nine Xantech keypads, nine custom wall panels, and much more. He has created many custom circuits and devices to automate devices that have no "factory" interfaces. His motto: "There's always a way to automate anything."
- Preston Gralla is the author of more than 30 books about computers and the Internet, including *Windows XP Hacks*, *Internet Annoyances*, and *Windows XP Power Hound*. He's been writing about technology since the dawn of the PC, was a founding editor of both *PC Week* and *PC/Computing*, an executive editor of ZDNet and CNet, and has contributed to dozens of publications, including *PC Magazine*, *The Los Angeles Times*, *USA Today*, and *Computerworld*, among others.

- Nick Jushchyhsyn (*http://www.jushhome.com*) has over a decade of experience in computer programming and software design. Over the course of the last several years, he has used these skills to branch into post-production and visual effects for film and video. Currently, Nick studies, teaches, and practices visual effects as an artisan in the Pixel-Corps (*http://www.pixelcorps.com*).
- Marc Loy is a trainer and media specialist in Madison, WI. When he's not working with digital video and DVDs, he's programming in Java. He can still be found teaching the odd Perl and Java course out in corporate America, but even on the road, he has his PowerBook and a video project with him. Marc is the author of O'Reilly's *DVD Studio Pro 3: In the Studio* and can be found online at *http://www.loyinc.com*.
- Ilya Lyudmirsky has worked on independent film projects as a cinematographer and producer for over 15 years. His background is extensive—he has worked practically every production job possible, with the exception of hair stylist. The projects that Ilya has been involved in have received numerous festival awards. Ilya also runs his own production company, Blacklist Productions, which specializes in cutting-edge independent productions.
- Don Marquardt (*kyham@k9soa.net*) has been playing around with home automation for almost 30 years. Each move to a new home just brought more toys to play with. His current home has just about everything he could think of. In total, it has almost eight miles of various types of wire to handle just about anything. Now, even his car talks to the Internet. Don's web site (*http://www.k9soa.net*) is open to visitors to take a little tour of Jeannie, The House That Listens. When you visit, be sure to sign his guestbook.
- Todd Ogasawara focuses on two distinct topics: Mobile Workforce and Mobile Lifestyle technology, with special attention paid to the Microsoft Windows Mobile platform (Pocket PC and Smartphone). Microsoft has recognized his demonstrated practical expertise and willingness to share his experience by honoring him as a Microsoft Most Valuable Professional (MVP) in the Mobile Devices category. His other technology focus is in the effort to bring commercial (especially Microsoft-related products) and GNU/open source software together in a synergistic and productive way. For lack of a better term, Todd calls this concept Eccentric Technology. Todd has written several articles related to mobile devices, including camera phones, for the O'Reilly Network Wireless DevCenter (*http://www.oreillynet.com/wireless/*). He previously worked as a technology analyst for GTE/Verizon. He also served as the contracted Forum Manager for the MSN (and later ZDNet) Telephony Forum and Windows CE Forum. More recently, he served as project

lead to develop an intranet portal for the state of Hawaii using open source tools. You can find his Mobile Workforce and Lifestyle commentary at *http://www.MobileViews.com*. You can learn more about Eccentric Technology at *http://www.OgasaWalrus.com*. For comments related to camera phones, you can reach Todd by email at *PhoneCam@OgasaWalrus.com*.

- Derrick Story is the Managing Editor for O'Reilly Network (*http://www.oreillynet.com*) and Mac DevCenter (*http://www.macdevcenter.com*), the latter of which he created in December 2000 for O'Reilly Media. His focus is on Mac OS X, digital media, and mobile computing. Derrick is the author of several O'Reilly books, including the *Digital Video Pocket Guide*, the *Digital Photography Pocket Guide*, *Digital Photography Hacks*, the *iBook Fan Book*, and the *PowerBook Fan Book*. He also coauthored *iPhoto 4: the Missing Manual* (Pogue Press/O'Reilly). Derrick's professional experience includes more than 15 years as a photojournalist, former managing editor of Web Review, and a speaker for O'Reilly, CMP, and IDG conferences. He manages his online photo business, Story Photography (*http://www.storyphoto.com*), which specializes in digital photography and special events.
- Gene Sullivan is a writer, musician, Irish dancer, and all-around computer geek. He was introduced to the joys of filmmaking by his wife, who drafted him as a production assistant on a student film at UC Santa Barbara. He has worked on a variety of independent films and television shows and is currently IT Manager for Evolution Film & Tape, Inc., an independent television production company in Burbank, California.
- Phillip Torrone is Associate Editor of *Popular Science Magazine*, feature columnist for Engadget (*http://www.engadget.com*), and author of numerous books on mobile devices and design. Phillip is also an Associate Editor for *Make* magazine, and he writes the magazine's weblog at *http://www.makezine.com*. Phillip's work and projects can be viewed at *http://www.flashenabled.com*.
- Richard Wolf is a research programmer at the University of Illinois at Chicago. His chief role is to provide support for the Macintosh at every level, from helping end users solve everyday problems to developing custom solutions for the university. In his spare time, he is writing an introductory text on PowerPC assembly and the Mach-O runtime, which he hopes Mac OS X programmers will find useful.

Acknowledgments

This book would not have been possible without the love, support, and devotion of my wife, Carrie. I also couldn't have completed it without the enormous help I received from my family, my friends, the great staff at O'Reilly, the contributors, and the entire online community. I would like to thank everyone (offline and online) who has helped me along the way . . . but room doesn't permit it.

To my wife, Carrie: you have filled my life with everything that was missing. Without you I could never be the person I want to be. You are my best friend, my playmate, my confidant, and my partner for life. Words cannot express my adoration for you. Every day I am with you, I love you even more. All of my days.

To my son, Avery: since the day you were born, I've loved you. You are the most precious gift I have ever been given and you fill my days with inexplicable happiness. Your smile lights up my life and your laughter feeds my soul.

To my mom: thank you for everything you've done for me. You've supported every decision I've ever made in life (well, other than majoring in Philosophy) and I couldn't ask for anything more. Yet, you do more than I could ever ask. I love you.

To my dad: you are an inspiration to me, even though you don't know it. Although the miles and days separate us, you are with me constantly. I love you.

To Dave: thank you for your support, guidance, and help as I've become an adult. I am honored to say you are a part of my family.

To my family: thank you for your help and tremendous support throughout my life.

To Susan Nessanbaum-Goldberg: you've helped guide my career and always made yourself available when I've had questions. Who knows where I'd be had you not opened the door? Thank you.

To Michael Fitzsimmons, wherever you are: you noticed my knack for computers and provided me the opportunity to *know* post-production. Thank you for the education.

To Ted Steinberg: you are an inspiration, a role model, and a friend. Thank you for being who you are.

To the staff at Evolution Film and Tape: thank you for letting me take a little time to myself and concentrate on my family. Specifically, thank you Doug, Greg, Dean, and Kathleen for creating and running such a wonderful company. It is a model that others in the industry should follow.

To Gene Sullivan: thanks for looking over my shoulder and keeping my hacks technically sound. Your feedback and knowledge were invaluable.

To Derrick Story: thank you for passing along my name to your colleagues. Without your confidence, I would have never had this opportunity. Oh, and thanks for the mass of contributions, too.

To Brian Sawyer: we've spoken on the phone, emailed almost daily, and experienced our children's first illnesses together...not to mention created this book. Thank you for your guidance. It's been a learning experience for me, and you made it painless.

With the publication of this book, I have fulfilled a promise made to my grandfather during his last years with us. Hopefully, there will be more to come...

Preface

Digital Video Hacks is meant to provide everyone, from beginner to professional, a new way of perceiving digital video. Whether you are looking for a new technique to include in your next project, a solution to a common problem, or just some insight into the digital video medium, my hope is you will find something new and exciting within these pages. By using the techniques in this book, you should be able to successfully complete a project, from preproduction to final delivery, while avoiding common, costly mistakes.

Why Digital Video Hacks?

The term *hacking* has a bad reputation in the press. They use it to refer to someone who breaks into systems or wreaks havoc with computers as their weapon. Among people who write code, though, the term *hack* refers to a "quick-and-dirty" solution to a problem, or a clever way to get something done. And the term *hacker* is taken very much as a compliment, referring to someone as being *creative*, having the technical chops to get things done. The Hacks series is an attempt to reclaim the word, document the good ways people are hacking, and pass the hacker ethic of creative participation on to the uninitiated. Seeing how others approach systems and problems is often the quickest way to learn about a new technology.

Trying to pinpoint exactly what defines digital video is difficult, let alone defining what it means to hack it. The digital video medium involves many different phases, from preparing for a shoot to delivering and viewing a final product. Finding the boundaries is difficult. Video production is, in and of itself, a creative medium, so you should be encouraged to free your creative energy, embrace the hacker inside you, and use this book as a starting point to your journey into the digital video revolution.

How to Use This Book

You can read this book from cover to cover if you like, but each hack stands on its own, so feel free to browse and jump to the different sections that interest you most. If there's a prerequisite you need to know about, a cross-reference will guide you to the right hack.

How This Book Is Organized

The book is divided into several chapters, organized by subject:

Chapter 1, *Prepare*
: Although every aspect of digital video production is important, none can save you more time and money than preparation. Being prepared, whether with technical know-how or physical tools, enables you to calmly deal with the *real* problems that will inevitably occur. Throughout the various stages of producing a project, you should be prepared for each stage before you reach that stage. The hacks in this chapter prepare you, and your project, for production and editing.

Chapter 2, *Light*
: Lighting is important to video, especially if you are trying to create a more professional look. Even though you can effectively light a scene "naturally" by using what's available, if you use specific techniques, you can bring out details (or avoid distractions) in your footage. Best of all, you can light most scenes for very little money. This chapter covers various lighting solutions.

Chapter 3, *Acquire*
: Every aspect of creating a video revolves around the footage, so acquiring it is quite obviously the most important element. Without footage, you can't edit or distribute a video. The decisions you have to make are *what* you want to acquire and *how* you should do so. The hacks in this chapter help you collect footage from various sources, shoot footage in unusual ways, and show you how to bring all of your footage (no matter where it originates) together.

Chapter 4, *Edit*
: Editing can be one of the most exhilarating aspects of a digital video project. It is also one of the most technically involved and, therefore, most error prone. Draw on hacks from this chapter to weave footage from different sources together.

Chapter 5, *Audio*
: Audio is often an afterthought in video production, even though it's a vital part of it. There is something very special about audio and its effect

on human emotion. Music, sound effects, and even the clarity of dialogue, will all have an effect on your audience. Follow the techniques in this chapter to collect audio for use in editing.

Chapter 6, *Effects*
: Effects are where the editing process gets to be really fun. Even though you've already captured your footage, you can still manipulate it, change its look, and give it your own style. Special effects can be subtle, such as those in *Forrest Gump*, or blatant, such as those in *The Matrix*. Using effects, you can also fix problems that occurred while you were shooting. Implement hacks in this chapter when you want to change the look of your footage.

Chapter 7, *Distribute*
: Distributing your final project is the most gratifying, and yet nerve-racking, facet of digital video production. When you distribute your video, it means you've completed your project. That, in and of itself, is a huge accomplishment. Employ the hacks contained in this chapter to pass along your video to other people in various ways.

Chapter 8, *Random Fun*
: Although the hacks in this chapter can be used for real, functional, and even business purposes, they're primarily here to provide a little fun and inspiration. Need a break or some inspiration? Pull a hack from this chapter and see what happens.

Conventions Used in This Book

The following is a list of the typographical conventions used in this book:

Italics
: Used to indicate URLs, filenames, filename extensions, and directory/folder names. For example, a path in the filesystem will appear as */Developer/Applications*.

`Constant width`
: Used to show code examples, the contents of files, console output, as well as the names of variables, commands, and other code excerpts.

`Constant width bold`
: Used to highlight portions of code, typically new additions to old code.

`Constant width italic`
: Used in code examples and tables to show sample text to be replaced with your own values.

Color
: The second color is used to indicate a cross-reference within the text.

You should pay special attention to notes set apart from the text with the following icons:

This is a tip, suggestion, or general note. It contains useful supplementary information about the topic at hand.

This is a warning or note of caution, often indicating that your money or your privacy might be at risk.

The thermometer icons, found next to each hack, indicate the relative complexity of the hack:

Using Code Examples

This book is here to help you get your job done. In general, you may use the code in this book in your programs and documentation. You do not need to contact us for permission unless you're reproducing a significant portion of the code. For example, writing a program that uses several chunks of code from this book does not require permission. Selling or distributing a CD-ROM of examples from O'Reilly books *does* require permission. Answering a question by citing this book and quoting example code does not require permission. Incorporating a significant amount of example code from this book into your product's documentation *does* require permission.

We appreciate, but do not require, attribution. An attribution usually includes the title, author, publisher, and ISBN. For example: "*Digital Video Hacks* by Joshua Paul. Copyright 2005, O'Reilly Media, Inc., 0596009461."

If you feel your use of code examples falls outside fair use or the permission given above, feel free to contact us at *permissions@oreilly.com*.

How to Contact Us

We have tested and verified the information in this book to the best of our ability, but you may find that features have changed (or even that we have made mistakes!). As a reader of this book, you can help us to improve future editions by sending us your feedback. Please let us know about any errors, inaccuracies, bugs, misleading or confusing statements, and typos that you find anywhere in this book.

Please also let us know what we can do to make this book more useful to you. We take your comments seriously and will try to incorporate reasonable suggestions into future editions. You can write to us at:

O'Reilly Media, Inc.
1005 Gravenstein Hwy N.
Sebastopol, CA 95472
(800) 998-9938 (in the U.S. or Canada)
(707) 829-0515 (international/local)
(707) 829-0104 (fax)

To ask technical questions or to comment on the book, send email to:

bookquestions@oreilly.com

The web site for *Digital Video Hacks* lists examples, errata, and plans for future editions. You can find this page at:

http://www.oreilly.com/catalog/digitalvideohks/

For more information about this book and others, see the O'Reilly web site:

http://www.oreilly.com

Safari® Enabled

When you see a Safari® Enabled icon on the cover of your favorite technology book, that means the book is available online through the O'Reilly Network Safari Bookshelf.

Got a Hack?

To explore Hacks books online or to contribute a hack for future titles, visit:

http://hacks.oreilly.com

Prepare
Hacks 1–16

Although every aspect of digital video production is important, none can save you more time and money than preparation. Being prepared, whether with technical know-how or physical tools, enables you to calmly deal with the *real* problems that inevitably occur. Throughout the various stages of producing a project, you should be prepared for a stage *before* you reach it.

Always keep in mind the project as a whole, from preproduction to delivery, as opposed to seeing a bunch of separate little projects that come together in the end. Anything you do during a project affects it from that point forward, and if you make a mistake, it carries through and continues to cause you problems until it is fixed.

HACK #1 Successfully Complete a Project

Following a series of steps enables you to successfully complete a project, in the least amount of time and at the least expense.

When it comes to completing a project, post-production is often an area of trouble for people. Getting a project through the editing process can be challenging. It can be difficult because of a lack of experience and/or knowledge, a feeling of intimidation by the technical jargon and equipment, or simply burning out in the final phase of a project. Fortunately, an easy-to-follow process can help you get through it.

This hack is meant to be a quick *hit list* of steps you should follow at the *very least*. Cross-references to other hacks are for your use as you see fit. I've witnessed people (professionals, no less) needlessly spend tens of thousands of dollars because they didn't follow these guidelines *in order*.

Labeling Your Media

Putting labels on your tapes is a no-brainer. But how you number your physical media **[Hack #3]** can make a world of difference in post-production. A simple numbering mistake early on can translate into having to renumber your entire library or reedit a project because the numbers are *cut off* by creating an edit decision list (EDL).

Avoid the pain and follow a couple simple rules:

- Don't duplicate numbers.
- Keep the labels simple.

You can get creative with your labeling, but don't go overboard. If you create a system that people have to repeatedly ask questions about, you're costing yourself time and effort. Simply put, do it right the first time and put a little thought into it.

Tracking Your Media

Keep a running tally of your media. You should be able to reference something to know what your tape numbers are and what they contain. Whether you use a database, a spreadsheet, or a notebook is up to you.

If you don't know what, or how many, tapes you have, you'll never know if you're missing footage. If you number your tapes before you shoot on them, you should enter additional information into your tracking system *after* you've shot on them. Just having tapes with numbers on them isn't enough.

Logging Your Media

If you don't know what's on your tapes, how are you going to find the footage you need? There are a number of ways to log your footage, from taking simple notes on a piece of paper, to using a spreadsheet **[Hack #5]**, using a third-party application, and even using your editing application.

However you log your footage, you should always be aware of one thing: *regenerating timecode*. This occurs when your timecode jumps *backward* in time and is troublesome for editing systems. If you come across regenerating timecode, you should create a digital clone of the tape **[Hack #48]**, with new and continuous timecode.

Making a Paper Cut

A *paper cut* equates to your edited video on paper. Before you start editing, you should have a clear idea of what footage you are going to use and where that footage is located. You can use a two-column script [Hack #7] to create your paper cut, which will greatly help with the process.

If you are working on a scripted project, such as a dramatic movie, you should have lined script [Hack #11]. If you don't have a lined script, you will probably want to watch your footage and take notes on the script. Your notes should include the tape number and timecode for the various scenes, at the very least.

Creating an Offline Cut

Your *offline cut*, also known simply as *offline*, is essentially where you start to edit. In offline, you will assemble your story and gather all of the necessary elements to tell it. This is where you should know what footage, sound effects, graphics, and other elements are missing, and be able to determine whether or not there is a way around the problems. During offline, you shouldn't worry about mixing your audio unless it's absolutely necessary.

You should use your paper cut or lined script as a guideline to creating your timeline. However, you should be familiar with your footage (and use your logs!) to fill in where your paper cut or lined script is lacking. Remember, what is on paper will more than likely *not* translate in the editing room.

Create backups of your project often [Hack #2]; otherwise, you may discover you've lost days, weeks, or months worth of work. Additionally, when you have completed a cut, you should output a reference copy [Hack #75] so you can make notes *away* from the editing system.

Locking Your Offline Cut

When you have determined you are happy with your offline cut, you should stop editing, back up, and output a reference copy. Your reference copy is *not* for making more notes! At this point, you should not have any temporary video elements and be ready to accept your story as it is. You should not make any video changes from this point forward.

Onlining Your Locked Offline Cut

If you have been editing in a low-resolution, such as OfflineRT on Final Cut or 10:1 on Avid, you are going to need to redigitize your footage at an uncompressed resolution. When redigitizing, you should make sure you have at least 15 frame *handles* as a safety net, in case there are problems you didn't see in offline. After you have transferred all of your uncompressed footage, you should digitize in your *locked offline reference cut*.

Place the reference cut on the top layer of video in your timeline. You then want to make sure your online matches your offline. Here are a couple of approaches you can take:

Cropping
: You can crop the reference cut about 50%, either horizontally or vertically, so you can see the online cut at the same time. When you play your timeline, you'll be able to see if any of the footage doesn't match. If it doesn't, you'll notice because the full image on screen will be split.

Toggling
: If you just want to *spot check* your online, you can simply jump to various sections of your timeline and toggle the visibility of the reference cut. This will allow you to see the full-screen image of the reference cut and then the online cut. Although this method won't always alert you to minor differences, such as a one-frame difference, it will alert you of any major problems.

After you determine your online cut is correct, you should output it to a new piece of tape. This tape is your *online master*. Keep it safe and be proud; you've made it farther than most people ever will.

Color-Correcting and Mixing

The final two steps can be completed on the same system or, if you can afford to hire a colorist or an audio mixer, on different systems. If you will be working on different systems, you should know what types of files are expected, especially for your audio suite. You will also need to know how you will get both signals (your audio and your video) back out to tape in sync. In other words, what application is going to be used and what are its requirements?

Color-correcting your online master. Yes, believe it or not, you shouldn't color-correct until after you have your online master completed. The reason is simple enough: if you make a mistake, you can easily undo it. Plus, you will always have a master tape you can fall back on.

If you are color-correcting using the same system and application that you have used to online your footage, you can proceed to color-correct using the same timeline. Before you color-correct, make sure you back up your online version of the timeline first!

If you are going to be using a different system to color-correct, you should export an EDL or a copy of the timeline (depending on the system you'll be using). The EDL or timeline will provide a reference to where the cuts occur in your video. This will make the job of color-correcting *much* easier.

When you have finished color-correcting, you should output to yet another piece of videotape. This tape is your *color-corrected master*. Things are looking good!

Mixing your audio. The final step is to work with your audio. Every time I work with audio, it absolutely amazes me how much a simple *tweak*, such as removing the hum [Hack #56], can make a scene come to life. When working with audio, make sure you have a good set of external speakers, so you can clearly hear what is there.

If you have done the color-correction and audio mix using the same application, you can simply output. If you have used different applications, you have two options, based on the capabilities of your audio-mixing system.

If the system is capable of outputting audio-only to videotape, while not affecting the video, you can layback the audio directly to your *color-corrected master*. This means that you can use the same tape that is your color-corrected master, and place your final mixed audio onto it. By doing so, you *know* which physical tape is your final master tape.

If the system cannot layback audio-only, you should output your mixed audio at the highest quality possible to an audio file. After you have the mixed audio as an audio file, you should import it into your video-editing application. After importing it, you should place your audio file onto your timeline and sync it with the video. After it's in sync, you simply need to output.

That's it! You're done.

Putting It to Work

It is a lot of work to complete a project. If you can solve problems early in the process, then you can avoid problems later. After completing a few projects, you'll discover that the later you discover a problem, the more difficult and expensive it is to fix. By following the guidelines in this hack, you can save yourself a lot of work, a lot of time, a lot of frustration, and a lot of money.

HACK #2 Keep Your Project Organized

As you continue to edit your project, your timeline will inevitably become more complicated. By following a few organizational rules you will avoid confusing yourself.

Editing is an artistic process. The key word in that sentence is *process*. Unfortunately, many people, including professional editors, overlook the process only to discover they are confused by their own work. As a side effect, they tend to blame their editing systems for problems, instead of blaming themselves.

Organizing Tracks

Keeping your timeline clean will help you work more efficiently, and therefore faster. To keep your timeline clean, you need to place your audio and video on specific tracks, based upon the use of your media. For example, you should place all of your title media, such as credits, on video track 5 (V5). Figure 1-1 shows a complex but well-organized Avid timeline.

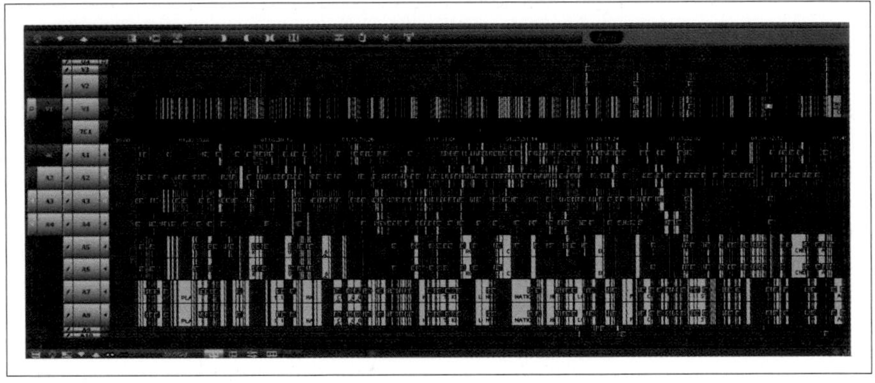

Figure 1-1. Keeping tracks organized in Avid

There are no rules you can apply to every project, because each one is different. By keeping your tracks organized, you will be able to easily spot media you want to work with by what layer it resides on.

As a starting point, consider organizing your video tracks like this:

Acquired
: V1 and V2

Mattes and Effects
: V3 and V4

Titles
: V5 and V6

And here are some suggestions for organizing your audio:

Dialogue and Narration
: A1 and A2

Natural Sound and Ambiance
: A3 and A4

Sound Effects
: A5 and A6

Music
: A7 and A8

Also, if you are using an external mixing board to monitor your audio, make sure you are hearing what is actually on your timeline. For example, if you have placed a sound effect of a kitchen blender on an audio track that should play only on your right speaker, and you hear it coming out of both speakers, your mixing board is not configured properly. More often than not, you will want your mixing board configured so that your right and left channels are panned right and left, respectively. Rarely will you want your channels *center panned*.

If you are limited in the number of audio tracks you can work with, it might be worthwhile to center pan your audio. For example, if you place your Narration audio on track A1, you should then center pan it. When using such an approach, make sure you are aware of it when mixing your audio.

Organizing Bins and Folders

Just as keeping a clean timeline will keep you working efficiently, so will keeping your bins organized. Just as there are near infinite ways to keep your computer organized, there are just as many ways to keep your bins organized. However, there are a few methods that have proven themselves on time-critical projects.

By tape number. Keeping your bins organized by tape is possibly the most common method of organization. It involves naming each bin as a tape number and only placing footage from that tape inside that bin. For example, if you have a tape numbered DV1031, only footage from tape DV1031 would be found inside that bin.

This method is extremely efficient, because most people look for footage based on the tape number and the timecode. Using this method, along with keeping logs of your tapes **[Hack #5]**, will make editing your project enjoyable, even as the days turn into weeks or even months.

By scene. If you are editing a scripted project, such as a dramatic movie, you might find it more efficient to organize your bins by scene. To use this method, create a bin and name it accordingly. Then, place footage that is a part of the scene inside the bin.

When using this method, you can name your bins by scene number, such as "Scene 3," or the action occurring during the scene, such as "Sam's Grand Jury Trial." Quite obviously, you can combine the two naming schemes, which would result in "Scene 3: Sam's Grand Jury Trial."

Taking the scene-labeling approach will allow you to concentrate and dig through only footage that's related to the scene you are editing. This is especially helpful when working with a producer or director who asks, "What other footage do we have for Sam's trial?" Instead of looking around your timeline to determine tape numbers and then locating those tape numbers in your bins, you can easily locate the requested footage.

By date shot. When working on a documentary or reality-style project, it is often more efficient to organize your footage by the date it was shot. When using this method, it is best to nest bins inside each other, so that you have a bin for a date and then within that bin are more bins labeled using tape numbers. For example, you would have a bin named 04Jan2000, and then inside that bin you would have additional bins for individual tapes, such as JVa039, JVb039, and JVa040, as shown in Figure 1-2.

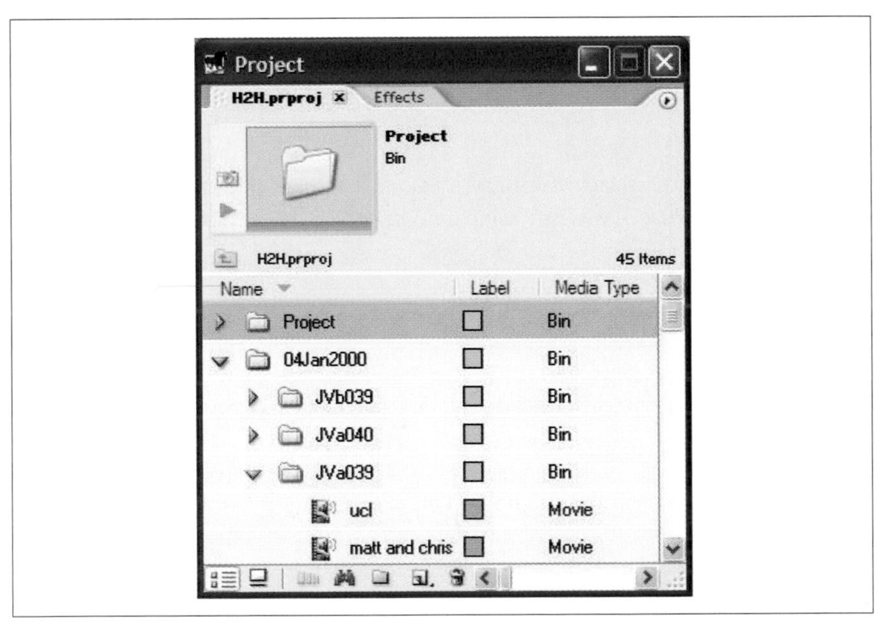

Figure 1-2. Organizing by the date shot

This method is efficient for documentaries because people recall events that occurred on a certain day, as opposed to on a certain tape. For example, a producer is much more likely to say, "Did you see the footage of John and

Vanessa kissing? I think it was on Tuesday, the fourth." He's less likely to say, "I think it's on tape JVa039."

Additional bins. You should also have a few other bins in your project:

B-Roll
: Your B-Roll bin is where you should place any b-roll footage, such as shots you plan on using for establishing scenes or for transitions.

Music
: The Music bin should be fairly self-explanatory. It is where you should place any music you plan on using in your project.

SFX
: Your SFX bin is where you should place all of the sound effects you plan on using in your project.

EFX
: The EFX bin is where you should place any effects, such as footage you are going to use for compositing or titling.

Project
: Your Project bin is where you should save your timeline sequences. Some people like to create multiple bins inside their project bin, such as Archives (for keeping backups), Promos (for cutting short promotional material), and Scenes (for working on small sections of a project), among others.

The most important thing about organizing your bins is to discover how you, and the people you work with, locate the footage and elements you need to edit. If you want to take the extra time, you might even go so far as to combine a few of these methods.

Tracking Drive Space

One thing many inexperienced editors *don't* do is keep track of their drive space. Keeping track of your drive space means knowing how much space you have available. Although something like drive space seems like it's unimportant ("Heck," you might say, "I've got 240 gigabytes in my system. What do I have to worry about?"), it is amazing how quickly video can fill a drive.

Keeping track of your drive space is important for two reasons. First, it lets you know if you have enough space to import new footage, when required. Second, it keeps you from running into a situation in which your drives fail or you are unable to work because there is no more space available. Some experienced editors have, well, *experienced* this second situation and vow to never experience it again.

Somewhat related to drive space, and often overlooked, is the item count inside your directories. Most editing systems create separate files for each audio or video effect you create and render. Over time, these files gather into huge collections and can cause problems.

Item counts can be costly, as a good friend of mine found out firsthand. He was working on a network prime-time television show and their editing systems started to "act funny." After some investigating, he discovered over 500,000 files in a single directory. These files wreaked chaos with his company's systems, 19 Avids linked together using a fiber connection, and cost them tens of thousands of dollars in *downtime* (time people aren't working).

On a semi-regular basis, you should plan on removing your render files and rerendering your project.

Backing Up

A final method for keeping organized is more of a *cover-your-bases* solution. Every day, or every time you start to address notes, you should create a duplicate of your timeline and save it using some type of naming system. You can name your copies by date (14Jun2004), by cut (Offline-pass-5), or some coded system ([DVv5dc]: DV project, version 5, Director's Cut).

Jackson Anderer, an Emmy-nominated editor, names his sequences as Project-Act or Scene-Cut-Date-Initials. So, a sample sequence title would be:

```
EMHE212-Act1-Cut10-Feb9-JA
```

. . . and translate to:

```
Extreme Makeover: Home Edition, Season 2, Episode
12, Act 1, Cut 10, completed on February 9, by
Jackson Anderer.
```

His system works extremely well for organizing large projects that involve many people.

However you choose to name your cuts, do it religiously. Sooner or later, you will want to go back to a previous cut, and if you've continually worked on the same timeline, you won't be able to. I witnessed a director *break his hand* (as he hit the floor with his fist) after being told that there wasn't a backup copy of a previous timeline; he lost a full day's worth of work.

You should also back up your project (not your media) to an external drive, preferably one that is portable. The small USB keychain drives are perfect for this type of use because they're large enough to hold even long, complicated projects and they're small enough that you can easily take them with you.

HACK #3 Number Your Tapes

How you number your tapes can have long-term effects.

Just as you want continuous, nonrepeating timecode **[Hack #4]** on each of your tapes, you also want a unique number assigned to each tape. Although tape numbering is an easy task to accomplish, more often than not, I hear about people who have duplicate tape numbers or no tape numbers at all!

Sticking with Your Choice

You should choose a numbering scheme that makes sense to you and the people you work with. Your numbering scheme should be easy to understand and allow you to glance at a tape and understand what the number represents. In order to make any of the following numbering schemes work, you should have a spreadsheet or database to track additional information, including the date shot and a brief synopsis of the footage.

Imagine trying to physically look through 100 MiniDV tapes, while trying to read what you printed on a small 1"×2" label. I can guarantee you will not be able to put all of the information you would like in such a small space. A spreadsheet or database can become an invaluable resource, especially when you are trying to locate footage.

Tracking information in a system that can be searched and sorted will help you manage your library over time. It is not uncommon for an independent project to amass one hundred or more hours of raw footage. Some professional projects can amass tens of thousands! Save your sanity and take an extra few minutes to enter the information in a central location.

If you are working on a project with more than just a few people, I recommend using a centralized system. Whether you create a method of checking in/out a tracking spreadsheet, use WebDAV to lock/unlock a set of files, or even design a database and place it on a server, you need to create a central location for tracking. This is because you need to take extra precautions to make sure people don't change other people's information, or enter conflicting information. If you do not take such precautions, you could find yourself completely confused by your own project.

Numbering Safely

Numbering schemes are a personal endeavor. Based on my background, I still cringe when I see a tape number longer than six characters. Although nonlinear editing systems are making progress in the realm of cross-compatibility, through Advanced Authoring Format (AAF) and Media Exchange Format (MXF) files, the Edit Decision List (EDL) is still the default and most common method of moving a project from one system to another.

In order to keep peace of mind when using an EDL, a tape number should not exceed six alphanumeric characters (or eight if you don't mind taking a very small risk). A tape number should also not use any special characters—such as an underscore (_) or ampersand (&)—nor should it end in a *B*. It is safest to simply stick with the basic set of A–Z and 0–9.

If you would like to know the reasons behind these rules, point your favorite browser at *http://users.rcn.com/brooks/maxguide.html#reeln%* and keep in mind that many systems export or import GVG EDLs.

If you know that you will never produce an EDL, you can be more flexible in your numbering scheme. However, you should still attempt to keep the numbers as simple as possible.

Okay...on to the numbering! The following are simply suggestions, but they have all been used successfully on professional projects.

Numbering Based on Project

One choice for tape-tracking is to number your tapes by project, then camera number, and then a sequential number. Using such a design results in a number like *PWA032*, which represents the Peter's Wedding (PW) project, camera A, tape number 32. You should take notice of the *032*, as opposed to only *32*. The prepending of the *0* allows up to 1,000 tapes (000–999). If you expect to exceed 1,000 tapes, you can easily prepend two *0*s to your load number and reduce the project indicator to one letter (e.g., *PA0032*).

Numbering Based on Date

To number your tapes based on date, and still maintain the six-character limit, requires you to track only the load count, camera, and date in your number. For example, the fourth tape shot on March 15 in camera A would read as 4A0315. If you use this scheme, you will probably want to place supplementary information somewhere on your tape's label or within your tracking system.

If you choose to expand your numbering scheme to eight characters, you can gain a little more flexibility using the date numbering option.

One approach to adding information to your labels, without compromising your tape number is to simply write information, like the project's name, on the label. This information would be in addition to the tracking number. It is easy to tell that a tape with the label of "Going Home 4A0315" does not belong to the same project as "Sunset 3B0322." This is an effective solution, especially if you are producing multiple projects.

When logging and/or digitizing using this scheme, you should enter only the tracking number as the tape number. Do not include the project name as a part of the tape number. Otherwise, you will have problems when creating an EDL.

Using Barcodes

A third choice is to buy or print out a series of barcode labels along with a human-readable number. I especially recommend this solution if you will be dealing with large amounts of footage. An additional advantage to this solution is you can then use a barcode scanner to help track your tapes as they move from person to person.

Purchasing barcodes. If you are looking to purchase a series of preprinted barcodes, I can recommend Barcode Discounters (*http://www.barcodediscounters.com*). I have used their products in the past and usually purchase the 2" × 1" Poly/Permanent style label. I also have the project's name printed above the barcode and a human-readable number printed below it, as shown in Figure 1-3.

Figure 1-3. A barcoded tape label

If you order your labels, you will probably discover that a label of the 2" × 1" size fits onto the face of MiniDV tapes. Depending on your preference, you may want to order the barcodes in duplicate, so you can attach a barcode to both the physical tape and its plastic case.

Generating your own. If you would like to print your own labels, and you own Microsoft Office, you can link an Excel spreadsheet to a Word document and generate a series of labels:

1. Create a new Excel document.
2. Enter `Barcode` into cell A1.
3. Enter the formula `=sum(A2+1)` into cell A3.
4. Select cell A3 and copy it.
5. Paste the formula in as many cells under column A as you require:
   - There are 2,000 labels in an Avery 5267 package.
   - You can paste to cell A1002, if you want to label both tape and case.
6. Enter `10000` (or whatever number you desire) into cell A2.
7. Save your document.

I like to start my series at *10000*, because some editing systems will trim your tape numbers from *00001* to simply *1*; I prefer consistency.

A nice bonus with this method is you can set up your spreadsheet to track additional information about the tapes, such as the date shot and a brief description of its content. To do this, simply add your criteria to the first row of your spreadsheet. So, to the right of barcode could come Project Name, Date Shot, Description, Format, and so on.

Okay, now to pull your information into Word and print the labels. You are going to print out labels using the Avery 5267 template. These labels are technically Return Address labels, but they measure 0.5 × 1.75 inches and fit nicely on MiniDV tapes. If you are shooting on a larger format tape, feel free to substitute an appropriate template.

If you have chosen to add a Project Name, you should enter it before printing your labels. Otherwise, you will have only the tape number printed.

1. Create a new Word document.
2. Choose Tools → Labels....
3. Change the Label type to Avery 5267.
4. Click the Data Merge button.
5. Using the Data Merge Manager, choose Data Source → Get Data → Open Data Source....
6. Locate the saved Excel document.
7. Accept the default import method.
8. Add the barcode to your label. If you have added a Project Name column, add it as well.
9. Choose Edit → Select All.
10. Choose Format → Paragraph....
11. Choose Alignment → Center.
12. Save your document.

After you have completed all of these steps, you can print out your labels from your Word document. Because there is only one entry per tape in your spreadsheet, only one label will be printed per tape. If you would like to label both your tape and your case, you can simply print two copies of your labels.

Keeping Everything Up-to-Date

Over time, your tracking system will become the central location for an overview of your projects. Make sure you enter information for *each* tape you shoot on, as soon as possible. The more diligent you are about keeping up-to-date, the fewer problems you will have during the editing process.

HACK #4 Black and Code a Tape

You can avoid problems with digitizing your footage by prerecoding timecode on your tapes before you record footage on them.

Digital video provides you with a professional level of accuracy while editing through the use of *timecode*. A movie is simply a series of still images, or frames, displayed quickly and in succession. Timecode is a method of referring to, and tracking, each of those frames. The *frame rate* of the video is the measurement of how many frames are displayed per second.

Timecode is measured in hours, minutes, seconds, and frames. The accepted notation is HH:MM:SS:FF. So, if someone provides you with a timecode of 42:21:33:04, you know the image occurs at 42 hours, 21 minutes, 33 seconds, and 4 frames.

There are two types of timecode: drop frame (DF) and non-drop frame (NDF). The accepted notation is to use a semicolon (;) to indicate drop frame, and a colon (:) for non-drop frame. Some people also use a period (.) to indicate drop frame. Here are some examples:

```
01;00;00;00 – drop frame
01:00:00;00 – drop frame
01:00:00.00 – drop frame
01:00:00:00 – non-drop frame
```

The type of timecode is important for editing systems, and the vast a majority use drop frame as a default.

Timecode is used as a reference point for editing. If you have the same timecode occur on a tape more than once, you will have an exceptionally difficult task determining which image to use. Therefore, you want each frame of your video to be associated with a unique timecode.

Viewing Timecode

Many cameras generate timecode, starting at 00:00:00;00, whenever they are turned off and turned on again. This can become a problem during production, because you might be attempting to conserve battery power while recording and consequently turning the camera off and on frequently. Because of this, you could wind up with a tape with numerous duplicate timecodes.

An easy way to tell if you have regenerated timecode is to simply look at the current timecode of the tape in your camera's viewfinder. For example, if you have recorded 10 minutes of footage, turned off and on your camera, and then notice the timecode is at *or near* 00:00:00;00, then you have more than likely regenerated timecode. Figure 1-4 shows the timecode displayed on an LCD both before (left) and after (right) cycling the camera off and on.

Some video cameras and decks have a feature to generate timecode internally, therefore ignoring any timecode on your tape. This feature is sometimes referred to as *Regenerate Timecode*. The feature is *not* the same as the *occurrence* of regenerated timecode explained herein.

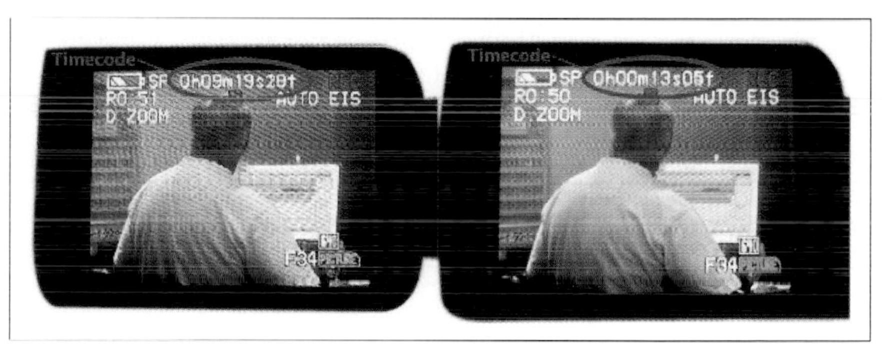

Figure 1-4. Keeping an eye out for timecode that "jumps" backward

In order to avoid a duplicate timecode fiasco, you can perform a simple task: black and code your tapes. *Black and code* refers to the process of recording a black image across your entire tape while allowing the camera, or deck, to assign a continuous timecode for every frame of video. After performing this task, you will be able to turn off your camera safely, because it *will not* overwrite timecode that is already present. However, it *will* record over the black image.

Determining the Frame Rate

The vast majority of time, you will be dealing with video that has a frame rate of 29.97 frames per second (fps) for National Television Systems Committee (NTSC) video or 25 fps for Phase Alternation by Line (PAL) video. There is a reason for the noninteger frame rate for NTSC, but I'll leave that for you to research. (Hint: television was not always broadcast in color.)

Here are the timecodes for the major digital video standards:

NTSC (29.97 fps)
: Used primarily in the Americas. Although the format is referred to as *drop frame*, *frames* are not really dropped. Rather, *frame numbers* are skipped. The calculation for drop-frame timecode is to drop one frame every minute, except when the minute is divisible by 10. When dropping a frame, the frame indicator will skip from ;29 to ;02.

PAL (25 fps)
: Used primarily in Europe. Running at a full 25 frames per second, PAL's timecode does not need much explanation, because it is straightforward.

24p (23.976 fps)
: Stems from High Definition video. Some DV cameras are now able to record in the 24p format, which runs at 23.976 fps. This seemingly odd frame rate provides a "film look" that many people prefer, as film traditionally is shot at 24fps. Most editing systems make using this frame rate fairly straightforward.

Panasonic has a 24p Advanced format which records 24p while applying a process, known as telecine, to make the video also work at 29.97fps. Some editing systems are able to recognize the 24p Advanced format and work with it seamlessly. If you plan on using this format, then you should make sure your editing system can deal with it correctly.

Creating a Black and Coded Tape

To black and code your tapes, you have a few options. Whichever option you choose, you should make sure you black and code your tapes *before* recording actual footage on them. Yes, people have actually attempted to do so after they have recorded footage...quite unsuccessfully.

Using your editing system. If your editing system provides a Black and Code option, you should use it. Apple's Final Cut Pro, Avid Xpress, and Adobe Premiere are a few of the mainstream editing systems that provide this functionality. If your editing system does not offer a Black and Code option, you can create a new timeline with a long black slug. Your slug should be as long as your tape, so if you are planning on recording to a 60 minute tape, your slug should be 60 minutes long as well.

After you have created the timeline, simply record it to tape. An advantage of this method, depending on the capabilities of your editing system, is that you can set the starting point of your timecode. For example, you can start the timecode of your first tape at hour one (01:00:00;00), your second tape at hour two (02:00:00;00), and so on.

If your editing system does not transfer the timecode you assign, you can still use it to send a black image while your deck or camera assigns timecode. But in such cases, you may just want to use your camera by itself, as it has a built in method of creating black.

Using your camera. You can create black with your camera by leaving the lens cap on and recording. Because the lens is covered, your camera will record a black image (because there's no image to record), while assigning continuous timecode to the tape. This will give you a functional black and coded tape.

Your camera's microphone might pick up audio and record it to tape. If you are able to do so, you should disconnect or mute the camera's microphone before using this technique. Just don't forget to reconnect or unmute the microphone when you are done.

Using your tape deck. You can save the wear on your camera's heads by black and coding your tapes using a digital video deck. If you are planning on acquiring a lot of footage, you should invest in a deck. When using a deck to black and code a tape, you can either use your editing system's built-in function (as discussed previously in "Using your editing system") or send a black signal to the tape deck's analog video input.

Purchasing precoded tapes. Many professional post-production facilities also offer precoded tapes. There is usually an additional charge, beyond the cost of the raw tape stock. Although most facilities use these tapes for their clients' editing sessions, you should be able to purchase them for use outside of their facility, as well.

Log Using a Spreadsheet

Logging your footage provides you with a searchable reference of your content.

Regardless of the size of your project, from a few hours of footage to a few thousand, you need to know what content you have acquired. Although there are many specialized applications to help you log your footage, you can just as easily use a simple spreadsheet.

If you do not have a spreadsheet program, you can create a tab-delimited text document that will accomplish the same goal. A tab-delimited file is a way to separate what you type into columns. In such a file, each tab is a column and each line is a row.

Double-Checking Your Tapes

Before you start, make sure all of your tapes are uniquely labeled and numbered **[Hack #3]**. You also need to be sure the timecode on your tapes does not regenerate **[Hack #4]**. Although you can catch these problems while logging and fix them **[Hack #48]** before you begin editing, it is easier to fix these problems early in the process.

Logging your footage is a time-consuming but worthwhile endeavor. As you progress, you will become intimate with your footage. While you're editing, you will discover that you recall seeing a particular shot, but you will not know on what tape or at what timecode the shot occurred. In such a situation, your logs will provide the necessary information.

Setting Up Your Spreadsheet

To set up your spreadsheet assign a column name for each of the following:

Avid
: Clipname, Tape, Start TC, End TC, Tracks

Final Cut
: Reel, Clipname, Media Start, Media End, Description

Premiere
: Tape, In, Out, Clipname, Log Notes, Description, Scene, Shot/Take

To log your footage, you need to enter the in-point, description, and out-point of the footage in a row on your spreadsheet. When entering information, you should enter unique information for Clipname for each log entry. You should also enter any additional information required by your editing system.

Setting Up to View Your Footage

How you view your footage while logging is completely up to you. Some people prefer to transfer their footage to VHS tapes with the timecode superimposed over the image, a process known as a window burn [Hack #75]. Others prefer to digitize all of their footage and view it on their computer.

However you choose to log your footage, I strongly recommend *not* using your master tapes. A simple mechanical failure with your playback machine can easily destroy your master footage, which cannot be replaced. You should treat your master tapes as prized possessions because they are irreplaceable. (Of course, you could always reshoot. But what if you are editing a wedding?)

Logging Your Footage

Logging your footage is a time-consuming process, but the payoff is well worth the investment. To successfully log your footage, there are three essential criteria: the tape number, the timecode of the in-point, and the description of the content. Beyond those three criteria, the out-point is helpful in determining the length of the log entry and is mandatory for most, if not all, editing systems.

If you've set up your logging spreadsheet properly, you can use it to create a *batch digitize list* to help digitize your footage into your editing system. Using the batch list will provide a breakdown of your footage, which will exactly match your logs. Therefore, you will have the same information in both your spreadsheet and your editing system.

Enter one log per row. For each new log entry, create a new row. Continue this process until the tape's content is completely captured in your own words. Each log entry will range from 1 minute to 60 minutes, depending on your attention to detail.

Most people I have worked with prefer to keep their log entries between 5 and 10 minutes long. If your footage is of an interview, you should log the dialog and *not* the action occurring on screen.

Check the format of your timecode. While logging, you need to make sure you enter your timecodes in the correct format. If you are working in PAL format, you can ignore the rest of this paragraph. For NTSC, the correct format for drop-frame timecode Hours:Minutes:Seconds;Frames (`HH:MM:SS;FF`). More often then not, you will be working in Drop-Frame mode, so you will need to use a semicolon before your frame number (;FF). If you are working in Non-Drop Frame mode, use a colon (:FF). If you do not know which mode you are working in, you should be able to safely assume you are working in Drop-Frame.

The colon/semicolon separator is important, because it indicates the type of time code with which your video has been recorded. The wrong indicator will cause problems in the long term, especially if you ever have to redigitize your footage.

If you are uncertain of which mode your tape is recorded on, you can tell by looking the timecode coming off the tape. If there is a semicolon or a period before the frame indicator, the timecode is Drop-Frame. You can think of it this way: one of the dots in the colon *drops* when using Drop-Frame.

Create a new file for each tape. You should create a new spreadsheet for each tape you log and save the spreadsheet as the name of the tape. For example, if you are logging tape number P001A, save the spreadsheet as P001A. Saving the spreadsheet with the same name as the tape's number will be essential for future reference.

Exporting a Batch Digitize List

When you have finished logging your footage, you can import your spreadsheet's information into your editing system. Since each system is different, there will be specific steps you need to follow. To begin, you should save your spreadsheet as a tab-delimited file for either Avid or Final Cut, or as a comma-separated file for Premiere. Once you have saved your log entries as a file, you will need to edit the file in a plain text editor.

The following text involves the Tab and Return keys. Since these don't translate well to the printed word, they are indicated in bold and **ALL CAPS**, surrounded by brackets.

Avid:

Delete the top line of headers, then type the following *above* your entries:

```
Heading
FIELD_DELIM[TAB]TABS
VIDEO_FORMAT[TAB]NTSC
AUDIO_FORMAT[TAB]44kHz
TAPE[TAB]TapeNumber
```

```
FPS[TAB]29.97
[RETURN]
Column
Name[TAB]Tracks[TAB]Start[TAB]End[TAB]Tape[TAB]Descript[TAB]DateShot
[TAB]Barcode
[RETURN]
Data
```

Final Cut
: Change Clipname to Name

Premiere
: Delete the top line of headers

Importing a Batch Digitize List

Once you have made the necessary changes, save the file. Next, you need to import the batch list into your editing system:

Avid
: File → Import

Final Cut
: File → Import → Batch List at 29.97 fps...

Premiere
: Project → Import Batch List

You will have to perform these steps for each tape you plan on using in your project. The effort, however, is well worth it. Upon successfully importing your batch list, you can easily locate shots either by referring to your spreadsheets or by searching within your editing project.

Search Your Project

If you have located a shot using your spreadsheet, and want to see the footage as it exists in your editing system, you can easily find it **[Hack #2]**, because the spreadsheet matches the information in your editing system. If you don't have access to your spreadsheet, you can simply search within your editing project and have the same result. Figure 1-5 shows the Find dialog box for Final Cut Pro, and Figure 1-6 shows the result.

To locate footage, you should make sure you search from your Project and not your Timeline. This is because most editing systems are context sensitive, so if you are currently working in the timeline and perform a search, you will only search footage currently in the timeline.

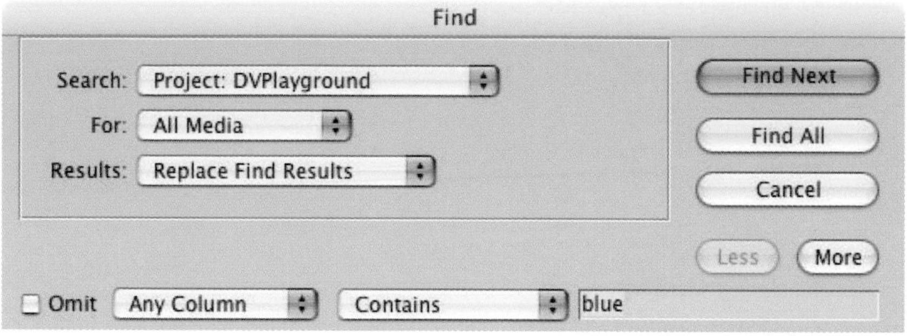

Figure 1-5. Finding media within Final Cut Pro

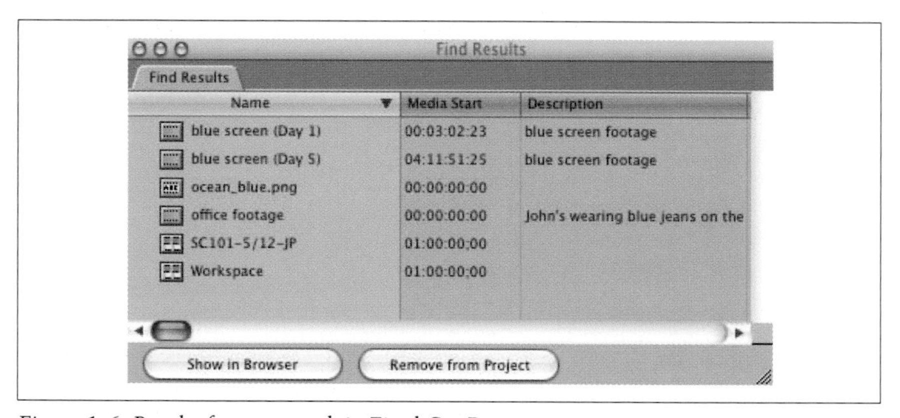

Figure 1-6. Results from a search in Final Cut Pro

To use the search function:

Avid
: Fast Menu → Custom Sift

Final Cut
: Edit → Find...

Premiere
: Edit → Find...

After locating the footage, open it in your application's source view, or just drop it onto your timeline. Having a detailed, and searchable, reference to your footage both in a spreadsheet and on your editing system, will save you hours of shuffling through your footage trying to find a shot.

Create a Digital Storyboard

Using free photo software, a digital camera, paper, and a pencil, you can create a digital storyboard.

Storyboarding is an important step in producing a movie. The process requires you to envision what images you plan on capturing to tell your story and put them to paper. Although it can be a tedious process, the positive effect it has when shooting is huge, because it helps communicate your vision with the people you are working with, including yourself.

Drawing the Story

Even if you are artistically challenged, you can still create a storyboard. To get started, you simply need paper and a pencil. If you have a script, break it out, because you'll need to refer to it quite often.

You can also use a computer program to create your images, or even professional storyboarding software. StoryBoard Quick! (*http://www.storyboardartist.com*; $279.99) offers a library of over 300 people and objects to help you design a storyboard.

Deciding what to draw. The process of storyboarding requires you to draw out each scene as you intend to capture it. For example, if you have a scene in which two people are talking in a car, you will probably want to start with an image of a car driving down a road. This would be your establishing shot. You would then draw a picture of two people talking, as viewed from the hood of the car, so your shot would capture the people from their shoulders up. Next, you would focus on one person, probably close up so you can see only his face. Figure 1-7 shows a sample, hand-drawn storyboard.

Figure 1-7. Drawing a storyboard by hand

Depending on how ambitious you are, you can use a full sheet of paper for each image, fold the paper in half, fold the paper in quarters, or use index cards. I prefer the simplicity and portability of index cards and a pencil. I also like the fact that I can write notes on the back of the cards.

Indicating camera movement. Quite obviously, drawings don't move. So, in order to demonstrate camera movement, you should draw arrows to indicate when the camera should pan, dolly, or zoom. If you want to indicate a zoom, you can also draw a square around the area you would like to zoom, as shown in Figure 1-8.

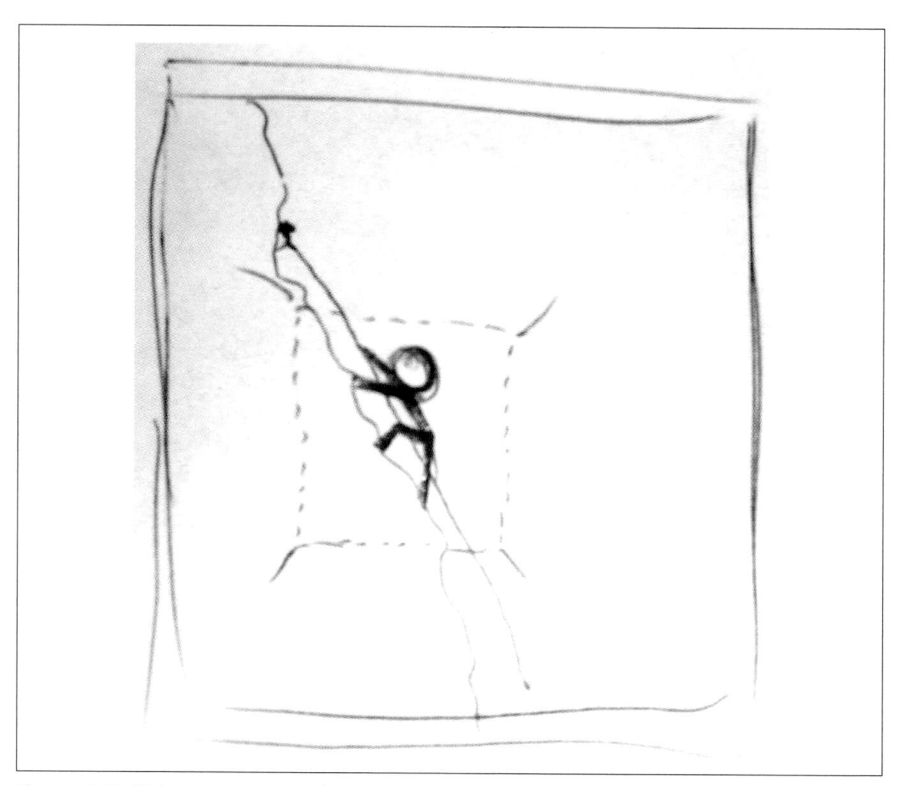

Figure 1-8. Using squares to indicate zooming

Since your storyboard is first and foremost a tool for communication, don't be shy about writing on your drawings. It is better to be clear than to discover there has been a miscommunication in the middle of shooting a scene.

Importing the Drawings

After you have drawn your storyboard, you have completed half of the process. Although you have a completely functional storyboard, going digital will give you a few advantages in the end.

Making analog digital. Your drawings are living in *meatspace* (the physical world) and not *cyberspace* (the digital world). To correct that problem, you can either use a scanner to scan all of the images or you can use a digital camera to take a picture of each image. Dr. Michael Johnson, of Pixar Animation Studios, has mentioned that some of the creative team at Pixar use the latter technique.

After you have bridged the analog-to-digital divide, you should import the images into your photo collection software of choice. For Macintosh users, this will probably be iPhoto; for Windows users, it will probably be Photo Story; and for Linux users, gPhoto. But, then again, there are numerous choices out there. The important thing is that you can import your drawings and arrange them.

Arranging the images. When you have successfully imported your images, you will need to arrange them accordingly. Some software allows you to annotate your images, so the annotation feature can be used to transfer any notes you may have made on specific images. You should look over your storyboard to see if there are any glaring omissions. You can also try moving scenes around, to see how they affect your story.

At this point, you should have a good sense of your movie and how it is going to turn out—and you haven't shot one frame of video yet!

Narrating the Story

Once you have arranged your storyboard, you can play your storyboard as a slideshow. While watching the slideshow, sit back and try to *follow the story* you are attempting to tell. This is not the time to be criticizing your artistic skill.

If your software supports it, you should narrate your storyboard using your script.

If you are working with other people, you should attempt to read your script along with the slideshow. You should be able to get a good idea of how your movie is going to come out, and where your weak spots are located. If necessary, rewrite portions of the script, and change the storyboard as necessary.

All of this might seem like a lot of work, but it is a lot less work, and a lot less expensive, than discovering problems in the middle of a shoot schedule, or worse...while editing.

Create a Two-Column Script

Two-column scripts clearly separate the audio and video portions of a scene. This approach provides more detail to help during the preproduction and editing processes.

Most people have seen a *script* at some point in their lifetime, whether for a play, a television show, or a movie. These are the traditional form of script, or *screenplay*, which focus on dialogue. They make mere mention of the actions the actors should perform.

A *two-column script* focuses on both the dialogue of the scene and the specific shots to be used. The two-column format provides visual instructions, allowing directors, editors, and everyone else on the crew to know what should appear on screen. Two-column scripts can be created before or after shooting and are especially helpful when working on a documentary or reality-style project.

Understanding the Two-Column Concept

Dialogue in a traditional screenplay typically looks like this:

> LAFEU.
>
> He cannot want the best.
>
> That shall attend his love.
>
> COUNTESS.
>
> Heaven bless him! Farewell, Bertram.
>
> [Exit COUNTESS.]

Although you can envision this scene in your mind, it is difficult to ensure we all imagine the same setting. *Storyboards* are one method of communicating what you "see" in your mind [Hack #6]. A two-column script provides another method. The dialogue in the previous example might look like Table 1-1 in a two-column script.

Table 1-1. Dialog in a two-column script

Video	Audio
TWO SHOT Lafeu is on a blue velvet couch and Countess is standing to his left.	Lafeu: He cannot want the best. That shall attend his love.
TRACKING SHOT Countess walks toward the door and leaves.	Countess: Heaven bless him! Farewell, Bertram.

As you can tell, there is more detailed information about the setting. But for this type of entertainment, a two-column script is probably more work than is necessary.

So, when should you use a two-column script? Well, there is no hard and fast rule, but if you do not have a script and you have captured your footage, a two-column script could become your most prized possession.

Setting Up a Two-Column Script

If you own Microsoft Word, setting up a two-column script is fairly straightforward:

1. Create a new document.
2. Choose Table → Insert → Table....
3. For the number of columns, enter `2`.
4. For the number of rows, enter `10`.
5. Click OK.

After the table has been inserted into your document, you should enter the headings of Video in the first column and Audio in the second column.

Writing with a Two-Column Script

If you are writing your script *before* shooting your scene, you will want to include descriptions of any action to occur on screen. In addition to descriptions of action, you will want to include information about camera movements. This type of script can be tedious to write, but it will effectively communicate your vision to everyone working on the project.

When working on a documentary, or anytime you are attempting to weave a story after footage acquisition, a two-column script might become your only roadmap to successful completion. With such projects, it is important to note both the tape number [Hack #3] and timecode [Hack #48] of the footage you would like to use. A good set of tape logs [Hack #5] will also prove invaluable during this process. Table 1-2 shows an example of what a two-column script looks like, when including tape and timecode information.

Table 1-2. An example two-column script

Video	Audio
DVH001 - 02:04:14	(voice-over - to be recorded)
David sitting at his computer, playing Solitaire.	Working aerial search and rescue takes years of training . . .
DVH032 - 00:25:15	DVH011 - 01:13:50
CU of walkie-talkie.	Air Five, we have a 920-Alpha at the Crest. Please respond.
DVH003 - 00:49:29	10-4. Responding now.
David running to his helicopter, speaking into the walkie-talkie.	
(con't)	(voice-over - to be recorded)
	. . . and the ability to remain calm under pressure.

Don't worry about being frame-accurate with your scripts. Frame accuracy is important only during the editing process. The preceding script uses the HH:MM:SS timecode format.

The previous script has a few interesting items. In the first row, notice the video is from footage that has already been shot but the audio is coming from a voice-over, which has yet to be recorded. We know the voice-over has not been recorded, because there is no tape number and no timecode associated with it.

In the second row, the video is a close-up shot of a walkie-talkie and the audio is coming from a different tape. Since there is no visual cue to indicate *what* is being said over the walkie-talkie, I have chosen to use audio from another tape that I feel moves my story in the direction I would like to take it. In fact, the audio from tape DVH011 is background audio that was obtained by accident.

Rows four and five are of the subject, David, running toward his helicopter while speaking into his walkie-talkie. Again, the voice-over portion has not yet been recorded.

Editing with a Two-Column Script

If you are editing a project and you have a two-column script, your job is going to be a lot easier than without one. You are provided both the tape number and timecode of the audio or video you need, so most of your initial decisions will be made for you. The majority of your early work will require you to locate the most suitable out-point for your edits, and then you will just have to make your edits smooth **[Hack #46]**. For the most part, however, you should be able to concentrate on the storyline of the project.

After you have digitized your footage, you simply need to locate the referenced audio or video and add it to your timeline appropriately. As you move down your script, you will move to the right on your timeline, continually layering in the audio and video together. If you've ever worked with a lot of footage and no script, you'll love the focus a two-column script provides . . . not to mention the amount of time it saves.

Build Your Own Apple Box

Apple boxes—simple wooden boxes, commonly found around television and film sets—can be used for a variety of purposes on any shoot. They're also easy to build.

Some common items you might find lying around your home can become indispensable tools when shooting your video. Items such as clothespins **[Hack #25]** are not only cheap, but highly functional as well. The same holds true for scrap pieces of wood.

Apple boxes are wood boxes that are used for everything from propping up cameras to being used as stepstools, and they are practical workhorses on any shoot. Using some scrap wood, or wood bought from a lumberyard, you can build your own. If you don't want to build your own apple boxes, they can be purchased from places such as FilmTools (*http://www.filmtools.com*; $31.00) or B&H (*http://www.bhphotovideo.com*; $31.95).

Gathering the Supplies

The supplies necessary to build an apple box are pretty limited. If you have some scrap wood lying around, use it. Otherwise, purchase the required wood at a local home improvement store or lumberyard. The price of the wood will vary depending on the type of wood and current market conditions. However you go about getting your wood, you will need six pieces per apple box, two pieces in each of the following dimensions:

- 1/2"×7"×11"
- 1/2"×7"×20"
- 1/2"×12"×20"

The resulting *full* apple box should measure 8 inches high, 12 inches wide, and 20 inches long. You could also build smaller *half* (4 inches high), *quarter* (2 inches high), and *eighth* apple (1 inch high) sizes. Figure 1-9 shows pieces of scrap wood, cut to size for a full apple box and placed together to check the dimensions.

Figure 1-9. Checking the pieces to ensure a proper fit

In addition to the wood, you will also need some nails or screws, along with a hammer or screwdriver. If you have access to other woodshop tools, and are comfortable using them, you can use them as you see fit.

Although it's not necessary, having a drill will help you to create handles in the sides of the box. To create a handle, drill two holes in the center of the 7" × 11" pieces, with the holes about 3 inches apart. Then saw across, from one hole to another, and punch out the bridge. Figure 1-10 shows a hole being drilled in one of the 7" × 11" pieces in order to create a handle.

If you plan on using the boxes heavily—and after needing one once, you'll probably never want to be without them—I recommend you use higher-quality wood.

Building the Apple Box

After you've gathered your supplies, you need to assemble your box. You might want to use some wood glue on the edges of your pieces. The glue will help keep the pieces in place, prior to nailing or screwing them together.

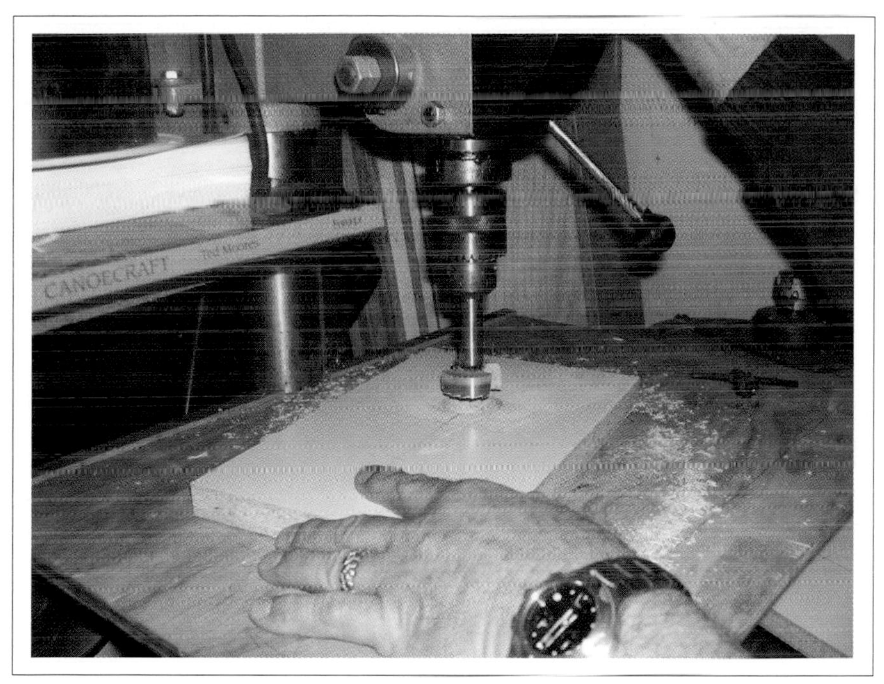

Figure 1-10. Drilling handles

To assemble the box, place a 12" × 20" piece as the base. Then, place the 7" × 20" pieces on each side, so that the 20-inch lengths line up, and the 7" × 11" pieces on the ends. Finally, place the second 12" × 20" piece on top and nail or screw the pieces together, to ensure the box doesn't fall apart. Figure 1-11 shows a completed full apple box.

When you are finished, file down any sharp edges and clean off the remnants of any glue, if necessary.

Using the Apple Box

There are no rules to how you should use apple boxes, so how you use them is up to you. You can stand on them, use them to prop up items, use them as small tables or chairs, and even staple or nail items to them. I've seen people run cable through the handles, and even place green fabric over one stood on end to make items "float" in front of a green screen [Hack #22].

Ultimately, you should abuse them for whatever purpose you need.

Figure 1-11. One full apple box, ready for duty

HACK #9

Make Your Own Slate

Using a small dry-erase board, you can keep track of your footage visually.

A *slate* is a record of information that is visually recorded onto video. In the physical sense, a slate is usually a plastic or wooden board with areas marked off for someone (usually a camera assistant) to write. When shooting on film, the slate will often have a small moving plank on top of it to make a *clack* sound and is sometimes referred to as a *clapboard*.

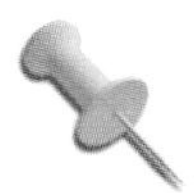

Clapboards are used so the audio and film images can be synchronized. Film captures only an image, so the audio is recorded separately. When you are watching the image, you can visually see the plank hit the slate, which indicates where the *clack* sound should be heard. When you synchronize the sound of the *clack* and the image of the plank hitting, the audio and film images are "in sync."

The reason to use a slate is simple. When you are watching footage, it is helpful to know what you are watching. For example, if you were to pick up a videotape, place it in a deck, and press the Play button, it would be helpful if something were to appear on screen to inform you what you were about to watch. This is why movies and television shows have opening title sequences and why your raw footage should have a slate as well.

Gathering Materials

There are a couple of ways to make your own slate.

Using household supplies. If you're looking to be crafty and save as much money as you can (maybe to feed your crew?), you can use a few common household items:

- One small piece of cardboard (around 8" × 11")
- One black marker
- One roll of wax paper
- One roll of tape

You will use the pen to draw sections onto the cardboard, almost like creating a grid. These sections will need to be labeled appropriately (as noted later). After creating and labeling your sections, cover the cardboard with the wax paper and then tape it so that it is taut. The wax paper should be tight enough against the cardboard so you can see the sections through it.

When using wax paper, you might discover that both dry-erase and permanent-ink markers can be used, and wiped off. However, the longer the ink sits on the paper, the harder it is to wipe off.

Using purchased supplies. If you're willing to spend a little money to make your slate, you will need these items:

- One small whiteboard (office supply store; $5–$10)
- One roll of 1/4-inch black electrical tape (home improvement store; $1–$3)
- One pair of scissors

You will need to cut the black tape into a few strips of the length and width of your whiteboard. For example, if you have an 8" × 11" whiteboard, you should cut piece of tape in 8-inch and 11-inch lengths. You will be using the strips to section off areas on the whiteboard. Figure 1-12 shows the items required to make your own slate.

Using Your Slate

The amount of information you want to keep on your slate will determine how you section it off. At the very least, you will want sections to write the project's title, the date, the scene, and the take number. You might to keep

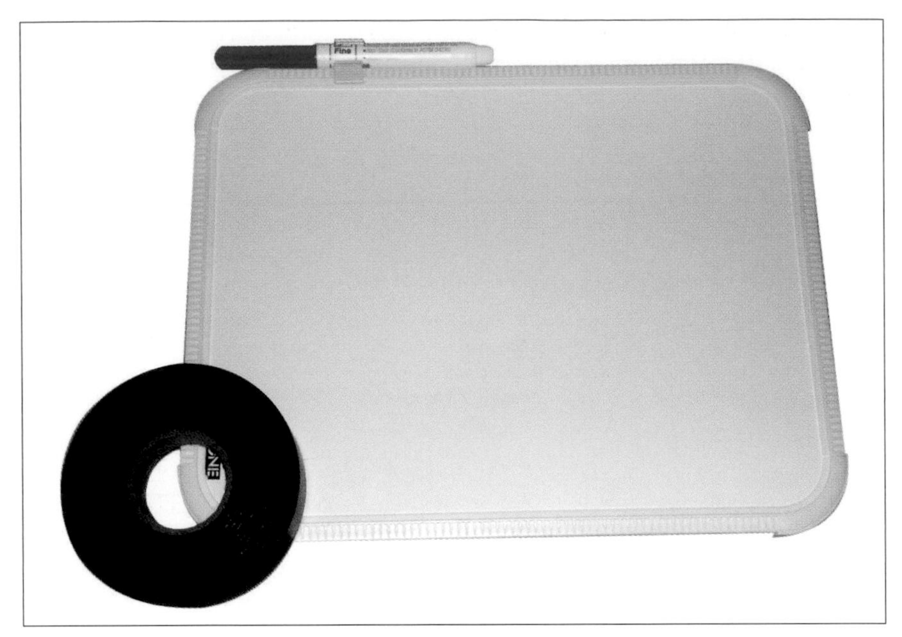

Figure 1-12. A small whiteboard, soon to be a slate

track of the camera (if you are shooting with more than one), the director's name, and the audio configuration (for example, Bob's microphone is on audio channel 1 and Sally's microphone is on audio channel 2). Figure 1-13 shows the slate, as completed and filled out for a shoot.

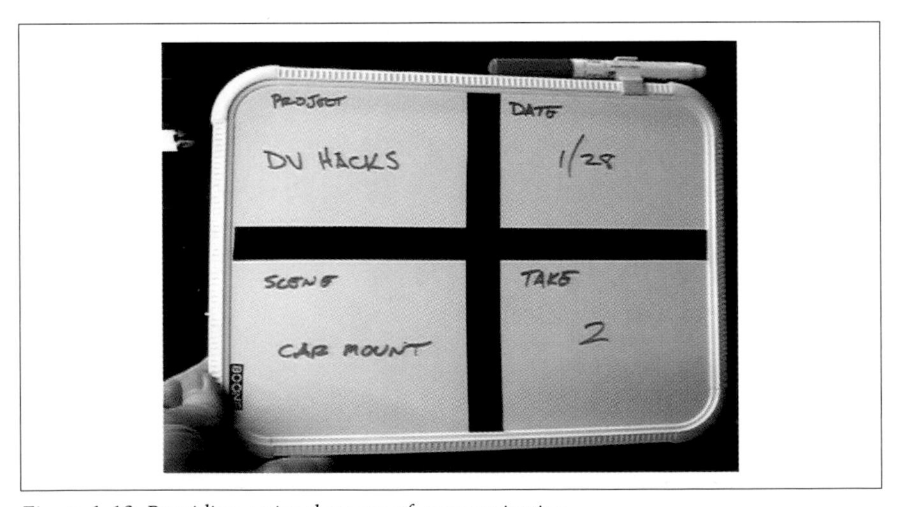

Figure 1-13. Providing a visual means of communicating

To use your slate, simply write the appropriate information on it and hold it in front of your camera before you begin to shoot. You should do this for every *take*. The nice feature about using either of these techniques is that you can wipe it clean when necessary. So, when shooting a lot of takes, or even a lot of scenes, it's easy to simply wipe the previous writing off and write something new. Because it's so easy to do, you won't become reluctant to slate your shots, which will pay dividends when you go to watch and log your footage [Hack #5].

If you are in dire need of a slate—if you're on set and forgot a slate, for example—use a piece of paper and a pen. Some information is better than none at all.

If you forget to use the slate before your take, you can still use it afterward. By holding the slate in front of the camera at the end of the take, you can capture the information. Even though it won't appear where you expect it, the information is still valuable. If you want to indicate that a slate is being used at the end of a take, you can hold it upside down.

HACK #10 Control Your Camcorder Remotely

It isn't difficult to control every aspect of your camcorder from some distance away. In fact, you can shoot video without having to touch your camcorder.

Sometimes, you need to take the camcorder out of your hand and stick it somewhere. For instance, you might be trying to film an animal that doesn't like people, or you might want to film a party from the far corner of a room, while remaining in the shot yourself. Whatever the reason, being able to control the camcorder remotely makes recording more flexible.

Replacing the Remote

Most camcorders come with a remote control that allows you to control many of the features they offer remotely. But these remote controls aren't generally that good; many of the ones I've used worked only from a few feet away. Fortunately, there are ways to get around this limitation: use a programmable remote control or use an infrared (IR) extender.

Programmable remotes. As the name implies, *programmable remote controls* are smarter than the average remote. They can learn the special codes that camcorder understands, allowing you to control the camcorder without the original remote. The more sophisticated models, such as the Philips Pronto range (*http://www.pronto.philips.com/*), can be programmed by a PC and configured to control several devices. These remotes can even string several commands together in a macro, so, you could, for instance, create a macro to switch the camcorder to low light mode and start recording from one touch of the remote—a big timesaver.

Programmable remotes can be a real boon if you've lost the remote, because many web sites offer downloadable configuration files for a huge range of devices. Check out sites such as Remote Central (*http://www.remotecentral.com*). The IR emitters in these remotes are also usually much stronger than the ones in the remote that came with the camcorder, which means that you can control the camcorder from further away.

IR remotes. IR remote controls work only where there is a line of sight, and that's not much use if you are in one room and the camcorder is in another. Fortunately, there is another simple solution: add an *IR extender*, such as the SmartHome 8220A (*http://www.smarthome.com/8220A.html*; $50).

IR extenders have two parts: a transmitter that picks up the IR signal from the remote and converts it to a radio signal, and a receiver that takes this radio signal and converts it back to infrared, which the camcorder picks up. By using radio signals, they can send remote control commands through walls or other obstacles.

If you are comfortable with a soldering iron, it's not difficult to build one yourself. You'll find plans for several different IR extenders at *http://www.mitedu.freeserve.co.uk/Circuits/Interface/candi.htm*.

Combining the best of both worlds. If you combine a programmable remote with an IR extender and a wireless video transmitter, you've got a remote control camcorder that you can easily put somewhere, and then retire to a nearby location to watch the video. And you can even control the camcorder, zooming in and out and stopping and starting the recording as required.

Adding a Remote-Controlled Pan and Tilt

While you're building the ultimate remote-controlled camcorder, why not add the ability to pan and tilt the camera? You can do this with a remote-controlled *pan and tilt head*: a device that fits under the camcorder and allows you to pan the camcorder left and right or tilt it up and down.

I picked up a Memorex Pan-O-Matic on eBay. It can pan 120 degrees left to right and up and down 20 degrees, and it comes with a remote control. When combined with the programmable remote (programmed to control the pan and tilt head, as well as the camcorder) and the IR extender, I have a fully controllable camcorder that could pan and tilt for maximum coverage. It also runs off batteries and fits onto a tripod to make it even easier to mount and use. Figure 1-14 shows my personal setup.

Figure 1-14. Sony Camcorder on a Memorex Pan-O-Matic base

Although the particular model I use is no longer manufactured, there are other wireless pan and tilt heads available, such as the X10 (yes, they are the people behind all of those annoying web ads) Ninja Pan Tilt head (*http://www.x10.com/products/x10_vk74a.htm*; $99.00). This one uses a radio remote control, so you won't need to use the IR extender to control it.

Pan and tilt heads aren't strong enough to move a large camcorder (such as a Canon XL1). They should be used only with smaller, palm-style camcorders. If you have a larger camcorder, you'll need to invest in a professional pan and tilt head that can handle the weight.

Secondhand pan and tilt heads from CCTV system also pop up on eBay a lot. You could pick up one for less than $40. However, most of them are wired models, in which you need to run a cable between the pan and tilt head and the controller. If you buy one, make sure that it comes with the controller and that you are confident about wiring it up.

—*Richard Baguley*

Monitor Your Camera

Knowing what you are recording while you shoot can save you hours, and even days, of frustration and money during the editing process.

Practically every video camera has audio and video outputs, but most people use them only when transferring video after it's been shot. By using these outputs while you shoot, you can monitor what is being recorded. Not only does this allow you to take care of troubles immediately, such as fixing a bad audio connection, but it also enables you to take field notes that you can refer to when editing.

Determining Your Cable Needs

There are a variety of connections for carrying an audio and/or video signal. If you plan on monitoring your camera while you're recording, you should run out to your local Radio Shack and grab the necessary cables and adaptors for your camera ASAP. While you're there, pick up a couple of spares, since you'll lose or break one at the most inopportune time.

Video. Monitoring your video requires you to send your camera's signal to a television or professional monitor. Some cameras have simple RCA jacks for audio and video, while others have either a combined A/V/Phones jack or something proprietary. Whatever your situation is, you will need to transmit the signal from your camera to your monitor.

Although the image of what's being recorded is important, you should also turn on the camera's display so that the current timecode appears superimposed over the image. By displaying the timecode, you will be able to reference it in your notes. Without it, your notes will be less useful. Figure 1-15 shows a Canon XL-1 and a JVC 3-inch LCD monitor connected together via an RCA cable, which allows for great mobility.

Unless you're setting up in a location for a long period of time, portability is key. A small, 13- to 15-inch television works well in most situations. However, if you plan on following a cameraman around—to shoot a documentary, for example—you should use a smaller, portable LCD monitor in the 3- to 7-inch range.

Figure 1-15. Connecting to a small monitor

Audio. When you are monitoring your audio, you will want to use headphones. The obvious reason for this is so that your microphone doesn't pick up what's being recorded, because your microphone will probably pick-up the audio emitting from your headphones. That would result in an echo and possibly even feedback.

The less obvious reason is that headphones allow you to hear what's being recorded, as opposed to what's being heard. The two are not the same; your microphone might pick up sounds you either can't hear or wouldn't normally pick up on.

Your camera will most likely have a jack that will allow you to plug in a pair of headphones. The most common connections for your audio will be a standard headphone jack (1/4") or a mini-headphone jack (3.5 mm), and adaptors are available to go from one to the other. Because adaptors are available, you are not limited to the type or style of headphone you choose. Figure 1-16 shows the male end of a 1/4" microphone adaptor and a 3.5 mm mini-adaptor, respectively.

If you are not personally working with the camera—if you've hired a cameraman, for example—you should run an audio line to where you are located. If you are shooting with multiple cameras, you should at least monitor the camera (or deck) where your primary audio, such as a boom microphone, is being routed.

Setting Up

Getting your equipment gathered is only half of the equation. The other half involves putting the equipment to use. To do so, you need to know about the location you will be shooting. Knowing about your location will help you determine how you will power your equipment and how to run the signal from the cameras, among other factors.

A good practice is to set up a rolling cart with your equipment on it. Using a three-shelf cart, you can use the top shelf to hold your monitor, place your

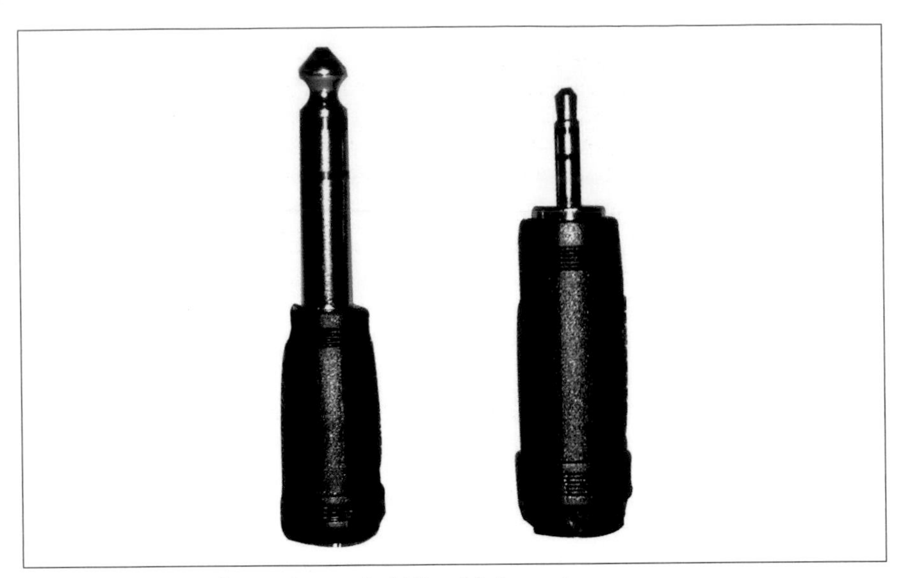

Figure 1-16. Microphone adaptors in 1/4"and 3.5 mm sizes

cup of coffee and other supplies on the middle shelf, and a bin with a variety of cables on the bottom shelf along with a surge protector. This basic setup allows you to move from one location to another quickly. If you want to be resourceful, you can even add a VCR to the mix and record a reference copy of your footage while you're shooting.

Wired. The easiest setup is to run power and the audio/video signal via cabling. If you are near a power outlet, you'll just plop your equipment down, plug it in, and go. Depending on how far you are from the camera or power outlet, you might need a long AV cable or extension cord, respectively.

Wireless. If you can't run cable from your camera to your monitor, you can still send the signal using a wireless transmitter and either display it on a television set or a compatible video receiver. There are many wireless transmitters on the market. Here are a few places to look:

X10
: VideoSENDER package (*http://www.x10.com*; $79.99)

Studio 1 Productions
: ShotWatcher video transmitter (*http://www.studio1productions.com/shotwatcher.htm*; $495)

The Spy Store
: A variety of video transmitters available at *http://www.thespystore.com/videotransmitter.htm*

NuTex Communications also offers a good, small, wireless video solution. NuTex is a Taiwan-based company, so if you want to purchase their products, you need to find a retailer who can import them. More information about their video sender is available at *http://www.nutex.com.tw/mini.htm*.

Going wireless has its advantages, but there are disadvantages as well. The most obvious is the fact that some wireless frequencies can interfere with your audio/video signal. In the extreme case, I have seen cell phones with built-in walkie-talkies, such as those from Nextel, *tear* a video signal, rendering the video useless. Studio 1 Productions is up front about how the ShotWatcher causes problems with Canon's XL-1. Whatever choice you make, you should run tests *before* you go into production.

Taking Notes

Being able to see and hear what's being recorded on tape will help you fix troubles before they become problems. But it also enables you to take notes while action is occurring, which you can use when editing to refresh your memory or communicate with your editor (kind of a "message in a bottle").

In addition to the current timecode, you want to be aware of what tape is in the camera. If you have prelabeled your tapes [Hack #3], you should note the tape number along with the timecode displayed in your monitor. If you have not prelabeled your tapes, you should be diligent about noting at least the date shot somewhere on the tape as soon as possible (such as when removing the tape from the camera).

You can take notes however you choose, but if you have a script to follow, you can use the same approach professional Script Supervisor's use: create a *lined script*. If you are shooting a documentary or unscripted video, you can write down the date and time of an occurrence in an ad hoc fashion. Both approaches will enable you to create a reminder of when something occurred that you found significant to your project.

Creating a lined script. To create a lined script, you need a physical copy of the script you are shooting. You also need a pencil, a sharpener, a three-ring binder (to hold the script), and a ruler.

When shooting a scene, mark the starting timecode just above the first line of dialogue being spoken. During the scene, follow along by drawing a line down the script, until the scene ends. At that point, write down the ending timecode. You can also write notes in the margins or on the back of the page. Such notes might include continuity notes, such as how full a glass of water is, or production notes of unexpected events, such as actors improvising. Figure 1-17 shows a lined script after a few takes have taken place.

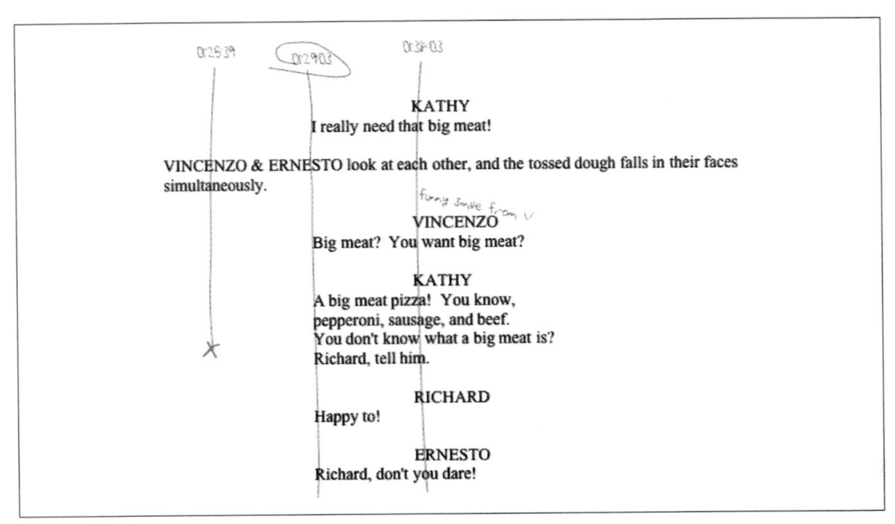

KATHY
I really need that big meat!

VINCENZO & ERNESTO look at each other, and the tossed dough falls in their faces simultaneously.

funny smile from V

VINCENZO
Big meat? You want big meat?

KATHY
A big meat pizza! You know,
pepperoni, sausage, and beef.
You don't know what a big meat is?
Richard, tell him.

RICHARD
Happy to!

ERNESTO
Richard, don't you dare!

Figure 1-17. A lined script, identifying that a good take for this scene occurred at 01:29:03

You continue the process of noting timecodes for every take, for the entire script. If you decide that a certain take is good, simply circle the timecode, as shown in Figure 1-17 This is known as a *circle take* and is usually an indicator of a director's preference. When you go to edit your project, you can refer to your lined script and immediately know where your good takes occur on tape.

Taking ad hoc notes. Even if you don't have a script, you can (and should) still take notes while shooting. The simple fact is, you won't remember everything that happens. This is especially true if you're shooting over a period of days or months.

For example, you might make a mental note that Bob and Sarah kissed for the first time at 3:15 p.m. on Tuesday. But when you start editing, their kiss will be one of probably hundreds of mental notes floating around your head. Taking notes early and often will free you to concentrate on other aspects of your project.

Using Your Notes

If you are under a serious time crunch, you might not have time to log your footage **[Hack #5]**. In such a circumstance, your field notes will prove to be invaluable. Their worth is compounded by each hour of footage you have.

With a lined script, you can log your footage quite quickly, because you have both in and out points for your takes. Moreover, you can enter a quick note in your log comments about which takes were deemed good when they were recorded. You can even try editing together your scene using your circle takes first and finesse it from there.

Using ad hoc notes isn't as straightforward as using a lined script, but they could prove even more valuable. Unscripted projects often amass much more footage than those that are scripted. In fact, some reality television shows acquire thousands of hours of footage. Imagine sifting through 10 hours of footage with and without notes. Which would you prefer to do? Now, compound that to 20, 200, or even 2,000 hours, and I'm sure you can begin to appreciate the value of field notes.

No matter how small your project is, you should take the to time to set up and monitor what your camera is recording.

Protect Outdoor Cameras

Build an inexpensive enclosure to protect your outdoor security cameras from the weather.

A few years ago, my home burned down while I was out of town. As a long-time home automation enthusiast, I used the opportunity to rebuild my home with as much automation equipment and provisions for future expansion as I could. One of the additions was to have security video cameras scattered all around the property and outside the home. With that in mind, I installed more than two miles of RG-6 cable and Category 5 wire, one length of each to every potential camera location so that I can expand my system simply by adding cameras.

My current video setup consists of eight cameras connected to a 16-camera video controller, a 960-hour time-lapse VHS recorder, and a video-to-TV modulator for each camera. With this system, I can view the video from the cameras on my television sets, as shown in Figure 1-18.

Each camera is connected to an input on the video controller, and each output from the controller is connected to a TV channel modulator. Each modulator rebroadcasts the camera picture on a different TV channel, so to view a camera I simply turn on a TV and tune in. The modulators are programmable and can broadcast on any channel from 14 through 64, so it's easy to set them to use unused channels on your cable system.

ChannelPlus (*http://www.channelplus.com*) makes a wide variety of TV modulators and camera controllers.

Figure 1-18. The video command center

The cameras I use don't have microphones, but the Category 5 wire that runs to each camera could carry sound back to the modulator if I decide to buy cameras that do have microphones. Currently, I use two wire pairs, of the four available in the Category 5 cable, to run electrical power to the cameras. The camera power supplies are plugged in near the video command center, as described earlier.

So far, everything I've described is standard equipment for video monitoring. The biggest challenge was the outside enclosures for the cameras. Instead of spending $50 to $100 each for commercial enclosures, I decided to build my own. These enclosures cost about $5 each and work just as well.

Here's what you'll need for this hack:

- One length of Schedule 40 PVC pipe cut to 4 inches longer than the camera and lens
- Two end caps that fit the PVC pipe
- One piece of standard window glass cut to fit inside the test cap

I use 4-inch diameter pipe, which should be big enough for almost any camera, but check to make sure your cameras will fit inside the pipe you select, as shown in Figure 1-19. You can cut the PVC pipe with just about any saw, but it might require two passes with the blade, given its size. Be careful, and make the cuts nice and even. Go slowly and make sure the pipe doesn't bind and shatter.

Carefully measure the location of the camera's mounting hole, and then drill a corresponding hole in the pipe that will result in the camera's lens being set about 1 inch inside one end of the pipe. You'll use this mounting hole with a bolt to securely mount the camera inside the pipe, during final assembly.

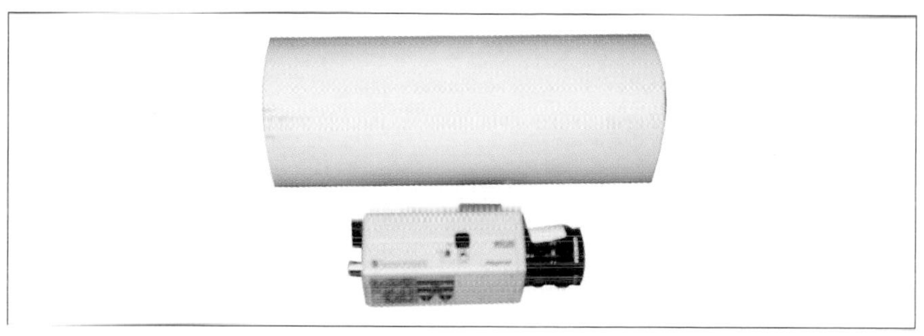

Figure 1-19. Appropriately sized PVC pipe

Drill another hole, just slightly larger than the coax cable you are using, 1 inch from the back end of the pipe. Add just one more hole, for the Category 5 wire that carries power to the camera.

Mount the camera inside the pipe and connect all the wires before you cap the end of the pipe. For the front cap, use a drill hole saw to cut a 2-inch hole in the center of the cap. Put a small bead of tub caulk around the test cap, place the glass inside, and press tightly. The caulk should spread around so that it is waterproof. Remove any caulk that has leaked onto the glass.

While the caulk is setting, which takes about an hour, connect the camera and adjust its iris and focus. A small piece of duct tape will hold the settings in place. After the lens cap has dried, slide it onto the end of the pipe. It helps to coat the edge of the cap with petroleum jelly so that it slides more easily.

Mount the case on an L-bracket, and you are in business, as shown in Figure 1-20.

Figure 1-20. The finished enclosure

—*Don Marquardt*

Digitize Lots of Footage Quickly

If you have 100 hours of footage to capture and only two days to capture it, using more than one computer to transfer the footage will buy you more time.

Although digital video has promised the ability to transfer footage from tape to disk faster than real time, no editing systems have included the feature. To overcome a shortage of time, many people use multiple editing systems to transfer their footage from tape to disk. The problem is, unless you have a shared disk array and the software to share it among your computers, there's no way for one computer to "see" the footage digitized on another machine. Fortunately, you can work around this limitation by following a few easy steps.

Setting Up Your Systems

Unfortunately, to use this hack, you can't overcome the fact that you need more than one computer and an equal number of video cameras or decks. So, gather as many as you can and set them up just as you would if they were going to be used to edit. There are two approaches to configuring your systems, depending on how you plan to edit. Both involve how you name the drives where you will be digitizing your footage.

- If you have a lot of external hard drives and plan on using them to edit, make sure each drive is named differently. All of the drives will be attached to your editing system when you are finished, so you do not want conflicting drive names.
- If you do not have a lot of external drives, or if you plan on keeping your footage on a single drive (or array), give your drives the same name. Taking this approach will allow you to transfer your footage to a new drive, while tricking your editing system into reading the footage off the new drive.

Digitizing Your Footage

On average, the time it takes to digitize your footage is 1.25 to 1.50 times longer than the length of footage you have acquired. For example, one hour of footage takes up to one and a half hours to digitize. This average takes into account the physical act of loading a tape, rewinding it if necessary, organizing bins, dealing with timecode breaks, and so on.

Determining the number of systems needed. If you have 100 hours of footage, you can safely assume it will take 125 to 150 hours to digitize. Therefore, if you have only two days before you are scheduled to edit, you can gather three systems and more than likely be done in time. The math breaks down like this:

3 (systems) × 48 (hours) = 144 available digitizing hours

Of course, to utilize every hour in the day, you'll either need a lot of caffeine or someone to work a couple of graveyard shifts.

Determining the amount of storage required. When digitizing your footage, you need to make sure that you have enough drive space available. If you have 100 hours of footage and three available systems, you will need at least 34 hours of available space per system. Considering that one hour of DV footage takes approximately 13 GB of space, you will need about 450 GB of storage per system. The math breaks down like this:

100 (hours) ÷ 3 (systems) = 33.33 hours per system
33.33 (hours) × 13 (GB) = 433.29 GB per system

Plan on being able to move the drives you use to digitize to your editing system, or to have an equivalent amount of storage available on your editing system.

If you don't have enough drive space, you might be able to capture your footage using compression, such as the OfflineRT setting in Final Cut Pro or the 10:1 setting on Avid.

Running a final check. After you have configured your systems and checked that you have enough drive space, run a final check to ensure your footage will all be digitized as expected. Basically, make sure each system is set to digitize the footage with identical settings. For example, each system should be set to digitize NTSC DV footage at 29.97 frames per second and 48kHz two-channel audio. If your systems have different settings, your footage will more than likely be incompatible with each other.

After running your final checks, start digitizing.

Consolidating Your Footage

After you have captured all of your footage, you need to consolidate it to your editing system. There are a few steps involved to accomplish this, some of which depend on how your drives are set up. In a nutshell, you need to move the footage from your digitizing systems to your editing system and import the references to your digitized footage into your project.

Moving your footage. Depending on how you set up to digitize your footage, you have to do one of two things to move your footage. If you have elected to digitize your footage to differently named drives, you simply need to detach the drives from their respective systems and attach them to your editing system. If you have digitized to identically named drives, you have a couple steps to follow.

First, you need to copy the footage from your digitize systems to your editing system. You can do this by setting up a network connection between the machines and copying the footage, or by using a large portable drive and copying the footage from the digitize systems to the portable drive. Either way, you should initially name the drive you are *copying to* something different than the drive you are *copying from*. Only *after* you have copied all of your footage should you rename the drive to its originally intended name. For example, if your digitize drives are named AVD, then name the drive you are *transferring to* XFER. Once all the media is transferred, rename the XFER drive to AVD.

Linking your footage. After all of your footage has been moved from your digitize systems to your editing system, you need to be able to access it. Because the footage was captured on your digitizing systems, you should go back to them and obtain the necessary information. Again, you have two options:

- Copy the projects from your digitizing systems to your editing system.
- Export a digitize list from each digitizing system and import it to your editing system.

The preferred and easiest method is to copy the projects. You need to open the digitize projects on your editing system and copy the bins from your digitizing projects to your editing project. When you have finished copying the bins, you should close your digitize projects and keep your editing project open. When using this solution, your media should automatically link.

If you choose to export digitize lists, you need to import the lists into your editing project and relink your media.

Export a digitize list:

Avid
    File → Export → Avid Log Exchange

Final Cut
    File → Export → Batch List

Premiere
    Project → Export Batch List...

Import a digitize list:

Avid
: File → Import

Final Cut
: File → Import Batch List at 29.97 fps...

Premiere
: Project → Import Batch List...

Relink media (you might need to select the clips that are offline):

Avid
: Avid → Bin → Relink

Final Cut
: File → Reconnect Media

Premiere
: Project → Link Media

After moving, importing, and linking your footage, you will be able to start editing your project—sooner than if you had pursued digitizing with only one system. When you're against a submission deadline for a film festival, editing time is a precious commodity. At times like those, a few systems to help you start editing sooner can be priceless.

Hacking the Hack

If you have a set of tape logs [Hack #5], you can import them into your digitizing systems. The best approach is to break up the logs by importing *sets* of tapes into each system. This approach effectively breaks up your digitizing task into manageable pieces and keeps you organized (and less confused) during what is usually a hectic time.

People who are lucky enough to have a shared disk among their systems can still use part of this approach to transfer information from one system to another. Since digitizing information isn't shared between systems automatically, you must either open one project in another or copy the bins between the two.

HACK #14 Build Your Own Blue Screen

You can build a large—in this case, 24'×8' removable, disposable blue screen for under $30.

As a child in the late 70s, I was first introduced to the concept of using a *blue screen* in film and video while watching *The Making of Star Wars*. The compositing technique of filming action against a solid blue background, which could be replaced by any environment or other action filmed later, opens endless possibilities to the filmmaker. Today, off-the-shelf software provides consumers with the ability to put this technique to use in amateur films.

Understanding Your Needs

Professional-grade compositing material can be costly, but almost any solid color material can serve as a small blue screen. One popular solution is to use fluorescent green or blue poster board to create small backdrops suitable for close-ups, but what about larger scenes?

Building your own blue screen is actually easy, and the whole project can be completed within a couple hours. The finished screen constructed in this hack is about 24 feet wide (including the wraparound to a second wall) by 8 feet high. Being made of cheap, common materials, the screen is also disposable. The materials are lightweight, making it possible to build a portable version if needed. In fact, I produced a martial arts instructional DVD that actually employed a portable, green screen version of this setup.

I strongly recommend attempting to shoot a few shots against a small piece of material and testing how well you can composite with it using your lights [Hack #22], camera, and software [Hack #70] before spending the time and energy to build a large screen. Once you're confident that it will work, go ask your parent or spouse for permission to cover one or two walls with the stuff. Most of all, have fun!

Gathering the Materials

In order to construct your blue screen, you'll obviously need a set of materials. This will require a run to your local strip mall to pick up the following supplies:

- One blue plastic tablecloth roll, around 100'×40" (party supply store; $13)
- Nine 1'×3' pine furring strips (home improvement store; $10)
- One hardware-style stapler (the kind used for stapling insulation) with staples (home improvement store; $20)
- One saw (home improvement store; $20)
- One level (home improvement store; $3–$10)
- Optional for mounting on basement walls: one can of Liquid-Nails (home improvement store; $4)

You should make sure you get a parent or spouse-approved room for the project. My spouse-approved room was an unused storage room in the basement. This was good for me, because it also happens to be the largest room in the house! Figure 1-21 shows me assessing the concrete wall.

Figure 1-21. Assessing the concrete wall

The strips used to mount the blue screen are removable, but the Liquid-Nails will leave a mark on (or strip the surface of) concrete. If you screw the strips into drywall, the screws will leave holes and the strips are likely to leave a crease. If you're mounting on concrete, you'll need to choose your poison.

Mounting the Pine Strips

The blue plastic is thin enough to see through if it lies against the wall, so you might want to use the strips even if you're mounting on plain drywall. I couldn't really mount the plastic against the brick wall; plus, the wall would show through. The strips leave a gap between the plastic and the wall, and they make it easy to mount the plastic by just using staples.

Before mounting the strips to the wall (as shown in Figure 1-22), make sure you measure for fit and cut the strips to size. Also, use a level to mark off a horizontal line for each set of strips. Make sure to leave room for overlap at the center, so if the plastic is 40 inches wide, space the about strips 36 inches apart.

Finally, mount the strips along your markings. If you're mounting to drywall, you could use drywall screws to mount the strips to the wall. Just remember that they will leave marks. If the drywall is a solid, light color, you might be able to staple the plastic directly to the wall. The staples will leave only small holes, but there will be lots of them.

Figure 1-22. Carefully mounting the pine strips

I was mounting to cinder blocks, so I used a can of Liquid-Nails. Liquid-Nails is like caulk, but it becomes much harder and stronger. It bonds extremely well to brick, concrete, and wood. I applied the glue to the wall and the wood before mounting, which yielded good results. The glue takes 10 to 30 minutes to set, so you might want to have some extra wood around to lean against the strips to hold them in place while the glue dries.

Adding the Blue Screen

Now, we can finally mount the blue screen itself, as shown in Figure 1-23. Roll out the length of material needed and start stapling! Work across from one side to the other. Use three or four staples per foot to prevent tearing as you pull tension on the material. After going a foot or two on the top end, work the bottom end too.

The strips are mounted horizontally for two reasons. First, this creates a single seam, horizontally, in the finished screen. This is much easier to address in post-production than several vertical seams. Second, it's easier to work from side-to-side along two strips of material, than up-and-down and up-and-down for several shorter vertical strips.

Make sure you pull the material as tight as possible without tearing for each staple to keep the finished screen flat. I goofed when placing the pine strips, and they were too far apart to completely overlap the two pieces. To fill the gap, I just used some blue, painter's masking tape. Figure 1-24 shows the completed blue screen, along with the painter's tape to cover the goof.

Figure 1-23. Mounting the blue screen

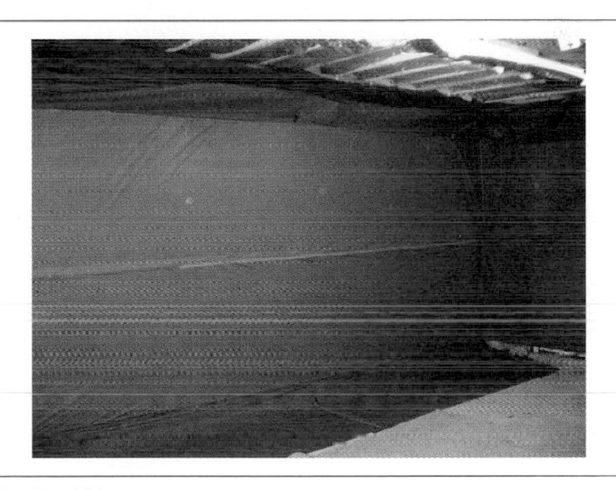

Figure 1-24. The final blue screen

Hopefully, you will learn from my mistake and measure it properly from the start.

Adding the Finishing Touches

To finish, I added some excess material to the top and bottom as well as to the adjacent wall. I did cover over the basement door, but the strips are screwed into the door frame rather than glued to the wall.

But does it work? Sure it does! Figures 1-25 and 1-26 show two examples of composites done using this blue screen room.

Figure 1-25. Shot on the home-built blue screen

Figure 1-26. A successfully composited image

If you would like to see some of the video created using this blue screen, you can check out the movie at the following URLs:

Windows Media
: *http://www.hiddenphantom.com/Scene3/03Share.WMV*

QuickTime
: *http://www.hiddenphantom.com/Scene3/03Share.mov*

You just have to love digital video technology!

—*Nick Jushchyshyn*

HACK #15 Stabilize Your Shots

Using a monopod, your video camera, and a 5- to 10-pound weight, you can create your own shot stabilizer.

If you have ever attempted to walk and record video at the same time, you have probably noticed the inevitable *jiggle* in you footage. In order to overcome this highly distracting result, a number of companies have brought products to market. Most notable is the SteadiCam, which is a professional solution to allow the cameraperson to move around freely while maintaining a smooth image. The problem is, SteadiCam gear is quite expensive and the knowledge of how to use it is not readily available.

Fortunately, you can couple a monopod and a standard weight (the type you might find in a gym) to create a solution to stabilize your shots. By attaching a weight to the bottom of a monopod, you will be able to counterbalance your camera when it is attached to the top. There are a great number of monopods available, but I prefer those by Bogen-Manfrotto (*http://www.bogenimaging.us/*) because you can really abuse them and the end cap is removable on certain models.

Assembling the Stabilizer

Once you have selected and purchased your monopod, you will want to attach a weight to the bottom of it. If your modopod has a removable end cap, you can remove the end-cap from the monopod, slide the weight onto the bottom, and reattach the end-cap. Figure 1-27 shows the result.

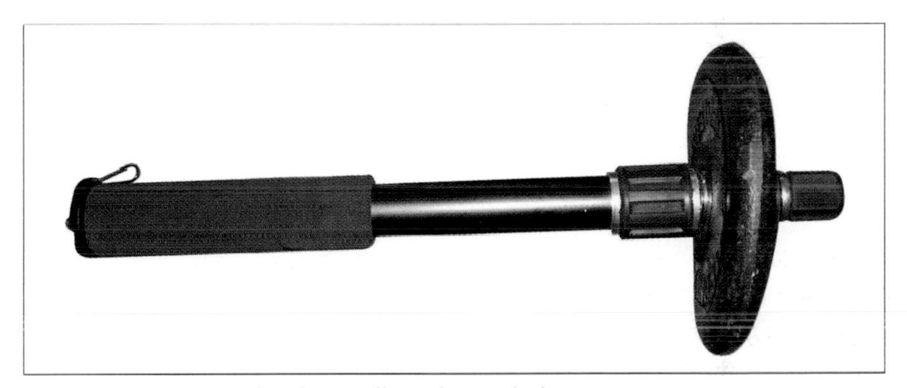

Figure 1-27. A monopod with a small weight attached

If your monopod doesn't have a removable cap on the bottom, then you can drill a hole through the bottom portion of the monopod, slide the weight on, and then place a bolt through your hole to keep the weight attached.

It is important that you attach the weight to the *bottom* of the monopod; otherwise, you will not counter the weight of your camera.

Once you have assembled your monopod and weight, place your camera on top of the monopod and secure it appropriately, as shown in Figure 1-28.

Using the Stabilizer

To use the stabilizer, lift and hold the monopod off the ground. As you walk around, your arm and shoulder will act as a type of shock absorber, allowing

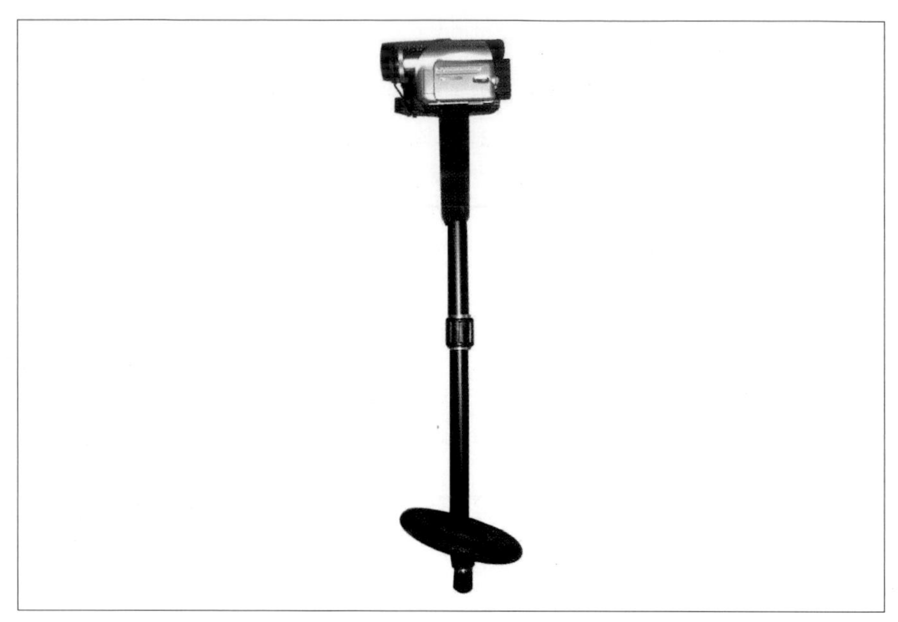

Figure 1-28. Counter-balanced monopod with camera

the shot to remain steady. You might even be able to jog alongside a subject while still capturing usable footage.

You should practice using this solution to understand how it affects the final look of your video. For example, try walking up a flight of stairs both while using the monopod and using just the camera by itself. You should notice a tremendous amount of difference between the two shots, but both could have their place in certain situations.

There are additional benefits of using the setup described here:

- You can use the monopod in the traditional manner, by resting the bottom of the rig on the ground. Doing so will allow you to maintain the "handheld" look of your video, while still steadying your shots.
- By extending or retracting the monopod, you can alter the counter-balance effect of the weight. This provides you the ability to refine the stabilization to your liking.
- You will find you get a bit of a workout if you use the monopod for an extended period of time. If you've got "chicken arms," make sure you switch off from right-to-left, so you can bulk up each arm equally.

Roll Your Own Dolly

A pair of wheels, and a way to roll around on them, can make a great dolly.

When attempting to capture a scene where your subjects are separated from the background, the camera needs to be moving. If you have ever tried to record footage while walking, you have probably noticed you get a very shaky shot. Your footage is probably the equivalent of *step-shake-step-shake*...

Most digital video cameras come with a feature to help stabilize your footage, which is especially useful for when you are standing in one place and holding the camera. When walking, however, you usually do not wind up with the shot you envisioned—even with the stabilization feature turned on.

Many professional camerapeople use a *dolly* to capture footage when they need to be moving. A dolly is basically a small, wheeled cart on which a cameraperson can sit while rolling along a given path. A variety of sporting goods, or even a wheelchair, can easily be substituted for the real thing.

Dusting Off Your Rollerblades

Many of us do not have the luxury of having a dolly, but we can make do with a pair of rollerblades and a trustworthy friend. Strapping on a pair of rollerblades and hitting the Record button on your video camera might seem like a simple feat, but there are a few caveats of which you should be aware.

Peter Smokler, an Emmy Award–winning cameraman, is known to throw on a pair of rollerblades, instead of using a dolly, to film some of his moving shots.

Trusting your friends. If you plan on skating and shooting, think long and hard about where you will be concentrating. Looking through the eyepiece of your camera will greatly restrict your vision. Even looking at an LCD monitor will probably distract you from noticing any obstacles in your path. Figure 1-29 shows a cameraperson on rollerblades being pulled backward by an assistant.

Skateboards, especially longboards, are good for low-angle shots. You can sit, lay, or simply attach the camera to one, in order to obtain the shot you want.

To be blunt, find a friend you can trust, and have her push, pull, and steer you around while you concentrate on the scene you are recording. Doing so will result in far fewer bruises, scraped knees, and broken bones. Whether

Figure 1-29. Rollerblades serving as a great dolly

you have a friend help you along, or you go it alone, make sure you use a broom and sweep the path you intend to take. A very small rock can cause a very big fall.

Moving backward. If your subjects are walking toward you, you will need to be skating away from them in order to keep them in frame and in focus. Because you will want to capture your subjects' faces, more often than not, you will be required to skate backward. Even though you will have your trusty friend pulling you from behind (you did read the previous paragraph, right?), you should be comfortable moving backward on skates.

Borrowing a Wheelchair

A wheelchair also makes a great dolly, because it is comfortable, stable, and requires no assembly. The use of a wheelchair as a dolly does not require much explanation: sit, point, shoot, and roll. As with rollerblades, it is best if you have someone push or pull you as you are shooting your scene.

You should avoid the temptation of a motorized wheelchair, because the motor's noise will most likely end up mixed in with your audio.

You will also find a wheelchair works especially well on more difficult surfaces, such as carpet, grass, and dirt.

Stealing from a Baby

If you have a child in the family, you more than likely also have a baby stroller. Depending on the type of stroller, you can alter it to hold your camera while you push the stroller along, recording your footage. Since each type and model of stroller is different, you will need to evaluate the situation to determine the best method to attach your camera. I have found jogging strollers to be the most versatile and easiest to work with. Figure 1-30 shows a Canon XL1 camera attached to a jogging stroller.

Figure 1-30. Using a jogging stroller as a dolly

There are a great number of other possibilities for creating your own dolly. Some people go so far as to create tracks out of PVC pipe to guide their dollies along a given path. Whatever your solution, you and your audience will appreciate the look a dolly shot provides.

CHAPTER TWO

Light

Hacks 17–25

Lighting is important to video, especially if you are trying to create a more professional look. Even though you can effectively light a scene "naturally" by using what's available, if you use specific techniques, you can bring out details (or avoid distractions) in your footage. Best of all, you can light most scenes for very little money.

HACK #17 Compile a Cheap Lighting Kit

Put together a bare-bones, super-cheap lighting package for digital video.

As most of us know, lighting is key to capturing a good-looking image. While preparing to shoot an independent film on location in Thailand and Cambodia, I realized that most of our gear had to be carried on our backs as we moved from location to location. I knew I was going to be lighting mostly with standard lights and using any available light source when possible, so I needed a portable and highly functional kit.

Gathering the Kit

Thailand and Cambodia both use 220-volt current, so I decided to buy most of my lighting elements in Bangkok. Before I left Los Angeles, however, I purchased the following items at a great place called The Expendables Recycler:

- 3/4 full roll of 12" black wrap (black tinfoil used to block out and shape light)
- CTO (color temperature orange gel)
- CTB (color temperature blue gel)
- ND (neutral density gel in 0.6 and 0.8)
- Full Blue (a vivid blue gel)

- Diffusion (both 216 and opal, with opal being more transparent)
- Gaffer tape in both black and white (a strong 1" cloth tape)
- A bag of clothespins (used to clip gels, diffusion, and black wrap to lights)

Here are the additional items I felt I needed to have in my kit, but didn't bring with me due to the difference in voltage:

- Two clamp lights with flexible necks (the kind you can clamp to your bedpost for reading)
- Two dimmers (to control the amount of light coming off the lights)
- A power strip
- About 100 feet of extension cords
- Additional dimmers to control any additional lights

You should be able to find all of these items at a home improvement store, such as Home Depot. Unfortunately, there wasn't a Home Depot in Bangkok, but I did manage to locate some in a small lighting shop. Figure 2-1 shows the lighting kit in its full glory. The blue containers on the left are extension cords (25 feet each).

I have found certain extension cords, designed where the cord coils up into the case of about 7 inches in diameter, are easy to transport and store. Everything in the kit can fit into a medium-sized backpack.

Using the Kit

Obviously, lighting is lighting and using a kit is a matter of placing lights as needed. But here's an example of the versatility of this kit.

During the shoot in Cambodia, I needed to light a shower that was located on the second story of a building. I was planning on beaming both clamp lights through a row of glass blocks running about shoulder high in the shower. The problem I had was that there was no way of clamping the lights to the wall outside the shower.

Gels are a tremendous help when lighting a scene. As a rule, you should remember that Color Temperature Orange (CTO) gels match tungsten lights, while Color Temperature Blue (CTB) gels match natural sunlight.

Placing the lights. When I looked around the location I noticed that there was an abundance of bamboo stalks lying on the ground. I decided that I would build myself a bamboo C-Stand (a stand to hold the lights) and clamp

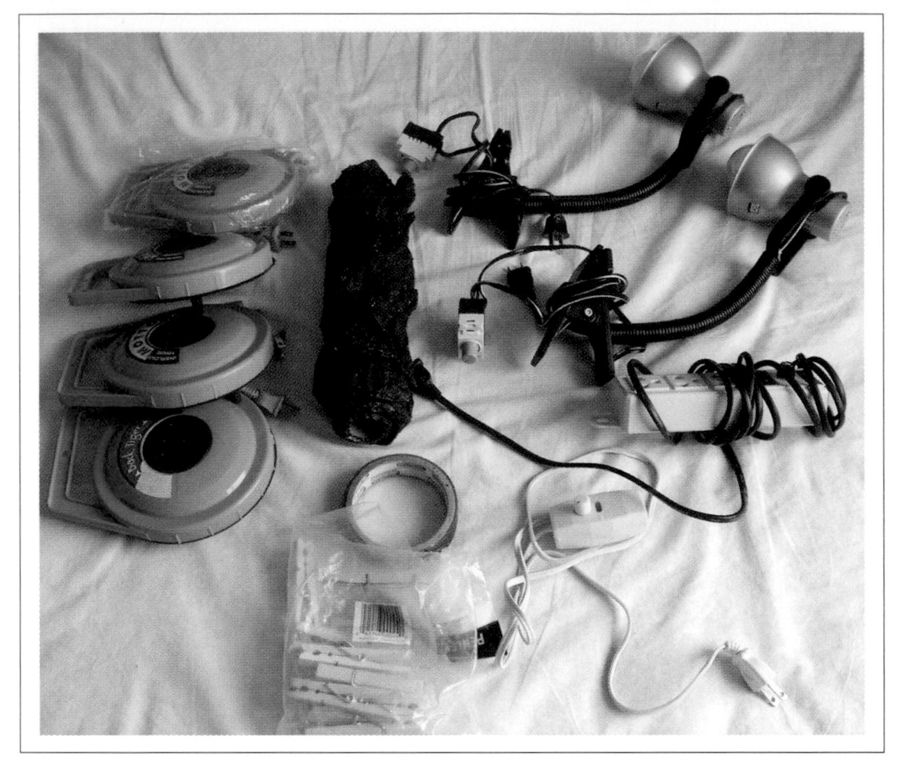

Figure 2-1. The lighting kit

the lights to the top of the stand. I found a solid-looking bamboo shaft about 10 feet long. About a foot down from the top of the shaft, I taped a three-foot bamboo cross member and clamped both lamps on left and right side of the cross member. I mounted my new C-Stand to the top of a nearby fence; the lights aligned just right with the glass. You just can't do that with the "professional" lights shown in Figure 2-2.

Using the dimmers. While I was setting up, it was late afternoon, but I was going to shoot the scene after sundown. I needed some way to control the level of light entering the shower without pulling the lights down every time and manipulating the dimmers that are hardwired to the lamp cord. This is where the extra dimmers come in handy, when you're placing lights that are out of reach but need to be adjusted periodically.

Staying Functional

This lighting kit is extremely versatile and functional, while staying affordable. Professional lighting kits usually start at $500 and go up dramatically from there. If I lose a light from my kit, it's no big deal.

Figure 2-2. Heavy-duty mounting required for professional lighting

The fact the kit is highly portable is simply another bonus. Shooting independent, low-budget (or no-budget!) movies often involves getting shots quickly and moving onto the next one. "Shoot and run" is often a reality in the truest sense, and being able to carry your lighting kit in a backpack is essential.

—Ilya Lyudmirsky

HACK #18 Light with Work Lights

"Work lights" are very bright lights, often 500 to 1000 watts, that are sold in home improvement stores. They make great video lights.

Unless you plan on a career that involves lighting, or you're really serious about making the next great independent movie, purchasing a professional set of lights just isn't worth the cost. If you're lucky enough to live somewhere where you can rent lighting packages, you always have that option. But if you don't, or if you're stuck somewhere between renting often (expensive) and purchasing (expensive), you might be looking for a way to light you video without a great expense.

Purchasing Work Lights

Practically every home improvement store sells very bright, portable lights, often called *work* or *shop* lights. The lights range in power, from small 50-watt lights to large 1000-watt lights. The range of power makes them a great choice for the aspiring moviemaker, because you can grab a few across the range. What's better is their price: you can often find them for less than $25 each. Figure 2-3 shows a 500W work light.

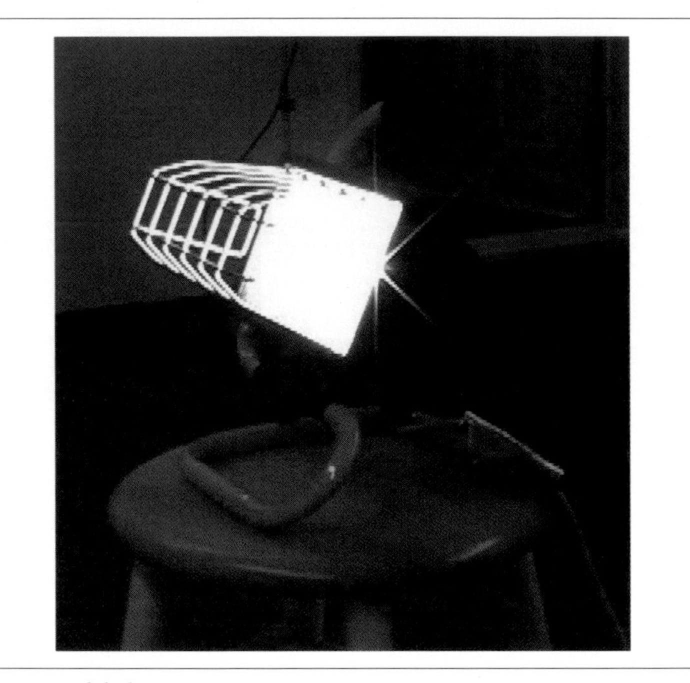

Figure 2-3. A work light

Using Work Lights

Using one really bright, powerful light, along with a small light kit **[Hack #17]**, can fulfill most of your needs. By using your bright light as your *key* light (the light most focused on your subject) and then using smaller lights as *fill* and *back* lights (lights to cut down shadows and balance out the key), you can effectively light a scene. Figure 2-4 shows a shot lit by a key light only, contrasted with one that uses key and a fill light. A key light can cause shadows; a fill can reduce them.

The technique of using a key light along with a fill light and a back light is called *three-point lighting*. Your key light should always be brighter than your other lights; a good rule of thumb is that it should be twice as bright.

Figure 2-4. A shot lit by a key light only (left) and both a key and a fill light (right)

Both your key and fill lights should be aimed from the same direction as your camera, somewhat opposite of each other and slightly above your subject. Your back light, as its name implies, should be aimed from behind and slightly above the subject you are shooting. Figure 2-5 shows a diagram of a three-point lighting scenario.

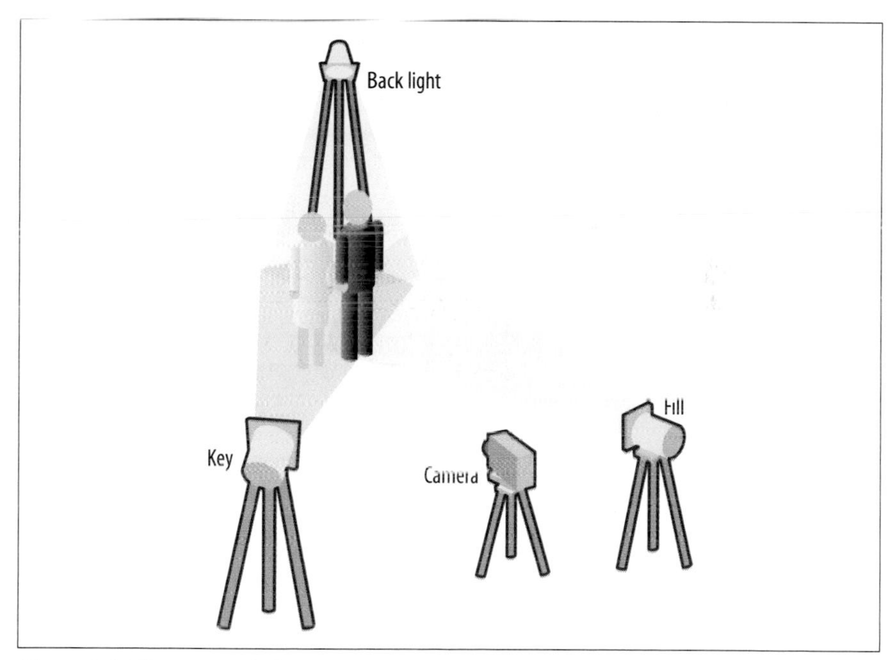

Figure 2-5. Three-point lighting

Controlling Work Lights

Work lights are cheap, functional, and work great for lighting video. The problem is, sometimes they are simply too bright. One solution to the problem is to attach a dimmer and bring down the intensity of the light. Another is to use diffusion. As a bonus, the two solutions can be applied together.

Using dimmers. Dimmers can be purchased at home improvement stores, lighting stores, and even some supermarkets. Use one dimmer per light, to gain the most control over the entire look of your scene. When purchasing a dimmer, be aware of the maximum wattage it can handle.

Using diffusion. Many photo and film supply stores sell sheets and rolls of diffusion. These sheets are often slightly transparent, white sheets of plastic that are very heat resistant. When diffusion is placed in front of a light, it *softens* the light.

A cheap alternative to purchasing diffusion is parchment paper (found at supermarkets; for $3–$5 per roll), which is used by people who bake. The paper is often white and slightly translucent—perfect for diffusion. Additionally, because it is designed to withstand heat, you can use with high-wattage lights. Figure 2-6 shows a 500W work light with parchment paper attached as diffusion.

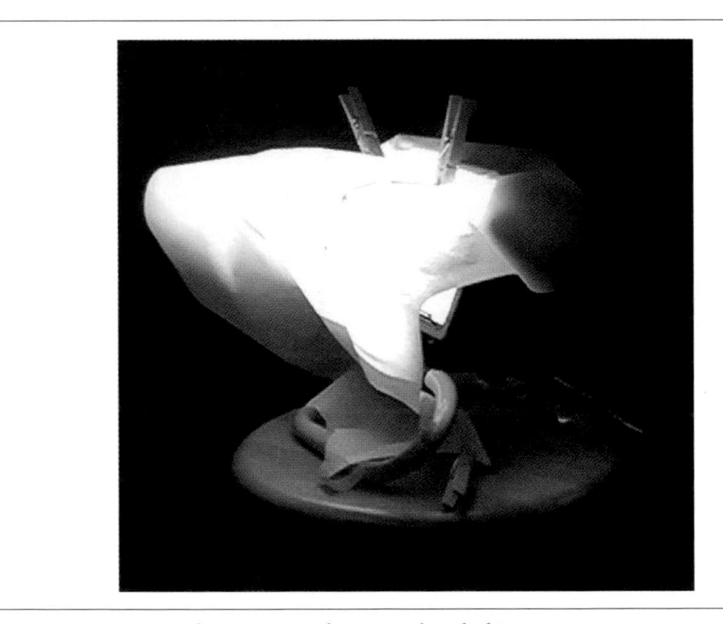

Figure 2-6. Using parchment paper for more than baking

As you can infer, work lights can be a great help in creating a more professional look to your video. But there's something else: they're designed for use in tough working environments. Because of their low cost (in comparison to professional lights) and their durability, work lights should be on your short list when shopping for video lighting supplies.

HACK #19 Use Paper Lamps for Lighting

Cheap paper lamps can be used to effectively light a scene.

Lighting a scene can make a dramatic difference in the look of your video. The problem for most people is that professional lighting kits are expensive, difficult to use, and not easy to transport. Fortunately, a few low-cost paper lamps can help bring a more professional look to your video. When purchased in sets of varying sizes, they are a perfect complement to a cheap, home-assembled lighting kit [Hack #17].

Asian paper lamps are often spherical in shape and come in a variety of sizes, up to 30 inches in diameter. They can be purchased from numerous places online, some home improvement stores, and even lighting stores. They cost anywhere from less than $3 to slightly more than $30, depending on the size and quality of the lamp. When illuminated from inside, paper lamps cast a soft, diffuse light. Figure 2-7 shows a frame of video without lighting and one lit by paper lamp (with a 100W bulb).

Figure 2-7. Lighting without (left) and with (right) a paper lamp

Most paper lamps do not include a cord and socket for a light bulb. Such assemblies cost around $5 and are almost always available when purchasing paper lamps. They practically go hand-in-hand and the distributors know this fact. An additional benefit of the cord/socket assembly shown in Figure 2-8 is that you can use it by itself, as well.

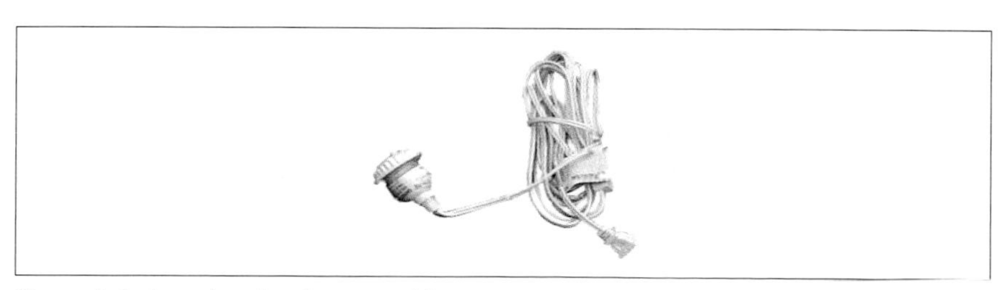

Figure 2-8. A cord and socket assembly

For additional control over the light, you can attach a dimmer. You can also use higher wattage bulbs or even use a colored bulb for effect. If you are trying to light an actor who is moving around, you can attach the light to a long pole and hang the light above him (much like a microphone boom **[Hack #53]**). In the end, paper lamps are highly functional, easy to transport, and best of all, cheap.

Add Diffusion to a Camcorder Light

Lighting is an important aspect of capturing good video images, but you need some control over it.

When shooting *on the go*—during a wedding reception, a nightclub party, or for a documentary, for example—it helps to have portable lighting. Many companies offer lights that attach to your camera and draw power from it. However, these lights offer little control, if any, over the beam of light provided.

Camcorder lights are usually very small, yet very bright, lights that sit atop your camera. Although some lights provide some control over the light, through the use of *barndoors*, most do not provide any control. It's light or no light.

Barndoors usually consist of four metal flags attached to the sides of the light, and are common on professional lights. By closing a flag, partially or completely, you can change the amount of light emitted. Barndoors still don't solve the problem of the sometimes harsh light emitted.

Herbert Wetherford, an independent producer in Los Angeles, introduced me to a trick he uses when shooting independent movies. Placing Scotch tape in front of the light will *knock down* the light, as well as soften it. You effectively diffuse the light with just a few strips of tape. Figure 2-9 shows the results of this effect.

Figure 2-9. Adding a little scotch tape to provide the needed diffusion

Scotch tape is not fireproof. Do not place it in close proximity to the light. If you have to fashion something to hold the tape off of the light, do it. The tape can, and will, melt.

If you need to, you can use clothespins **[Hack #25]** to position the tape away from the light.

Adding diffusion to a light causes it to become *softer*, thereby reducing harsh shadows and casting light evenly across an area. If you have control over your environment, using paper lanterns **[Hack #19]** or some work lights with diffusion **[Hack #18]** can provide excellent, soft light. But when you're on the move and can't take time to light a scene, a camcorder light is your best option.

Take Video in Total Darkness

All sorts of interesting things happen in the dark, but how do you take video of them? These tips will help you take video in even the darkest places.

It could be a mystery animal rustling in your garden or the phantom cookie thief who keeps sneaking into the kitchen in the dark of night. Whatever it is, you want to take some video for proof. But how do you do this without turning on the lights? Neither the mystery animal nor the cookie thief is likely to come out if you leave all the lights on and your camcorder running.

It might seem like an insurmountable problem, but there are a couple things you can do to take video in poorly lit or dark places. You can either make the most of the light you have, or you can add more light that only your camcorder can see.

Using Slow Shutter Speeds

By slowing down the shutter speed of your camera, you give it more time to collect the available light, making more of the light that is out there. In many situations, such as under streetlights or other indirect lighting, using this slower shutter speed mode can be enough to produce usable video, and most camcorders offer a special Slow Shutter Speed mode. The resulting video can often be rather jerky and blurry, though; much like a still camera, a slower shutter speed exaggerates both camera shake and the movement of the things that you are capturing. Figure 2-10 is a frame from a video captured using the Color Slow Shutter Speed mode of a Sony camcorder in low light.

Figure 2-10. A blurry image captured with a slow shutter speed

You can't do much about the blurring from moving objects, but you can minimize the camera shake by mounting the camcorder on a tripod or holding the camcorder in both hands and leaning against a wall or building.

Adding More (Invisible) Light

The other trick is to add more light. While this might seem illogical, there are types of light that your camcorder can see but that you (or your subjects) can't. Specifically, your camcorder can see infrared (IR) light, which is invisible to the human eye. Most animals can't see it either, or they aren't bothered by it.

To see what I mean, try this: take a remote control, point the end of the remote toward your eye, and press a button. You won't see anything, because your eyes can't see the IR light emitted by the remote. Now try the same thing, but pointing the end of the remote at your camcorder lens. On your camcorder screen, you'll see a brief flash of light from the end of the remote, because the camcorder can see the IR light. And this is what we can use to capture video in situations where there isn't enough light to record normal video: by using IR light, we can illuminate the subject without annoying or distracting it.

In fact, this hack has been adopted by some of the camcorder manufacturers themselves. Many camcorders (such as most Sony models) have a Night Vision mode that uses a built-in IR light emitting diode (LED) light (Sony calls this mode Super NightShot, while Panasonic calls it 0Lux MagicPix). But this single IR LED isn't all that powerful: it can illuminate things only to a few feet away. This is fine if you are filming another Blair Witch knockoff (and please don't; they may have been funny the first five times, but not anymore), but it's not much use for things that are further away.

If you plan on recording anything worthwhile, you'll need something a bit brighter, such as an IR illuminator. Some manufacturers also produce their own illuminators: Sony makes one called the HVL-IRH2 that works with any of their NightShot camcorders and also doubles as a conventional light. However, this only works when connected to the camcorder, and it uses the camcorder's battery. The fact that the illuminator is separate from the camcorder means that you can place it near the subject and leave it, while remaining at a distance with the camcorder. Figure 2-11 shows an IR illuminator purchased on eBay.

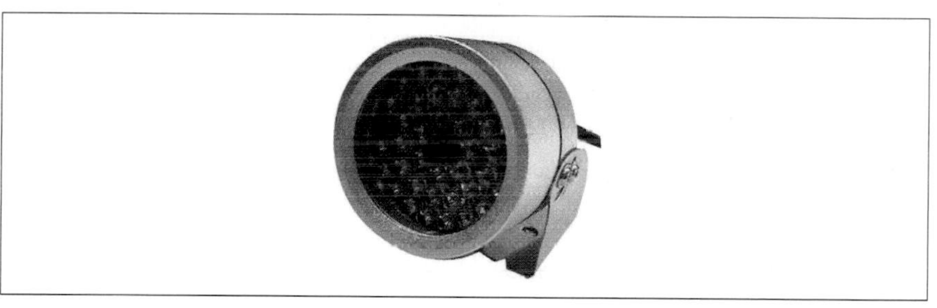

Figure 2-11. A 39 LED IR illuminator

The downside of using IR mode is that your video turns to black and white, because the camcorder is only capturing the IR light, not the full spectrum of color light. This does make your video look rather like the night vision goggles that the military uses. Figure 2-12 shows light emitting from the illuminator; notice the lack of color.

Getting the Best Results

If you are shooting through a window or have other reflective surfaces nearby, turn off the built-in IR light of the camcorder to avoid reflections. And don't get confused if you turn on the IR illuminator and you don't see any light coming from it: remember that it emits infrared light that you can't see, so the most you'll see with the naked eye is a very dim red light.

These techniques can be useful for keeping an eye on things in low light or total darkness. For example, my wife and I work with feral cats, and many of them don't like coming out when people are nearby. So, we use the IR illuminator and camcorder in night vision mode to watch them without disturbing them. Figure 2-13 is a frame of video taken with a Sony camcorder in very low light, using the Super NightShot Plus mode and the IR illuminator.

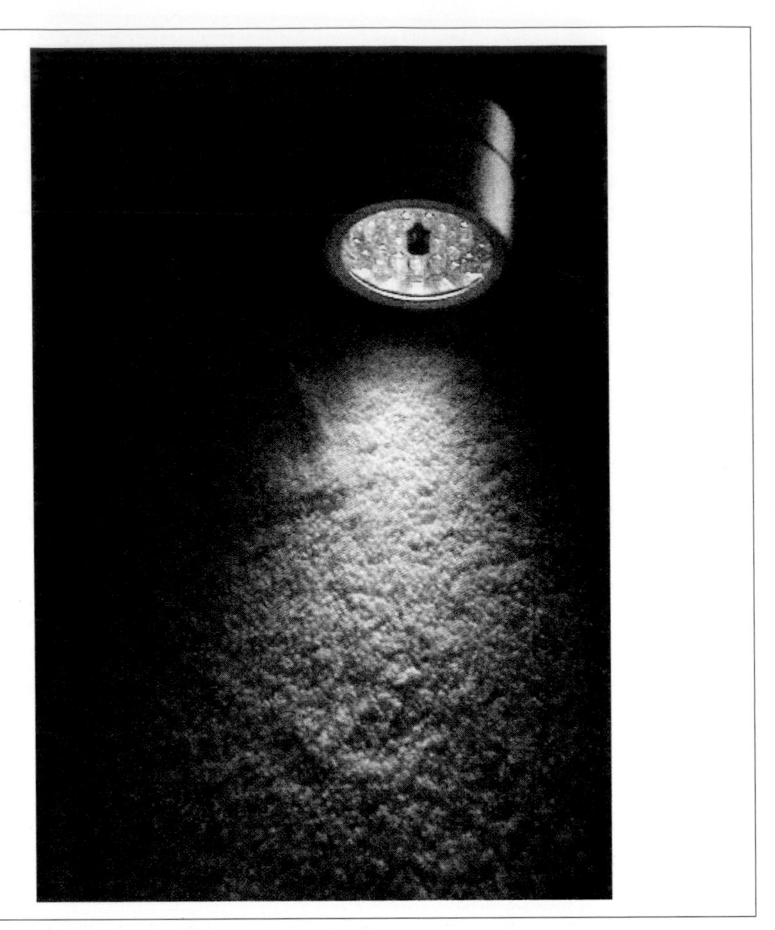

Figure 2-12. IR light captured using a digital camera

Figure 2-13. The ability to see at night, using IR light

If you're trying to video that elusive animal in the bushes, the moon can be your friend. Time your shooting with the full moon, because a full moon can provide enough light to take video if you use the night vision mode on your camcorder. You can check the phases of the moon at *http://aa.usno.navy.mil/idltemp/current_moon.html*.

Shoot a Green Screen Image

Successfully shooting a green screen (or blue screen) requires attention to detail.

Creating a believable composite image requires you to capture a usable green screen image. Although most software applications are capable of dealing with poorly acquired footage, you shouldn't count on your application's ability. Simply put, if you shoot a bad green screen, don't be surprised if you wind up with a bad composite. Garbage in, garbage out.

For the purpose of this hack, *green screen* refers to any background that consists of a single, solid color, whether it's green, blue, or some other color you decide to use. When I refer to *green*, you can substitute whatever color you are using.

Locating a Green Screen

If you're lucky, you'll be able to find and rent a stage with a complete green screen *cyclorama*. A cyclorama is often referred to as a *cyc* (pronounced "syke"), so if you are going to call around for a stage, you should use that terminology. Most of us, however, aren't lucky enough to either live near a stage or afford one.

Fortunately, there are a number of methods you can use to create your own green screen. If you have a room or office you can use temporarily, you can assemble your own green screen using home-improvement supplies [Hack #14]. Or, if you don't mind something semipermanent, you can paint a wall a solid color, being careful to keep the color consistent across the wall. Finally, there is always the option of simply using a large section of green fabric and hanging it from a rod.

Many amateurs and professionals shop for green screen materials at Film Tools (*http://shop.store.yahoo.com/cinemasupplies/*) in Burbank, California. Film Tools carries a wide range of video- and film-related items, from director's chairs to video carts.

Lighting a Green Screen

The primary key to capturing a good green screen image is the lighting. If you light your green screen poorly, you'll spend time in post-production trying to fix it. Save yourself the headache and just do it right while you have the chance.

Your actors should not wear anything green. When you go to matte out the color green from your scene, you don't want to matte out portions of your actors! This same principal applies to your props, as well.

Lighting the background. When lighting, concentrate on lighting the green screen first. This means you should not have any of your actors or props within the area of the green screen. You will want to light the green screen evenly, meaning you should light to remove any shadows, light or dark spots, or glare. You might also find it useful to use green gels over your lights to help even the hue across your green screen. Figure 2-14 shows a poorly lit green screen, opposed with a well-lit one. The goal is to have an even color across the screen.

Figure 2-14. A poorly lit green screen (left) and a well-lit one (right)

Lighting the subject. Only after lighting your green screen should you bring in your actors and props. Light your subject just as you would normally, keeping in mind the background you intend to use in the composition. While lighting your subject, make sure no shadows are cast onto your green screen, and if they are, remove them (by adjusting or adding lighting).

Also be aware of any *spill* occurring from the green screen onto your subject. Spill refers to the reflection of green off of an object. You can remove spill by, yes, more lighting. So, if you notice a green hue coming off of your lead actor, attempt to reduce it by lighting your lead actor from behind. Additionally, gels can be helpful when trying to reduce the occurrence of spill, as they can counteract the reflection.

Shooting a Green Screen

Although shooting a green screen seems easy enough, I recommend practicing a few times before taking other people's time to shoot your scenes. There is a lot you can learn by recording just a prop in front of a green screen and then attempting to composite it onto a background. You might even want to practice on a smaller scale, just to get a feel for what it takes to light and composite a scene successfully.

When shooting, you should keep your camera in a static position. If you plan on having a moving background, it will be extremely difficult to match the camera movement to the movement of your background (if not impossible). Even if you're able to view your composite in real time, it will take a *long* time for you to get the camera movement to match the background in a believable manner.

I once worked with a director who insisted on using a moving background for a shot. Even though we were able to composite the green screen and background in real time, it took us 43 takes to get one shot that was somewhat usable in the final project. Unless it is truly needed for your story, stick with a static background and save yourself the headache.

If you have lit your green screen and your subject well, then actually recording the footage will be as easy as recording anything else. When you're finished and you go to composite your footage, it should be almost effortless. The less time you spend working with your composite, the more time you'll have to make it look believable by adding shadows [Hack #71] and editing your scene.

Shoot Clearly Through a Window

When shooting through a window, images are often blurred or overexposed. But they don't need to be.

Sunlight provides great natural light. But when you point your camera toward a window during the day, you'll notice the view outside is blurry and possibly even *blown out*. By covering the windows you are shooting toward with a neutral density (ND) gel, you can capture sharp images outside the window while keeping your subject in focus.

Gels are not cheap, considering they are small sheets of colored plastic. Small sheets (21" × 24"), primarily used to affect lights, can be purchased for $5–$10 from most photography or film supply stores, such as Film Tools (*http://www.filmtools.com*) or B&H Photo (*http://www.bhphotovideo.com*). Larger rolls (4' × 25') can cost anywhere between $100 and $200. Fortunately, gels are fairly durable, when not abused, and can last a long time.

When covering a window, you should place the gel inside the window. In addition to keeping the gel free of debris, you will easily be able to make quick adjustments, should the need arise. Figure 2-15 shows a neutral density gel in a standard 21" × 24" sheet.

Figure 2-15. A standard size sheet of neutral density gel

Depending on the physical aspects of the window, you might be able to clip the gel to hold it in place **[Hack #25]**. If you can't clip your gel, you can resort to taping your gel in place. You should consider using a special adhesive tape made by Permacel, commonly called *gaffer tape* or *camera tape* (depending on the width). What makes gaffer tape special is that it doesn't leave a sticky residue when it's removed. This is especially important when you're shooting in a location that you're "borrowing," because you probably want cause the least amount of damage possible.

After placing your gel, you can go ahead and shoot your scene with a clear view of the outside world. Figure 2-16 shows a window half-covered with an ND gel to demonstrate the difference realized by using a gel.

Although your viewers will never notice the gel or the difference it has made (because they have nothing to compare it to), you will have made a difference in their experience of your scene.

Figure 2-16. Shooting through a window using a neutral density gel

HACK #24 Reflect Light from a Shade

Improve your lighting by using a car windshield shade.

When lighting a scene, every so often you need to *bounce* light in order to overcome slight shadows. Professionals often use *bounce boards*, which are large sections of heavy, white paperboard, or light discs, which are flexible frames, covered with a reflective material. Luckily, both are made of materials similar to common windshield shades.

Cardboard shades are the cheaper of the two types of shades and can usually be purchased for $5.00 or less. They are perfectly good substitutes for bounce boards, especially when they are white. A nice feature is that they can be folded. This allows you to alter the size of the shade and therefore the amount of light reflection it provides.

Fabric shades are almost identical to professional light discs, except that they are designed to fit a car windshield. Because of this, they are much more affordable. Figure 2-17 shows a reflective fabric windshield shade.

To use the shade for additional light, angle it so that it faces the subject you would like to light. If your subject has a few shadows on his face, attempt to reflect light off of the shade up and onto his face, in order to *fill* the shadows. This requires you to hold the shade at about waist level and angled toward the subject. Figure 2-18 demonstrates the result of using a shade as a reflector.

If you already own a windshield shade, or can borrow one, then you don't need to hunt one down. Otherwise, you will probably have to make a run out to the local auto supply store, or purchase one online. Some grocery stores also carry them. However you obtain one, a windshield shade can

Figure 2-17. A windshield shade made of reflective fabric

Figure 2-18. Reflecting light from the left side to help reduce shadows

possibly make the little difference needed to light your shot, while pulling double-duty to keep your car cool.

Use Clothespins Like a Professional

Clothespins are used throughout the television and film industry to aid in lighting scenarios.

Seeing clothespins on a television or film set is so common that it's easy to forget that they are intended to hang clothes. In fact, in the film industry they are often referred to as *C47s*, instead of *clothespins*. Figure 2-19 shows a bucket full of clothespins, waiting to be used as C47s.

Figure 2-19. Clothespins, or C47s as they're known "in the biz"

There are a few explanations why clothespins are referred to as C47s, but no one knows for sure. Some say that C47 was the part number used by the military, others claim it was a bucket number where they were stored in a local lighting shop, and still others claim it's an old accounting code. We'll probably never know, so at this point, you can make up your own story . . .

Lighting people most commonly use clothespins to attach gels to lights. When near hot lights for an extended period of time, the wood doesn't retain the heat in the same way as metal clips do. Additionally, although they might burn, they won't catch fire.

To use a C47 for lighting, simply place the gel or diffusion near the light you would like to effect. Then, clip a few C47s to hold the gel or diffusion in place, as shown in Figure 2-20.

Be careful not to put gels too close to a light source, because they can melt.

Figure 2-20. A blue gel attached to a light using clothespins

In addition to lighting people, production assistants use C47s to keep papers together, wardrobe people use them to alter clothing when time is of the essence, and producers...well producers pay for them.

CHAPTER THREE

Acquire

Hacks 26–38

Every aspect of creating a video revolves around the footage, so acquiring it is quite obviously the most important element. Without footage, you can't edit or distribute a video. The decisions you have to make determine *what* you want to acquire and *how* you should do so. The hacks in this chapter will help you collect footage from various sources, shoot footage in unusual ways, and help you bring all of your footage (no matter where it originates) together.

Create a Time-Lapse Video of a Sunset

Compressing a long event into a few seconds can create a wonderful transition.

You have probably seen a time-lapse video in educational movies, such as the ones that show a rose as it progresses from a bud, to full bloom, to its leaves falling off. This type of imagery can be quite powerful. Unlike film cameras, most digital video cameras do not allow you to change the frame rate at which they capture.

Although the technique somewhat limited, this hack should provide an effect similar to what you can accomplish with a film camera, at a fraction of the cost.

If your camera has the ability to capture time-lapse sequences, consider yourself lucky. However, you can still experiment using the following information to fine-tune your final sequence.

Researching Your Shot

When you plan on capturing a sunset, or any time-lapse sequence, you should determine an ideal location at which to shoot. The following points might be obvious, but I am compelled to mention them anyway:

- Limit the possibility of someone walking in front of the camera.
- Locate an area with minimal vibration from passing traffic.
- Make sure you are able to stay in the location (i.e., get permission).
- Check your local weather report.

Many DV cameras will record only one speed, so you might have only 60 minutes of coverage. Therefore, you should know what time sunset will occur and begin recording about 30 to 40 minutes beforehand. If your camera has an LP option, you should be able to capture 120 minutes of footage. Either way, you should plan on being at your location at least an hour ahead of when you plan on pressing that little Record button.

Setting Up Your Shot

In order to capture the sunset, you want to place your camera on a stable surface; a tripod is the most effective option. In order to discourage people from walking in front of your camera and ruining your shot, you might want to rope off the area you are going to shoot. You don't want anyone, or anything, coming in between your camera's lens and the scene you want to capture. You should also make your best effort to not move the camera after you've set your shot.

Once you have your shot set up, start recording. Then, sit...and...wait...

If all goes well, you will have a long, boring shot of a sunset. You can use this footage for a cool time-lapse transition in your movie or for when you have guests over and you would like them to leave: "Hey, do you guys want to see the sunset I shot?"

Making Time Fly

As usual, digitize the footage into your editing system. Time lapse is essentially the compression of time, so you will need to add a speed effect to your footage. You should experiment with the speed setting to accomplish to final look you prefer.

If you are using NTSC footage, you should de-interlace your footage before changing the speed settings. Some editing systems will perform this function for you when altering the speed of footage. If you or your editing system do not de-interlace the footage, you might find your footage to be jittery.

Here's how to change the speed of your footage in some popular editing systems:

Avid
: Click the Motion Effect button and then check Variable Speed.

Final Cut
: Choose Modify → Speed...

Movie Maker
: Choose Add Video → Speed Up X2.

Premiere
: Choose Clip → Speed/Duration.

iMovie
: Select the timeline viewer and adjust the speed using the Faster/Slower slider.

Some systems require you to apply the speed effect to your footage multiple times until you reach the desired effect. Others allow you to speed up your footage by entering a percentage.

After you have applied your settings, render your sequence. Voilà! You have your desired time-lapse effect; a sample is shown in Figure 3-1.

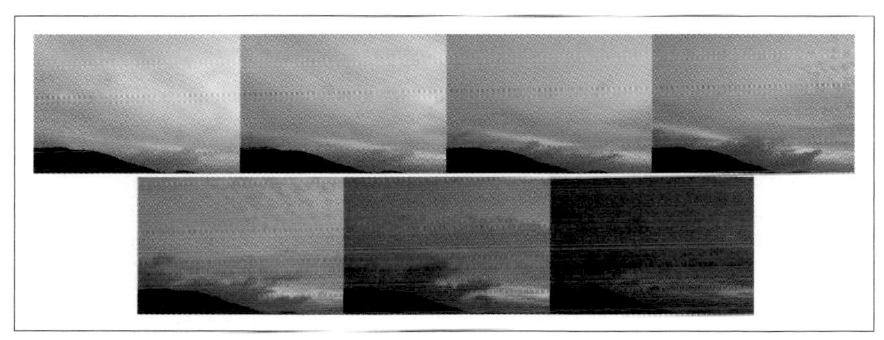

Figure 3-1. Frames of video from a sunset

Using Stop-Motion Software

Stop-motion movies are used to assemble a series of images, taken at intervals greater than 1/24 second, into what appears to be motion. People usually use stop-motion techniques to create claymation or brick movies [Hack #91]. However, the same technique can also be used to create great time-lapse sequences.

A tremendous advantage of using a software solution is that you can capture images over a much longer period of time, because you will be limited only by available disk space. This means you can create time-lapse sequences of subjects such as plants growing or even of the seasons changing.

A couple companies have created software to aid in the creation of stop-motion and time-lapse movies:

Boinx
: Publishers of iStopMotion for Mac OS X (*http://www.istopmotion.com*; $39.95)

Stop Motion Pro
: Publishers of Stop Motion Pro for Windows 98 SE or ME, XP, and 2000 (*http://www.stopmotionpro.com*; $249)

Both of these applications are available in demo versions and both provide a tutorial for creating time-lapse movies. To use the applications, you need to attach your video camera to your computer using an IEEE-1394 (a.k.a. FireWire and i.LINK) cable. Once attached, run your motion-control software, configure it appropriately, and create your magic!

Mount Your Camera to Your Car

There are some elaborate rigs for attaching a camera to a car. But when it comes to "build versus buy," one product makes building a solution a tough sell.

There are certain shots that are difficult to acquire, particularly those that are shot from a moving vehicle. Instead of building some elaborate apparatus to mount your camera to a car, you can purchase a heavy-duty suction cup, with camera mount, called The Cleat. It is manufactured by PowrGrip (*http://www.powrgrip.com*) and can be purchased online at *http://shop.store.yahoo.com/cinemasupplies/cleatsuccamm.html* for $75. Using The Cleat, and a camera that weighs less than 10 pounds, you can capture great shots from a moving vehicle, or from just about anywhere there's a smooth surface.

Mounting The Cleat

Essentially, The Cleat is a well-designed suction cup, with a particular set of uses in mind. To attach it to a surface requires the surface to be smooth and nonporous. The area you are attaching to should also be clean.

When you have located and prepared the area on which you want to mount The Cleat, place it and then pump the grey plunger. Continue pumping until the red indicator no longer shows. The plunger is spring-balanced, so you can expect a little resistance while using it.

Quite possibly the best feature of The Cleat is the red indicator on the plunger. When you can see the red indicator, it means that your suction is weak. To make it stronger, you simply need to give the plunger a few good pumps. Figure 3-2 shows The Cleat without a good vacuum seal, while Figure 3-3 demonstrates a good seal. If you see red, don't attach your camera.

Figure 3-2. A bad seal

Knowing that you have a good vacuum seal will provide you peace of mind that your expensive digital video camera isn't going to take a nasty tumble.

Figure 3-3. A good seal

Attaching Your Camera

After mounting The Cleat, attach your camera using the threaded screw on the top of the unit. For security, you should also tighten the knob just below where you've mounted your camera. You can then proceed to adjust the arm of The Cleat, so that you can frame your shot as you'd like to record it. Figure 3-4 shows a camera mounted to a car using The Cleat.

Recording Your Shot

How you physically get your camera to record will depend on the make and model of your camera. If you have a remote control, the process will be easy. If do not have a remote control, you will need to start your camera manually. This will probably require you to capture more footage than required, but it shouldn't cause too many problems. Figure 3-5 shows a frame of video, as captured from the car-mounted camera and then rotated **[Hack #42]**.

If you are not taking part in the scene, or driving the vehicle, you should attempt to view what's being recorded **[Hack #11]**. If you are shooting the interior of the car, you might be able to hide in the back seat and provide direction from there. Otherwise, you can attempt to do so from another car using wireless communication.

Figure 3-4. A camera mounted to a car

Figure 3-5. Video captured at 35mph

Removing the Camera

When you have completed your shot, first remove your camera from the mount. This is purely to be safe and ensure your camera doesn't topple when you release the suction cup's vacuum. Then, to remove The Cleat, all you need to do is firmly pull on the tab on the suction cup. Figure 3-6 shows the pull tab on the suction cup.

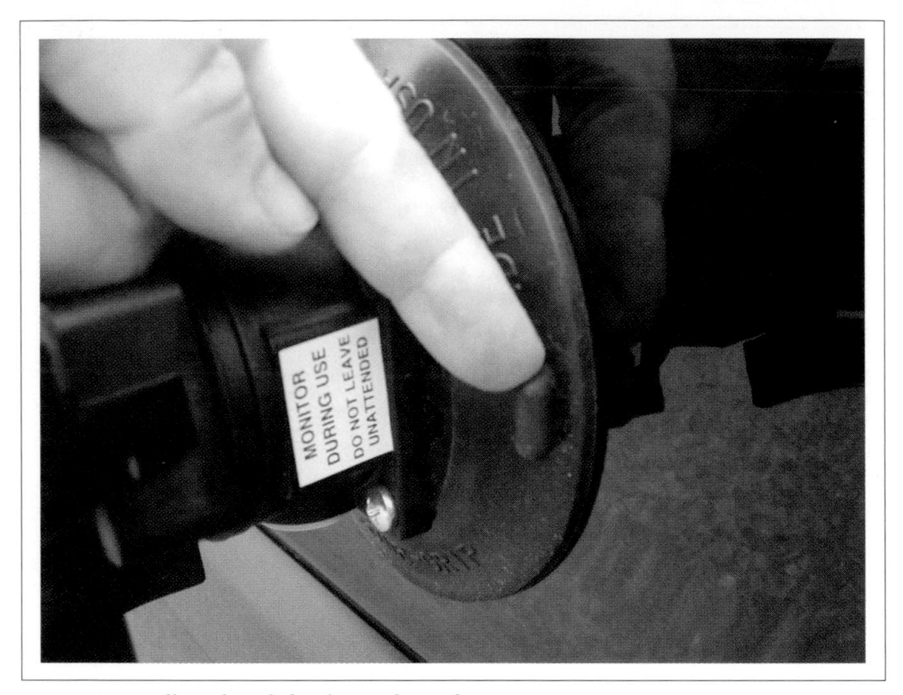

Figure 3-6. Pull on the tab firmly to release the vacuum

Hacking the Hack

Due to it's small size, The Cleat can be used much like a tripod. Additionally, it can be mounted in unusual or difficult locations, providing for similarly unusual shots. For example, you could mount your camera to the outside of a second story window and shoot into your house.

You can also use multiple mounts and shoot a multicamera scene using a limited crew. Imaging trying to shoot using three cameras inside a car *without* The Cleat; it would be quite difficult. There are a vast amount of opportunities for creative use, and by purchasing The Cleat, you'll free up a lot of time to discover them.

Slate Your Cuts

Visually track each version of your video

Every editing system includes a way to create text and place it on screen. You can use this feature to add a visual tracking system to your cuts. Much like you use a slate with raw footage [Hack #9], you can use your system's ability to add titles.

A slate is easy to create and takes only a couple of minutes. Yet, many people don't take the small amount of time to do so, only to discover they are completely confused, after a few weeks of editing, as to which tape has which cut. To build a slate, you need to place text on screen.

Avid
: Clip → New Title...

Final Cut
: Effects tab → Video Generators → Text → Text

iMovie
: Titles → Centered → Centered Multiple

Movie Maker
: Edit Movie → Make titles or credits → title at the beginning

Premiere
: File → New → Title...

Once you are able to type, add the information that is pertinent to your project and save the slate. At a minimum, you should include the project's title and the date of the cut. You might also want to include the version of the cut (e.g., Cut 4), the total running time (TRT), the audio configuration (e.g., Channels 1 & 2—Stereo Mix), and the editor's name.

Some editors and producers like to add an artistic flare to their slates. If that's your cup of tea, feel free to go wild and express yourself. Just don't forget to include the necessary information. Figure 3-7 shows a basic slate over a black background, as well as a more creative version.

After finishing your slate, place it at the head of your timeline. Although some editing systems allow you to add a slate on output, I recommend placing the slate, as it provides a secondary means of determining the version of a cut (the first being the title of the timeline). Figure 3-8 shows an Avid timeline with a slate at the head.

If you get in the habit of slating every cut you output from your editing system, as well as creating a copy of the timeline associated with it, you will always have the opportunity to go back to any cut of a project.

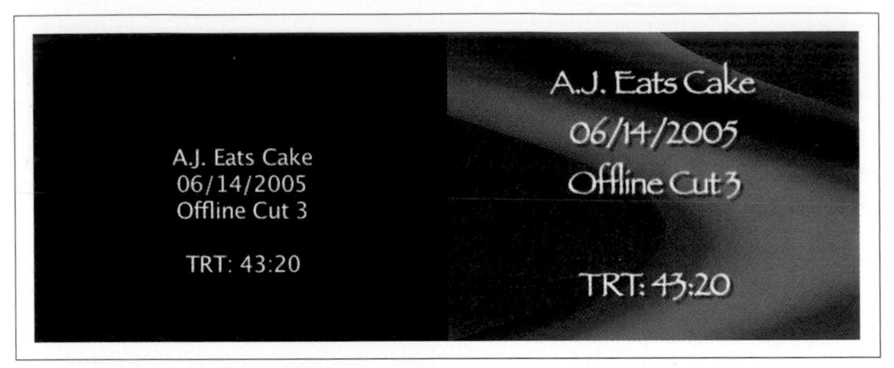

Figure 3-7. No rules, just a question of how much time is available

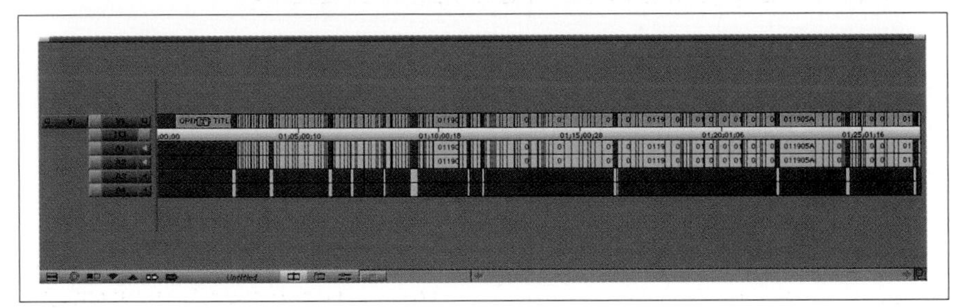

Figure 3-8. A slate at the left side of the timeline

Transcode a Movie's Codec

Just because you are using one digital video format, that doesn't mean you can't distribute others. By converting from one format to another, you can reach a new audience.

You might initially intend on distributing your video on DVD, only to discover there is a market for online distribution. Moving digital video from one format to another is easy, if you know which tools to use. There are a large number of applications, both commercial and free, that will enable you to change your footage from one codec to another.

Of the applications available, Discreet's Cleaner (*http://www.discreet.com/products/cleaner/*; $549) and FFmpeg (*http://ffmpeg.sourceforge.net/index.php*; free, open source) are two solid options. Both programs are available for Macintosh and Windows operating systems. FFmpeg is also available on Linux.

Using Cleaner

Cleaner is capable of doing much more than simply transcoding a movie file; for example, it can enable a movie to open a URL [Hack #04]. However, it is known for being one of the best video encoders available. Additionally, it makes encoding video easy for beginners, because it has a wide variety of presets, while also allowing advanced users to fine-tune the encoding process.

Here's how to create an MPEG-4-compliant, streaming video, suitable for a 1Mbps connection:

1. Launch Cleaner.
2. Drag and drop your movie file onto the Batch window.
3. Choose Batch → Specify Setting.
4. Choose MPEG → MPEG-4 → IMSA Profile 1 1 Mbps streaming.
5. Click the Apply button.
6. Choose Batch → Encode.

Before encoding, Cleaner asks you to provide a name for the file and a location on your hard drive to store it. Once the encoding process begins, Cleaner displays a Progress window. The Progress window updates throughout the process, providing both visual feedback in the video portion of the window, as well as in the Status, Data Rate, and Statistics tabs. Figure 3-9 shows the Progress window for a movie after it has been encoded.

Cleaner will *ding* like an egg timer when the process has completed. You can then proceed to distribute the resulting movie as you see fit.

Using FFmpeg

FFmpeg is an open source audio and video tool that can record, convert, and stream media files. Although it is developed and targeted for Linux, it can also be used on Macintosh and Windows. You can find instructions for downloading the current release at *http://ffmpeg.sourceforge.net/download.php*. FFmpeg uses a command-line interface (CLI), so you use the program by typing and not using a mouse.

After you have either built or obtained a binary of FFmpeg, locate the application on your hard drive from a command-line prompt. Using the `-i argument` followed by the video file you want to convert and the video file you would like to convert to, you can easily convert your video from one format to another:

```
ffmpeg -i mymovie.dv mymovie.mpg
```

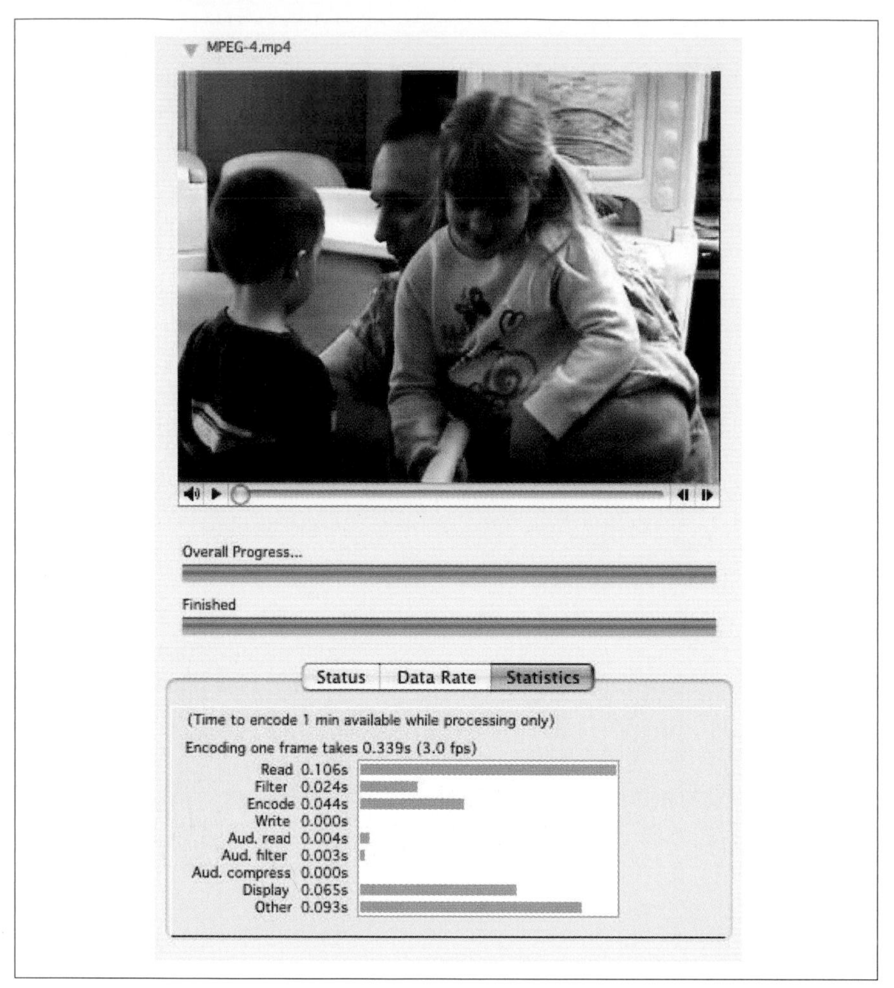

Figure 3-9. Displaying statistics for a completely encoded movie

Replace `mymovie.dv` with the movie you're converting and `mymovie.mpg` with the name you would like to save the converted movie. FFmpeg is capable of encoding many different formats, including asf, mpeg-1, mpeg-2, mpeg-4, dv, mjpeg, and many more. Documentation is extensive and available on the FFmpeg web site (*http://ffmpeg.sourceforge.net*).

Figure 3-10 shows FFmpeg converting an MJPEG-encoded movie named *Promo.mov* to an ASF-encoded movie named *Promo.asf* using the command `ffmpeg -i /tmp/Promo.mov /tmp/Promo.asf`.

Running FFmpeg From the Command Line

To run a command line program on:

Mac OS X
: Open the Terminal application, located in the *Utilities* folder.

Windows
: Open the Run... item from the Start menu, and then type `command`.

If you are not familiar with using the command line, or if you've never compiled a software program, you will probably achieve more success if you can locate either a *binary* or *GUI* release of FFmpeg for your particular operating system:.

Mac OS X
: FFmpegX (*http://homepage.mac.com/major1/*; $15.00)

Windows
: MeWiG (*http://mewig.sourceforge.net/*; free, open source)

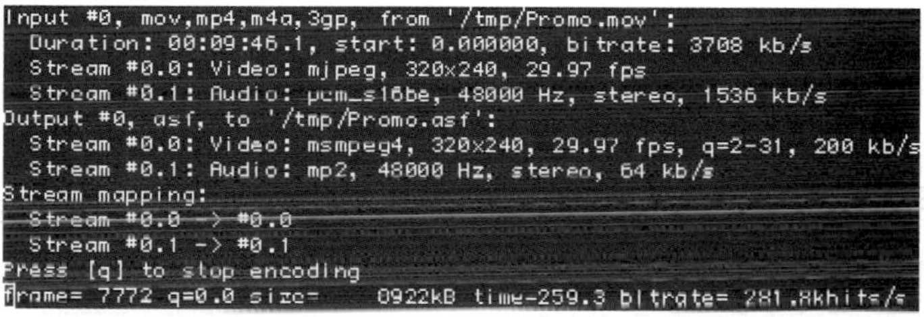

```
Input #0, mov,mp4,m4a,3gp, from '/tmp/Promo.mov':
  Duration: 00:09:46.1, start: 0.000000, bitrate: 3708 kb/s
  Stream #0.0: Video: mjpeg, 320x240, 29.97 fps
  Stream #0.1: Audio: pcm_s16be, 48000 Hz, stereo, 1536 kb/s
Output #0, asf, to '/tmp/Promo.asf':
  Stream #0.0: Video: msmpeg4, 320x240, 29.97 fps, q=2-31, 200 kb/s
  Stream #0.1: Audio: mp2, 48000 Hz, stereo, 64 kb/s
Stream mapping:
  Stream #0.0 -> #0.0
  Stream #0.1 -> #0.1
Press [q] to stop encoding
frame= 7772 q=0.0 size=    0922kB time-259.3 bitrate= 281.8kbits/s
```

Figure 3-10. FFmpeg in action

Distributing Your Movie

Every distribution method has a preferred codec, or set of codecs, associated with it. There are numerous ways you can distribute your movie, including on video tape, on DVD **[Hack #79]**, on VCD **[Hack #78]**, streaming on the Internet **[Hack #83]**, via peer-to-peer network **[Hack #86]**, and even via cell phone **[Hack #89]**.

In the end, you should always make sure your distribution method and your encoding match up **[Hack #77]**. There is a slight "chicken and egg" paradox with distribution and encoding, but once you know which comes first (e.g., you are distributing via DVD), then you can easily figure out which comes second (e.g., use the MPEG-2 codec).

See Also

- QuickTime Pro (*http://www.apple.com/quicktime/*) can convert to and from a wide variety of file formats. See *http://www.apple.com/quicktime/products/qt/specifications.html* for a complete list.
- Windows Media Encoder (*http://www.microsoft.com/windows/windowsmedia/9series/encoder/default.aspx*) is able to convert various Windows Media-compatible formats (*.wav*, *.wma*, *.wmv*, *.asf*, *.avi*, *.mpg*, *.mp3*, *.bmp*, and *.jpg*) to the Windows Media Video format.

HACK #30 Pantyhose Diffusion Filter for Flattering Portraits

Razor sharp optics are great—unless, that is, you're photographing the love of your life. In those instances, you might want to borrow her pantyhose.

A flattering portrait is often praised for its soft lighting, good angle, and natural expression. You'll rarely hear a subject rave about a picture that highlights her pores, wrinkles, and blemishes. Sometimes, modern camera lenses can be too sharp!

A popular solution used by pros is what's known as a *softening* or *diffusion* filter. Simply put, these accessories attach to the front of the camera lens and downplay the appearance of texture on the face. The wrinkles don't go away; you simply don't notice them as much.

These specialized filters can cost as much as $200 and are difficult to find for less than $20. Plus, if you use a variety of lenses for your portrait photography, you might have to buy more than one filter to fit the different lens diameters. That's fine if you shoot portraits for a living. But what if you just want to take a nice shot of your sweetie?

Ask her for her pantyhose.

That's right, by stretching a piece of light beige pantyhose over the front of your lens and securing it with a strong rubber band, you can create the same flattering effect achieved in professional portraits. The more tightly you stretch the material, the milder the effect—the looser the material, the softer the image.

You can capture good portraits without filtration, as shown in Figure 3-11, if you use good technique. But there will be situations in which you'll want to add a little softening effect, as shown in Figure 3-12 using a *pantyhose filter*. Be sure to keep a knee-high stocking, along with a couple sturdy rubber bands, in your camera bag for just these occasions.

Figure 3-11. A portrait without filtration

Figure 3-12. A portrait using a pantyhose filter

I actually prefer knee-highs to pantyhose, because I don't have to cut the material. One knee-high fits nicely in my accessory pouch. And it doesn't run or unravel, because I haven't had to trim it.

For best results with this technique, I recommend the following camera setup:

- Position the subject at least 10 feet from a background that has few distracting elements. A big green bush, wood fence, or even the side of a house work well.
- Look for diffused lighting, such as an overcast day. If the sun is too harsh, you can also place the subject in the shade of a tree and use a bounce board **[Hack #24]**. The best lighting is usually before 10 a.m. or after 4 p.m.
- Take lots of shots using different tension levels of pantyhose stretched over the front of the lens. You won't be able to pick your favorite by looking at the image on the camera's LCD monitor. So, having lots of shots to choose from once they've been uploaded to the computer will ensure success.

If you don't get the results you like with one pair of pantyhose, try another with a different weave or thread count. You'll be amazed by the results.

—Derrick Story

Convert PAL to NTSC

You can shoot on PAL format and then convert to NTSC for richer colors.

DV video comes in two flavors: PAL, which is used in Europe, Asia, Australia, and most of the world, and NTSC, which is used in North America. PAL has better resolution and richer colors. You can shoot on PAL then convert to NTSC and have a better-looking NTSC project.

Getting to Know PAL

PAL is actually a superior format, having more natural colors, a higher resolution, and a more film-like frame rate (25 frames per second much closer to film, which is 24 fps). NTSC runs at 30 frames per second (well, 29.97 when in color). PAL also has a bigger frame size (720×576 pixels, compared to NTSC's 720×480), so it captures a larger image.

Most commercial DV cameras area also available as a PAL model rather than an NTSC model. Video shot on a PAL model has a much better look than NTSC. But video shot on PAL gear will not play on NTSC DVD and

VHS players. It won't usually even display properly on TVs in America. However, you can shoot on PAL and then convert to NTSC if you need to, and it will still look better.

Affordably Converting PAL to NTSC

Commercial post-production facilities often charge an arm and a leg for this service—up to $300 for a one-hour tape. And the result often looks jerky or slightly distorted. However, there is an easy way to do it at home, in the form of an inexpensive utility called DVFilm Atlantis 2 (*http://www.dvfilm.com*; $195), available for both PC and Mac.

Atlantis requires QuickTime 6 (included on the Atlantis CD-ROM) or QuickTime 6.5.1. You should not use QuickTime 6.5.

Atlantis is also useful if you're making an NTSC project and having people in other countries shooting for you on PAL. I did this on my new film, *Hubert Selby, Jr.: It/ll be better tomorrow* (*http://www.cubbymovie.com*). A lot of the interviews were shot in Europe by friends, which saved me the problem and expense of having to travel there.

Preparing for Conversion

First, you have to have your footage in a data format on your hard drive. You want to export it as an uncompressed file. Atlantis can deal with either QuickTime or AVI files. You have this option from almost all video-editing programs. It's usually under File → Export Timeline/Movie, or something similar. You want to export as an uncompressed video. Figure 3-13 shows the initial steps to exporting a movie from Adobe Premiere.

I like to use Microsoft AVI. It looks better to me than QuickTime, for some reason, even though theoretically there should be no difference. It also exports more quickly from some programs.

You will also need to configure the format, codec, frame rate, and audio rate of your movie. Figure 3-14 shows the Export Movie dialog box for Adobe Premiere. Notice the Settings... button, as well as the current configuration for the export settings on the bottom part of the window.

You'll want to set up the options in your editing program to make sure it's exporting your PAL footage in PAL format. The default is usually to export

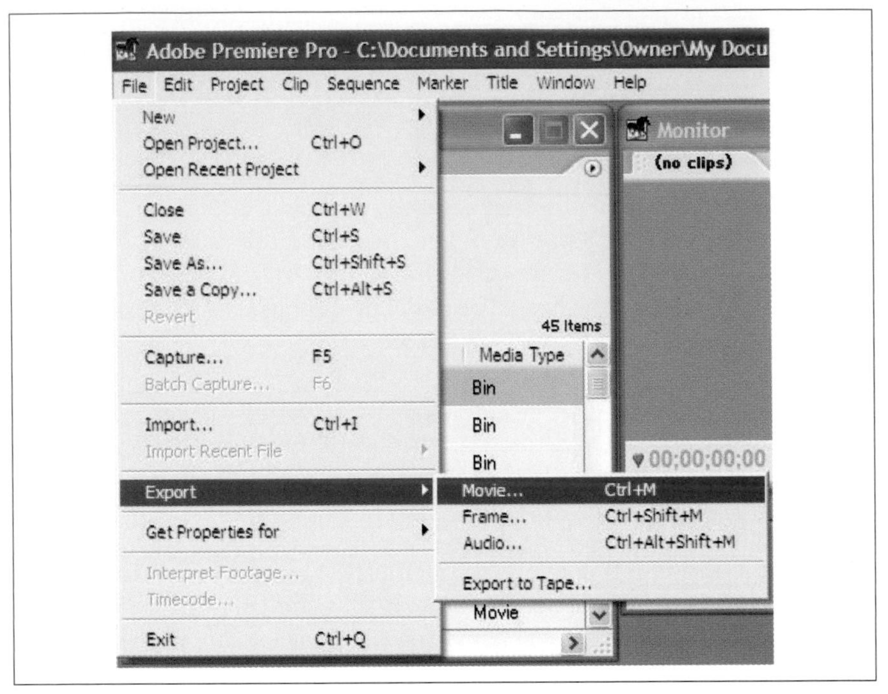

Figure 3-13. Exporting a completed timeline from Adobe Premiere

as NTSC. If you do change it, and you normally work in NTSC, then don't forget to change it back when you're done. Figure 3-15 shows the export settings for Adobe Premiere. Notice the Pixel Aspect Ratio is set for D1/DV NTSC and needs to be changed.

It will take a while to render (probably about 10 minutes per minute of video, perhaps more, depending on the speed of your computer). When it is done, drag and drop your file onto Atlantis.

Using Atlantis for Conversion

Atlantis is a tiny utility (350KB zipped for PC, 219KB for Mac) and a model of purity and simplicity. Unlike most programs that are packed with features you'll never use, Atlantis does only one thing, and does it better (and more simply) than any other program or process. It converts PAL footage to NTSC. (Actually, it does two things well: it also can convert NTSC to PAL.)

When you double-click on the utility, you'll get a drag-and-drop welcome screen dialog box, as shown in Figure 3-16.

To use Atlantis, drag your video onto the screen. You'll get a pop-up menu to configure your options. In addition to doing PAL to NTSC conversion,

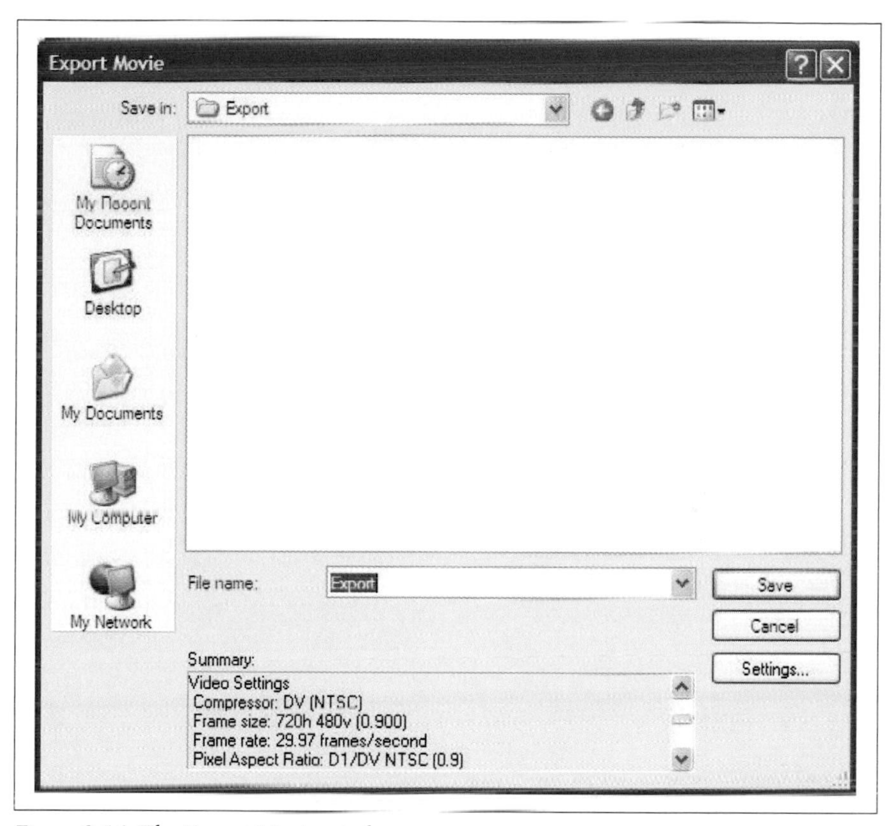

Figure 3-14. The Export Movie window

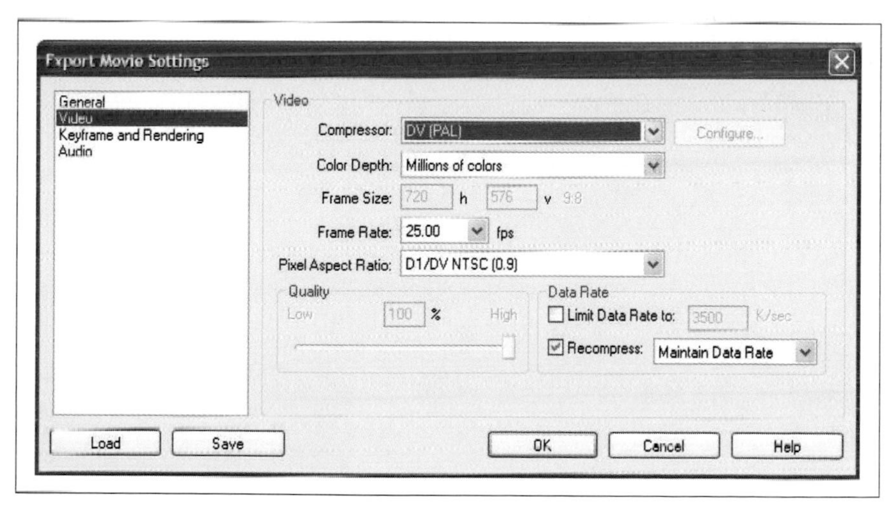

Figure 3-15. Exporting a DV AVI file in PAL

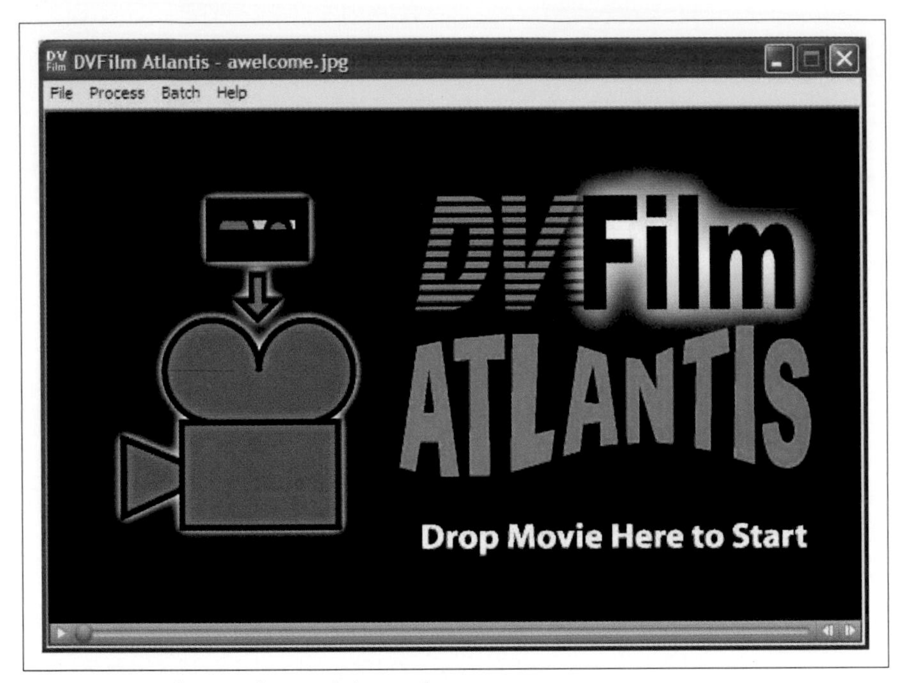

Figure 3-16. Atlantis's drag-and-drop welcome screen

you also have options to experiment with letterboxing, adding film grain, and adjusting colors, as shown in Figure 3-17.

I recommend experimenting with the letterboxing, film grain, and advanced options after you've done a simple conversion without them. To perform the actual conversion, click the Start button. Once Atlantis has started your conversion, you will get a pop-up status bar showing how long it takes to finish, as shown in Figure 3-18.

From there, the program does the rest. It does all the nifty background math it takes to make the conversion and spit out a new file, called *new movie*. When Atlantis is done, it sounds a tone and gives you a "Processing completed OK" box.

You can then import your footage into an NTSC project and output to tape, or make a DVD from it.

Hacking the Hack

You can also use Atlantis, if need be, to convert NTSC to PAL. This is great if you are making a film as an NTSC project but need to make a copy for the European DVD, or if you need to send your movie to a film festival abroad.

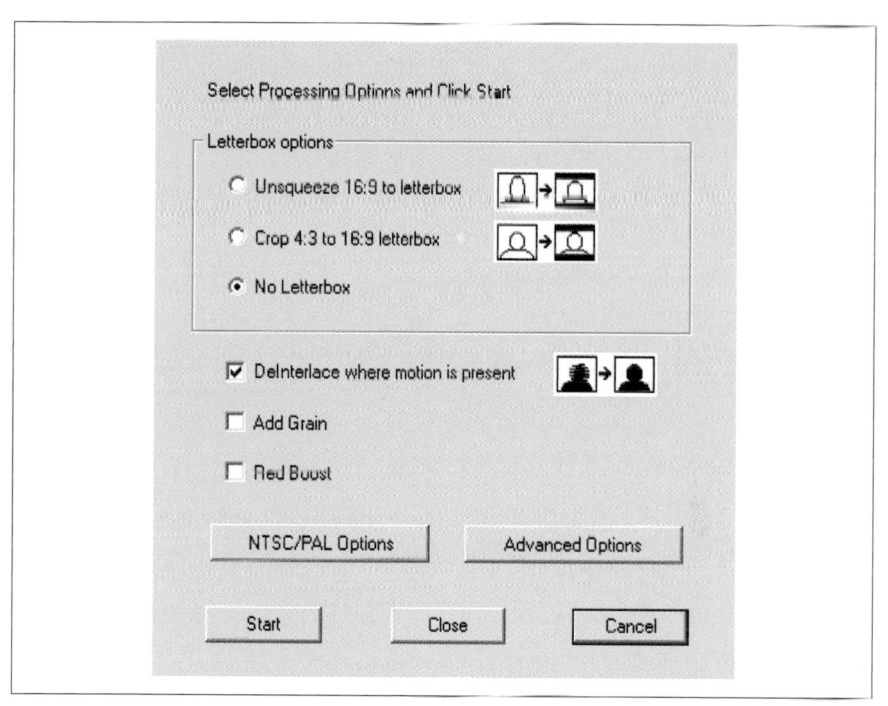

Figure 3-17. Atlantis conversion options

Figure 3-18. The conversion status bar

Atlantis can tell which type of file you are dragging in. If you drag an NTSC file, it will convert it to PAL. If you drag in a PAL file, it will convert it to NTSC.

Keep in mind that if there is already a *new movie* file in the same folder, it will overwrite it, unless you're using the batch processing command.

You can also batch process a number of files, simply by using the *batch* command from the interface. I often do this before I go to sleep; Atlantis does all the work and has it ready for me in the morning.

You don't need to worry about the audio; Atlantis takes care of that, too.

All in all, I have to say "bravo" for Atlantis. There should be more things out there that solve a problem this elegantly.

—*Michael W. Dean*

Record a Television Screen Without Flicker

If you are shooting a scene in which a television or computer monitor is visible, you might notice a flicker occurring on the screen. You can remove, or at least reduce, the flicker by making a few minor adjustments.

It's not uncommon to be watching a video and notice a television screen or computer monitor in the background *flickering*. The flicker is caused by the screen's presentation of moving images, through a process called *refreshing*, being at a different rate than you are recording. Depending on your situation, you might be able to remove the flicker completely, or at least minimize it so that it isn't a distraction for your viewers.

Recording a Television Screen

Shooting a television screen is actually easier than shooting a computer monitor. Because your television and video camera are recording at the same frame rate, you'll probably notice a slow, crawling flicker. The flicker occurs because of the slight difference between when the television displays a frame and when your camera captures it.

Within North America, a 60Hz current powers each television set, while Europe uses a 50Hz current. If you do a quick calculation, you'll notice that NTSC runs at 30 fps and PAL runs at 25 fps. Coincidence? No. The frame rate is directly tied to the electrical current. For a technical explanation, visit *http://en.wikipedia.org/wiki/Interlace*.

In simpler terms, when you turn on your camera, it begins its frame rate. This frame rate begins at a different time than the frame rate of your television. The fraction of a second difference causes the flicker to appear. Figure 3-19 shows a television screen mid flicker, as recorded using a Panasonic PV GS-120 MiniDV camera. Notice the large grayish band running across the top of screen; it is part of what causes the flicker to occur.

Figure 3-19. A flickering television screen

The solution to reducing, and usually removing, the flicker is to turn off your camera's image stabilization. For example, the Panasonic PV GS-120 has a feature called Electronic Image Stabilization (EIS). By turning EIS off, the camera will record a flicker-free television screen. Figure 3-20 shows a television screen as recorded using a Panasonic PV GS-120 with EIS turned off.

Figure 3-20. Turning off image stabilization to remove the flicker

When using a camera that records 24p, such as the Panasonic AG-DVX100, you need to change your frame rate from 24 fps to the appropriate rate (e.g., 29.97 fps for NTSC) in order to record a flicker-free image. This is in addition to turning off any image stabilization.

Recording a Computer Monitor

Capturing a screen of a computer monitor is only slightly more difficult, because televisions *refresh* based on the electrical current, which is the same for every television. Computer monitors, on the other hand, refresh at different rates, based on the user's preferences and the computer's capability.

The refresh rate of the monitor will reflect in the visible speed of flicker. Put another way, the higher the refresh rate, the faster the flicker will appear. Figure 3-21 shows a frame of video, recorded of a monitor set to refresh at 85Hz.

Figure 3-21. A monitor refreshing at 85Hz, causing a fast flicker

To reduce the speed of the flicker, you should change the monitor to a lower refresh rate that is a multiple of your frame rate. So, if you are recording at 29.97 fps, you should set the monitor to refresh at 60Hz, 90Hz, or 120Hz. (Yes, I'm rounding; monitors don't offer rates like 59.94Hz.) Figure 3-22 shows a frame of video, recorded from a monitor set to refresh at 60Hz.

As when dealing with the flicker from a television screen, if you turn off the camera's image stabilization feature, you will dramatically reduce the presence of the flicker. However, because the refresh rates are slightly different than you need, you will still see a slight flicker, which I feel is more of a slow *scan*. Figure 3-23 shows a monitor at 60Hz with the video camera's image stabilization feature turned off. At 60Hz and image stabilization off, there is still a slight flicker. Notice the small light blue line about two-thirds of the way down the screen.

Even though you (more than likely) can't completely remove the flicker while shooting a computer monitor, being able to reduce it to an almost imperceptible level will help keep your viewer's attention.

Figure 3-22. A monitor refreshing at 60Hz, causing a slow flicker

Figure 3-23. A slight flicker with image stabilization turned off

Import Footage from a DVD

Have you ever dreamed of interviewing a Harrison Ford? By importing footage from a DVD, you can have a celebrity take a role in your next movie.

Perhaps you would like to have Julia Roberts appear in your daughter's movie, cheering for her soccer team?

> "Hi. Can I speak with Julia Roberts, please?"
>
> "Who, may I ask, is calling?"
>
> "A little fat girl, from Ohio."*
>
> (click)

* This is a reference to Francis Ford Coppola's famous quote from *Hearts of Darkness*.

So goes your daughter's efforts to get Julia Roberts to appear in her movie. Okay, the scenario is a little far-fetched, but it demonstrates the fact that most of us can't get a celebrity to appear in our home movies—that is, unless we use scenes of them from other movies. Luckily, there are ways you can import video from your DVDs and use the video in your project.

Determining Your Needs

The signal from your DVD player might pass to your computer without any problems. If this is the case, you simply need to convert the analog signal to digital [Hack #37] and almost any breakout box should suffice. However, if you notice the image is fluctuating from bright to dark while you are digitizing, your DVD is probably copy-protected using a feature called Macrovision. If it is, you will have to take a few extra steps to digitize the footage.

For more information on DVDs and Macrovision, visit the DVD FAQ at *http://www.dvddemystified.com/dvdfaq.html*.

In order to import footage from a DVD with Macrovision, you need a way to pass the signal from your DVD player to your computer in a usable state. To do so, you will need a time base corrector (TBC), a breakout box, or a PCI video input card. All of these are different than the IEEE-1394 (a.k.a. FireWire or i.LINK) input you probably use under normal circumstances.

Copyright law protects the vast majority of major motion pictures produced. If you plan on using footage from a movie, you should make sure your use falls under *fair use* or that you have permission to use it. Stanford University has a good web site that covers copyright and fair use at *http://fairuse.stanford.edu/*.

Using a Time Base Corrector

There are also other types of hardware (in addition to breakout boxes, discussed in the next section) that can enable you to import your footage from a DVD. *Time base correctors* (TBCs) help to stabilize video signals while they are passed from one machine to another. Therefore, they are able to *normalize* a Macrovision signal.

TBCs tend to be expensive, because they are often considered *professional* tools. Some of the cheaper models, such as DataVision's TBC-1000 or AVToolbox's AVT-8710, start around $300. Sima Products Corporation (*http://www.simacorp.com*) manufactures a line of Video Enhancement and

Duplication products. Although they are not technically time base correctors, the products are capable of removing Macrovision from a video signal. You can also find decent deals on TBCs through eBay.

If you have old VHS footage you recorded and the image has degraded, some TBCs also have features that can help improve the degraded image. These features include controls to adjust the contrast, brightness, and sharpness—two birds, one stone.

If you choose to use a TBC, you will still need a way to digitize the signal coming out of the device. So, you'll need either a breakout box or a PCI card. The nice thing is, you won't have to worry about which breakout box or PCI card you buy. You can simply choose the one that works with your system.

Using a Breakout Box

If you are at all concerned about hardware compatibility with your computer, a *breakout box* might be your best choice. Most breakout boxes are designed for "plug and play" compatibility, so you can just plug it into your computer and convert your footage easily. A breakout box is a piece of hardware that has multiple connections to take various signals of video and audio and then convert those signals to another type.

Some breakout boxes, such as Miglia's Director's Cut (*http://www.miglia.com*; $269), ignore the Macrovision signal, while others, such as the Canopus ADVC-100 (*http://www.canopus.com*; $249), need to be *instructed* to allow the signal to pass through. For example, to instruct the Canopus ADVC-100 to pass the video signal, do the following:

1. Turn off all of the dipswitches on the bottom of the breakout box.
2. Hook up the DVD player to the box, using either an S-Video or RCA cable.
3. Hook up the box to your computer using an IEEE-1394 connection.
4. Turn on the power to the DVD player, your computer, and the breakout box.
5. Launch the program you plan to use to capture the video.
6. Set the program to manually capture, if applicable.
7. Start playing the DVD.
8. If the DVD is protected by Macrovision, the red status light will blink intermittently.

9. Press and hold down the Input Select button until the status light stays off.
10. Release the Input Select button and make sure the Analog In light is on.
11. If the Analog In light is off, simply press the Input Select button to change from Digital In to Analog In.
12. Begin digitizing.

Newer versions of the Canopus ADVC-100 sold in the United States have removed this feature.

Using a PCI Card

As with breakout boxes, some video input cards for your computer will ignore the Macrovision signal. The Happauge WinTV-GO is a great deal at around $50 and it seems to ignore the Macrovision signal.

In addition to being able to capture your DVD footage, WinTV-GO includes a 125-channel TV tuner, so you can hook up your cable signal and watch, or digitize, your favorite shows. Other Happauge cards should work just as well, so if you want you can get a card with an FM-tuner or Dolby Surround audio.

Using Software

There are a lot of software applications available that can copy footage off of a DVD, and a search for "`dvd copy`" or "`dvd rip`" on Google (*http://www.google.com*) will yield a long list of them. Unfortunately, most require you to copy an entire disc, instead of just the shots you want to use. Additionally, the applications usually encode the footage using the MPEG-2 codec.

So, if you plan on using software to get the DVD footage into your project, you will likely have to copy an entire disc, and then transcode the footage [Hack #29] using the same codec as your project's footage (most likely DV). The entire process can be very time consuming, especially if you are just trying to get one shot!

The following web sites provide a lot of information and tremendous peer support (through user forums):

http://www.afterdawn.com
http://www.doom9.net
http://www.videohelp.com

The information goes well beyond just DVD-related topics, so when you have time, it's worthwhile to visit and see what topic each community is currently focusing on

Knowing What You're Using

Whatever approach you take, you should always make sure you import your DVD footage in the same format you will be editing. All of these approaches, except the software one, are taking the DVD signal over an analog cable, so you need to make sure you digitize your DVD footage with the same setting with which you are editing your project. So, if you are editing a DV project, digitize your DVD footage with a DV setting.

Use HDV for Better DV Quality

Even if you don't plan to release a high-definition project, HDV can provide great-looking results.

You often see High Definition (HD) video associated with High Definition Television (HDTV). While that's certainly a reasonable association, remember that HD is not one specific format or resolution. HDTV, on the other hand, is indeed a specific format and resolution. (Well, okay, it is several specific, competing resolutions, but they are clearly defined.)

You could use an HD camera to produce content destined for HDTV, but you could also use an HD camera to produce content destined for a digital movie theater, standard broadcast television, or even the Internet.

Defining High Definition

We should start by tackling what defines HD video. *HD video* is basically anything that has a higher quality than a broadcast D-1 NTSC signal, with a resolution of 720×486. (For the record, that's not hard to do.) Therefore, HD is not a standard, just a category.

DV and DVCPro record video at a resolution of 720×480 (NTSC).

There are several HD cameras out there, such as the Thomson Viper and the Sony HDW-950, and several editing applications, such as Final Cut Pro HD and Adobe Premiere Pro, that support various HD formats. The formats include the most popular 720p and 1080i options. Some cameras have better color. Some formats have better resolution. Some are just different. Whatever the particulars, it's safe to say that HD is a pretty loose term.

Understanding HDV

HDV is the hi-def version of DV. HDV is itself a clever hack. It uses real-time MPEG-2 encoding to squeeze a big picture (up to 1920×1080, interlaced) onto MiniDV tapes. HDV is an emerging standard, but with Sony's HDR-FX1 (*http://www.sonystyle.com*; $3,699), it's likely to be embraced by the independent video and film producers, as well as companies that produce for cable television.

You'll primarily see two variations when looking at HDV cameras and formats: 1080i and 720p. The 1080i format is an interlaced image of 1080 lines with 1920 pixels/line. The 720p format is a progressive scan format with 720 lines of 1280 pixels (see Figure 3-24 for a size comparison). Because of the heavy-duty compression that is used, the data rate for these much larger pictures is comparable to standard DV. Capturing video from HDV cameras can still be done easily over FireWire. (In fact, FireWire even handles fancier formats, such as Panasonic's DVCPRO HD.)

Figure 3-24. Relative resolutions for DV, HDV 720p, and HDV 1080i

Down-Converting HDV

So what's the hack? Well, you can use an HDV camera to record superior-quality content that you eventually *down-convert* to DV for editing, recording to DVD **[Hack #79]**, or distributing on the Web **[Hack #82]**. Down-converting is the process of reducing a high-resolution image (like HDV) to a lower-resolution image (such as DVD). The HDV cameras are as affordable as most other prosumer DV cameras, such as Canon's XL2 (*http://www.canon.com*, $4,999). The tape stock is the same, so why not try it out?

In order to down-convert, you don't need to do anything special. By acquiring your footage on HDV and then distributing it on another medium, such as a VCD **[Hack #78]**. Yes, it's that easy to get the absolute best quality for any distribution avenue you choose.

Sony has released a new tape specifically for use with its HDV camera. While the company claims the tape is superior in quality (i.e. lower drop-out rates), and optimized for use with Sony's HDV camera, you can still use regular MiniDV tapes.

Now, you might be thinking, "But DV is DV. If I down-convert, I'll lose any original quality gains I had with the HDV." You certainly will lose quality, but even with that loss, your end result will be better than simply starting with normal DV.

Consider your favorite television show. You are usually watching a program that was filmed—really filmed on 35mm—and then down-converted for broadcast. You can clearly see a difference in the quality between your show and the commercial for the local car dealership shot on DV and then broadcast. Colors are richer and the image has more depth. You're watching both sources on your TV set as sent to you by a TV station or a cable provider, but you can definitely tell when you're looking at a film source versus a video source.

Or consider something closer to home. Rent and watch your favorite big-budget Hollywood blockbuster on DVD. Then, watch your family vacation that you transferred to DVD using your cousin's computer. They both have their high points and special moments for you, but unless you travel with your own 70mm crew, you'll notice a difference in the quality of the two images.

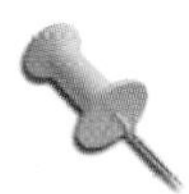

We won't discuss the quality of the *content*. As indie film fans, we're entirely prepared to believe that your vacation video is a better overall experience than the blockbuster.

The same holds true for your DV projects. Starting with a high-quality source will improve the overall quality of your project, regardless of the final destination. Consider renting some HDV equipment, or picking up an HDV camera if you're already thinking about a high-end purchase. It can't hurt your current projects and you'll be all set to go when the next generation of hi-def DVDs hit the streets!

—*Marc Loy*

Make a Movie Without a Camcorder

Even if you don't have a video camera, you can still make a movie by downloading video footage from the Internet. The Internet Archive web site provides free video you can use to create your next masterpiece.

Every once in a while, you might be in the mood to edit together a movie, but you don't have a video camera. The Internet Archive (*http://www.archive.org*) is a nonprofit organization, which was founded to build an "Internet Library." A vast array of media is available through the site, including audio and video files.

You can also use footage obtained from the Internet Archive for establishing shots and transitions within your movies.

Using the Archives

There are two good, free archives available through the Internet Archive site. The first of the two archives is the Open Source movies section (*http://www.archive.org/movies/opensource_movies.php*), which contains movies that have been contributed by the public. The second is the Prelinger Archive (*http://www.archive.org/movies/prelinger.php*), which dates back to 1983 and was founded by Rick Prelinger. The Prelinger archive contains close to 2,000 movies online.

Browsing the site. Both the Open Source and the Prelinger archives are part of the larger Moving Images collection. Within the collection of Moving Images are a variety of sections, including Independent News, Computer Chronicles, and Feature Films. Of particular interest is the Feature Films section, which contains genres ranging from Action to Western. Do you want to make a Mystery movie? Then browse the Mystery section, gather your footage, and edit to your heart's content.

All of the archive sections are set up the same. The homepage contains some quick links to the top movies, movies in the "spotlight," and most recently reviewed movies. Clicking on any of the links returns more information about the specific movie, as well as a way to download it. Figure 3-25 shows a list of recently updated reviews and links to their respective movies within the Open Source section.

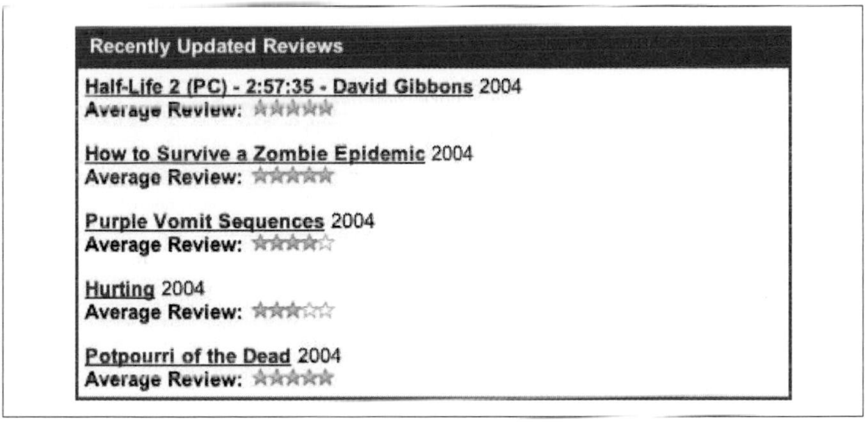

Recently Updated Reviews

Half-Life 2 (PC) - 2:57:35 - David Gibbons 2004
Average Review:

How to Survive a Zombie Epidemic 2004
Average Review:

Purple Vomit Sequences 2004
Average Review:

Hurting 2004
Average Review:

Potpourri of the Dead 2004
Average Review:

Figure 3-25. Taking an interest in "How to Survive a Zombie Epidemic"

One movie that caught my attention while I was browsing was *How to Survive a Zombie Epidemic*, shown in Figure 3-25. So, I clicked the link and retrieved the details page. There I discovered it is a movie made from *Night of the Living Dead* and *Amid the Dead* footage, both of which are available on the site.

On the details page, you will also find information about how long the movie is, which formats it is available in, and who was involved in the production of the movie. You can also surf to other movies that are listed under the same keywords as the movie you are looking at by clicking on one of the relevant links.

Finally, the license restrictions (a.k.a. what you are legally allowed to do with the footage) are listed just below the Keyword links. Make sure you understand the license and what you are allowed to use the footage for before you download and use it. Figure 3-26 shows the details page about the movie *How to Survive a Zombie Epidemic*.

In the upper-left corner of the details page you will see stills from the movie that play in succession. This will give you an idea of what the footage looks like: black and white, color, animation, and so on. If you click on the View movie scenes link, you will be shown images from each minute of the movie.

Downloading a Movie

After finding out more information, if you would like to download the video, simply click on the respective link listed under Download on the right side of the page. You will have the most success if you right-click (or

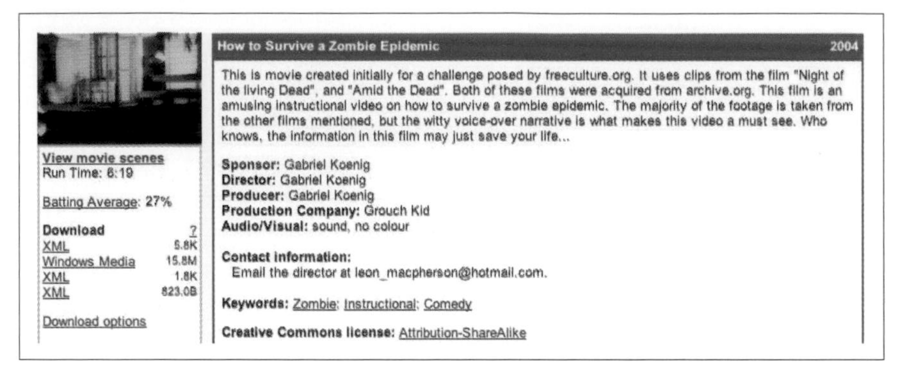

Figure 3-26. Evidence that Gabriel Koenig is a very busy man

Control-click on Mac) and choose to download the link as a file. Figure 3-27 shows the pop-up menu selected to download the movie.

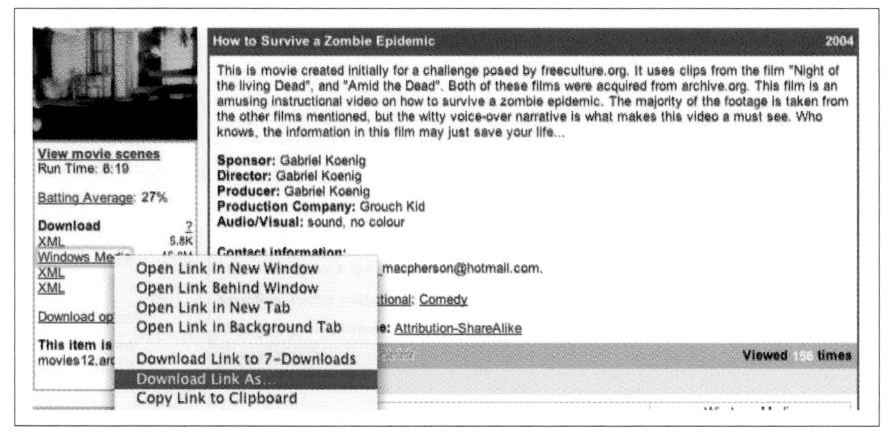

Figure 3-27. Downloading "How to Survive a Zombie Epidemic"

Downloading the link as a file will allow you to save the movie to your hard drive. After you have downloaded the entire video, you can import it into your editing system and get to work on your new movie.

Some of the videos you download from the Internet Archive might not be in the format you would like to use. If this is the case, you will need to convert the video **[Hack #29]** so you can use it.

Depending on your approach, you can mix and match footage from many movies, reedit a movie, or dream something up like *Mystery Science Theater 3000*. You might want to log your new footage **[Hack #5]** and write a script **[Hack #7]** before you start editing, though.

HACK #36 Freeze Time

Roll your own "bullet time," a la The Matrix

Some time ago, I discovered how to make my own version of the popular *bullet-time* effect seen in *The Matrix*. What I'll call *freeze time* is visual effect known by several names (detailed later in this hack), in which some, presumably dramatic, action comes to a halt while the camera shooting the action appears to move through space. The effect has been around for over 20 years, but it was recently made very popular by *The Matrix* and its sequels.

Discovering the Past

I'm quite amused by this whole concept and method, since it has been considered so "revolutionary" for the last several years. Why am I amused? Well, the very first *motion picture* was shot over 100 years ago, using an array of still cameras set along a racehorse track, connected to trip wires. Each camera took a photo in turn as a horse ran past the trip wires.

Lore has it that the rig was set up to settle a bet about whether a horse ever has all four feet off the ground when running (turns out it does). The side effect, when flipping through all the photos in succession, was a moving picture of a horse running. Now, almost the same rig is being used a full century later, to make motion *stop*!

Designing the first solution. According to an article in *Videomaker* magazine (June 2001; "Magic Morphing" by Scott Anderson), the first example of bullet time was in 1980 by an art student named Tim Macmillian. He was busy making collages of photos taken at several angles, usually random, at the same instant. As he progressed to adding more and more cameras per shot, he got to a point where he wanted more cameras than he could afford, so he built one to suit his special need.

This camera was just a big, long piece of wood with a slot, a window, a strip of film, and a shutter with lots of individual round holes. By tripping the shutter, the entire strip of film would be exposed through the holes. The result: when quickly flipping through the images, the action remained frozen, but it would appear that a single camera was in motion rather than several cameras sitting still.

Using hardware and software. Fast-forward several years and along comes *The Matrix*, and special effects guru John Gaeta, to refine this effect to new levels. For *The Matrix*, Gaeta pioneered the concept of computer-controlled still cameras that shot a fraction of a second after one another to create both a camera sweep and a slow progression of time rather than a frozen moment. He also made extensive use of interpolation software to smooth the transition from one photo to the next and shot the subject in a green screen room to be composited into a CGI environment.

In the United States, the invention of the effect is credited to Dayton Taylor. His company, DigitalAir (*http://www.digitalair.com*), owns patents on not just the cameras used, but also the concept of the effect itself.

So, what exactly is this effect called? Tim Macmillan called it *time slice*. Manes Studios (a.k.a. *The Matrix* guys) called it *flow-mo*, but Gaeta called it *bullet time*. The effect is also called *temps mort* (dead time), *timetrack*, *virtual camera*, *multicam*, and (viewers of the Super Bowl) *EyeVision*. Whatever you call it, I think it's a great effect.

Approximating the Effect

Here's how you can approximate this effect with some friends and a six video camera setup.

Setting up for the shoot. For the shoot, you should collect six video cameras and set pair of them at the 90-degree points. If all else fails, you'll at least have good video from these two angles. Between the two cameras, fill out the arc with the other four cameras.

You should place a light stand at the focal point of the arc. This will be your target point. Then use a string, with a bubble level on it, to measure out a common distance and height for the front/center for all of the camera lenses. The measurement will ensure that all of your cameras are the same distance and height from the target point. Figures 3-28 and 3-29 show the initial setup.

To calibrate pan, tilt, and focal length, you need to connect the first camera to a video monitor [Hack #11] and aim it at the light stand. Then mark the screen position of the target point on the tripod directly onto the monitor with a dry-erase marker.

Finally, connect each of the remaining cameras, in turn, to the same monitor and adjust them to position the target point directly behind the marker spots on the monitor glass. Figure 3-30 shows a cameraman adjusting one camera while using an external monitor for reference.

Figure 3-28. Placing the main cameras on the outside of the arc

Figure 3-29. Measuring the distance to each camera, an important step

You should do your best to eyeball the entirety of the scene, with the expectation that any necessary corrections can be done in post-production.

Capturing the scene. With all the cameras running, either slate your shot [Hack #9] or fire an still-camera flash just prior to each *event*, in order to mark the same instant in time on all the tapes. From this marker, it will be easy to identify the starting frame on all cameras. You should then be able to step forward to a frame during the event while editing, and jump to the same event instantly on any camera.

Finishing the scene. You can finish the shot using practically any video editing application. You should use a paint tool, or other frame of reference, to

Figure 3-30. Adjusting a camera using an external monitor

capture the synchronized still frame needed from each camera. From there, you can use the timeline to make a template that properly aligns and matches-up the six cameras. The template can then be used to create as many different shots as you want.

You can see how the effect turns out at *http://www.hiddenphantom.com/NBBF/PalmHeel.wmv* (Windows Media Player 9 or higher, or VLC, required).

—*Nick Jushchyshyn*

HACK #37 Convert Analog Video to Digital Video

A media converter will help transfer your old Hi8, VHS, and Beta footage to digital video.

During last couple decades, families have accumulated numerous hours of video. Whether from their children's birthdays and school events, or from making independent movies, all of this analog footage can now see new light, thanks to the advances in digital video. Now is the time to reedit those movies you cut together using two VCRs and your two index fingers.

> If you have professional decks, like a Sony UVW-1800 (Beta) and a Sony DSR-80 (DV), you can connect the two decks using component video and the RS-422 connection to create analog to digital copies.

Knowing Your Needs

It is important to know whether you will be capturing your converted footage to digital videotape or to computer hard disk. For editing purposes, you should record to videotape, because editing systems rely on timecode **[Hack #4]**. Videotape is also a more reliable medium when it comes to long-term video storage. Converting analog video directly to a hard disk will not provide a timecode reference.

If you plan on using your converted footage within a project, make sure you use the right codec **[Hack #77]**. For example, if you will be editing a DV project, then you should convert your footage using the DV codec.

Keeping your timecode. There are many conversion hardware options, ranging in price from $100 to more than $3,000, so you should do some research and figure out what your *real* needs are. Unless you have a library of timecoded analog footage, such as BetaSP, you will not have to spend a lot of money on analog-to-digital conversion hardware.

Most converters provide two-way conversion, which means they will convert both analog-to-digital and digital-to-analog. This will enable you to dub your analog tapes to digital video, transfer your digital videos to VHS, and even convert a digital signal to display directly on a television screen.

Some video cameras can convert an analog input to a digital output; check the manual for your camera to see if it has such a feature. Converting an analog input to a digital output using a camera is sometimes referred to as signal *pass through* or *bounce*.

If you plan on converting from an analog source that has timecode, such as BetaSP or 3/4", make sure your converter can transfer the timecode correctly. Some converters, such as the ProMax ProMedia Converter (*http://www.promax.com/Products/Detail/27187*; $1,495), will translate the analog timecode signal to digital, while others, such as the Canopus ADVC-500 (*http://www.canopus.us/US/products/ADVC500/pm_ADVC500.asp*; $1,649) will not carry any timecode signal.

If you are planning transferring some old VHS or Hi-8 footage, you can save well over $1,000 by purchasing a basic converter like the Canopus ADVC-55 (*http://www.canopus.us/US/products/ADVC110/pm_advc55.asp*; $229).

Dealing with the lingo. When shopping for a technology product, it is nearly impossible to not get caught in lingoland. Here's some of the lingo you might encounter while shopping for a converter:

Composite Video
: A connection in which luminance and chrominance are combined; most often carried over an RCA connector

S-Video
: A connection in which luminance and chrominance are separate

Component Video
: A connection in which luminance and chrominance are separate; often used to refer to a more detailed signal than S-Video

SDI
: A serial digital connection for passing a video signal; used professionally

Locked Audio
: Assures both audio and video are in synchronization with each other

AES/EBU
: A digital audio standard used in the U.S. and Europe; used professionally

Unless you have a real need for Component Video, SDI or AES/EBU, you should feel comfortable using S-Video or Composite (RCA) to transfer your footage. If you've never heard of these terms before, you can probably keep avoiding them. In other words, if you are planning on converting some old VHS tapes, don't buy a machine with an SDI connection. You won't use it; don't pay for it.

Recording to Videotape

If you are planning on transferring to digital videotape, find out what options your camera or deck offers to record video. Most digital videotape is 60 minutes in length. However, some machines allow you to record at either standard (SP:60 minute) or extended (LP:90 minute) play on the same type of tape.

Recording a digital tape in LP mode will reduce the quality of your audio. Some editing systems will not allow you to mix SP and LP footage, while others will require you to render your audio if the formats are different than expected. To avoid problems, always check what mode you are using to record and stay consistent from one tape to another.

Sony has a method of recording digital video called DVCAM. This format will record on a standard MiniDV tape, but at a speed that is about 33% faster than standard. Because of the speed difference, you will only be able to record about 40 minutes on a 60-minute tape. If you own a Sony camera or deck, check to see what method you are using to record your footage to know how much time you *really* have available.

Also take note of how long the footage you will be converting is. This will allow you to determine whether or not your camera or deck can be used in the conversion process. Obviously if you have a 120-minute VHS filled with footage, it will not be able to fit onto a 60-minute MiniDV, even if you record using the LP mode. In such an instance, you will need to split your 120-minute VHS across two 60-minute MiniDV tapes.

Recording to Hard Disk

If you will be capturing your converted footage directly to a hard disk, you'll need at least a few large disk drives. On average, digital video will require 13GB of storage per hour of footage. Fortunately, drive capacity is becoming cheaper by the day (at least it seems that way!) and portable terabyte (1,000GB) drives are not uncommon.

In order to record footage to hard disk, you will need some software to save the video to a file. Your editing system will more than likely be able to digitize your footage, or your converter may include such software. When digitizing a tape that does not have timecode, such as VHS, your system might warn you there is no timecode reference. This is to be expected.

After you've converted your footage, you can use it within a project, archive it to DVD [Hack #79], put it up on the Web [Hack #82], or simply keep it on your computer to watch whenever you want.

HACK #38 Create a 3D Video

You don't need really expensive equipment to create your own 3D video.

During the 1950s, there was a 3D craze for moviemakers. People flocked to movie theaters to put on cheap, funny-looking, paperboard glasses with different-colored lenses. Well, the craze is over, but you can still create a movie that requires your viewers to don those silly glasses.

To create a 3D movie, you need to choose a technique and become familiar with its strengths and weaknesses. Although there are four common techniques, only two can be used at a reasonable cost: *anaglyph* and *Pulfrich*. The other types, *polarized* and *alternating field*, are both complicated and expensive to produce (and the glasses aren't cheap, either).

A 3D technique called ChromaDepth® does not require you to apply any effects to your footage. However, you must follow numerous guidelines in order to create a successful video. The guidelines can be obtained from the ChromaTek web site (*http://www.chromatek.com/ChromaDepth_Primer/chromadepth_primer.html*). Viewers will need special glasses to see the movie in 3D (also available from ChromaTek).

Creating an Anaglyph 3D Movie

Traditional three-dimensional movies, such as *Jaws 3D* and *SpyKids 3D*, are created as *anaglyphs*. Anaglyphs are composed of two images, taken at the same time, at slightly different angles. Basically, the images are composed as if taken from the eyes of a viewer—one camera on the left and one camera on the right.

Unfortunately, anaglyphs work best in black and white, because the lenses of the glasses used to view the images are blue and red. Therefore, if you plan on creating a 3D movie, you should plan on working in a black-and-white medium, or accept the fact that the colors of your video might be off when you are finished with it.

When composed together, color-corrected, and viewed through glasses with one red lens and one blue (or cyan) lens, anaglyphs appear three-dimensional due to the perception of depth. Figure 3-31 shows a pair of glasses for viewing anaglyph images.

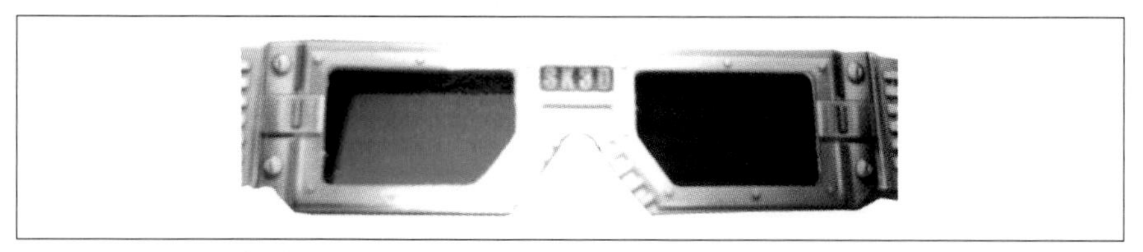

Figure 3-31. Anaglyph 3D glasses

Recording for a 3D effect. To create an anaglyph, you need to capture your video using two cameras. If you are planning on shooting a multicamera-style video, you will need twice as many cameras as you intend to use. For example, if you are planning on doing a three-camera shoot, you will need six cameras total to capture your video to process it as an anaglyph.

You will have better results if all of your cameras are of the same make and model. This is because lenses, charge coupled devices (CCDs; a.k.a. *chips*), and additional features vary from one manufacturer to the next and can affect the color and quality of the images you acquire.

If you don't have two videotape cameras, you can still create a 3D video by using two web cameras. Depending on the speed of your computer and the software you use to capture your video, you might need two separate computers to capture each video independently. Figure 3-32 shows a setup using two iSight cameras.

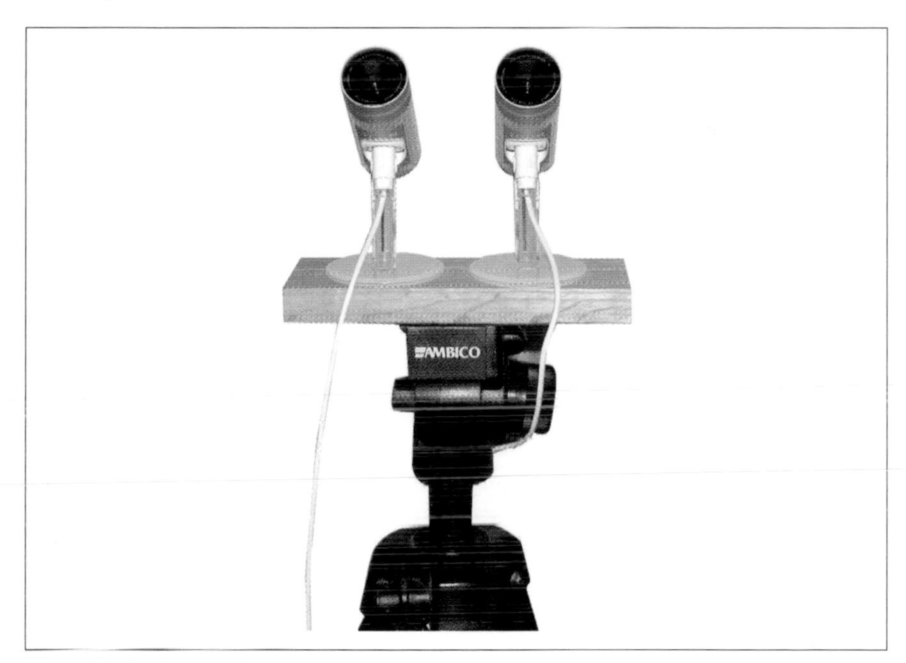

Figure 3-32. A two-web-camera setup

When setting up your cameras, attempt to place them about three inches apart from each other. This is intended to mimic the distance between a person's eyes. You can either set up two tripods as closely together as possible, purchase a mounting bar like one offered by Jasper Engineering (*http://www.stereoscopy.com/jasper/heavyduty-bar.html*; $199–$289), or construct your own mounting bracket. If you plan on capturing close-ups, you might have to angle the cameras inward, in an almost cross-eyed fashion.

To make the editing process easier, find a way to synchronize your cameras. You can do this either by setting the cameras to record timecode as the current time of day, or by using a slate [Hack #9]. Also carefully label your tapes, so you can determine the left from right camera angels.

Creating the effect. After you have acquired your footage, you need to import it into your editing system and place your footage on your timeline. Place both videos within the same space, so that the videos are *stacked* in your timeline. Although how you place your footage is up to you, make sure you know which image is from the left camera and which is from the right camera.

If you are planning to make your video black and white, completely desaturate both video layers (or simply make them black and white, if your system offers such an easy solution). Also apply any effects you plan on using to your scenes *before* applying any of the anaglyph effects. Otherwise, you might not get the results you desire.

The images for both the left and right cameras need to be composed together, so the top layer needs to be slightly transparent; about 50% should suffice. Otherwise, the footage on the lower layer will never be seen. Here's how to change the opacity of your footage:

Avid
: Effects → Blend → Superimpose

Final Cut
: Motion Tab → Opacity

Premiere
: Effect Controls Tab → Opacity

Once you have changed the opacity, you need to alter the RGB (Red, Green, Blue) values. To create the 3D effect, you need to remove red from the right camera and remove green/blue from the left camera. Doing so will cause the viewers, when wearing anaglyph glasses, to see the image in a stereo pair thereby creating the illusion of depth. Here's how to alter RGB values:

Avid
: Effects → Image → Color Effect

Final Cut
: Effects → Video Filters → Color Correction → RGB Balance

Premiere
: Effects → Video Effects → Image Control → Color Balance (RGB)

After you change the RGB values, your image will look a little odd. This look is expected, because you have removed colors from the composition. To finalize your 3D look, attempt to line up your two images so that they appear as closely as possible to be one image. Figure 3-33 shows an image before it is lined up and after.

Figure 3-33. Lining up two images to make the effect work

Viewing the video. To see the 3D effect, you need a pair of glasses with one red lens and one blue (or cyan) lens. If you don't have a pair lying around the house and you need a pair ASAP, you can make a run to the local bookstore and pick up a 3D children's book (which should include a pair of glasses). If you are willing to wait for them, get a pair for free from *http://www.rainbowsymphony.com/freestuff.html*. You can also purchase them in bulk from *http://www.3dglassesonline.com*.

Once you have a pair of glasses, viewing your video in 3D is as easy as putting on your glasses and pressing (or clicking) the Play button.

Creating an Pulfrich 3D Movie

You can create a 3D using a single camera by using the Pulfrich effect. Using the Pulfrich effect to create a 3D movie does not require the images to be effected in any way and involves only the movement of the camera or the objects it is recording. In fact, when creating a 3D movie using the Pulfrich effect, you can view your raw footage in 3D! Additionally, viewers who do not wear 3D glasses can still enjoy the video without realizing they are missing anything.

The Pulfrich effect was explained by German physicist Carl Pulfrich in 1922. It occurs because it takes your brain longer to process a low-light image. Therefore, when covering one eye with a grey filter, your brain processes the left and right images at different times. It was used successfully in an episode of *3rdRock from the Sun* in 1997.

Recording the video. Making a video using the Pulfrich effect is easy. You simply need to move the camera horizontally left-to-right or right-to-left. This is essentially a dolly move; however, for the effect to be of use, the movement must be smooth. Figure 3-34 diagrams how a camera should move to create the Pulfrich effect.

Figure 3-34. The Pulfrich effect, caused by horizontal camera movement

The fundamental ingredient to the effect is an object's movement in relation to the camera. A simple and fun experiment to experience the Pulfrich effect is to sit in the passenger seat of a car and record the scenery out of your window. Driving at various speeds will provide ample insight to how momentum can affect the Pulfrich effect. If you plan on using this technique, I highly recommend experimenting with it first.

Viewing the effect. To view a Pulfrich effect video, you need a pair of glasses with one lens of transparent grey and the other clear. You can create a pair of Pulfrich effect-friendly glasses by simply removing one lens from a pair of sunglasses. Plus, if you wear them while watching a sporting event, such as football, you will probably experience some 3D images.

CHAPTER FOUR

Edit

Hacks 39–50

Editing can be one of the most exhilarating aspects of a digital video project. It is also one of the most technically involved and, therefore, most error prone.

Digicam Movie Editing Made Easy

Almost all point-and-shoot digicams capture video footage in addition to still photos. But how do your turn those short snippets into your own personal movie?

If it comes down to a choice between bringing along a camcorder or a digital camera, I'm going to choose the camera. Even though I enjoy shooting video, I like still photos even more. Plus, digital cameras are smaller, and managing pictures is easier than dealing with hours of video.

But there are times when I also want to capture a few snippets of video. Certain special events—such as a speech at a wedding, a greeting from an old friend, or a child's first steps—are communicated better with moving pictures and sound. Fortunately for us, most digital cameras have a respectable Movie mode, and some even have great ones.

The problem is, once you have captured the footage in your camera, what do you do with it? Before I answer that question outright—and I will—let me explain the difference between the video your digicam captures and the footage from a digital camcorder.

These days, most people shoot video with a digital video (DV) camcorder. DV is becoming the format of choice. After you record your movie clips, you can plug the camcorder into your personal computer and an application on your computer launches, ready to download and edit your footage. On the Windows platform, you might use Microsoft's Movie Maker; on the Mac, you could use Apple's iMovie or Final Cut Express.

But when you plug in your digital still camera, these applications don't seem to recognize it, even when it's full of video. One of the reasons for this is that you're probably plugging your still camera into an USB port, while your camcorder uses the FireWire port. Most video-editing applications look for devices connected via FireWire. But that's not the only difference.

Digital camcorders typically record in the DV format. Your digital still camera is using a completely different format, such as Motion JPEG OpenDML, which is an extension of the AVI file format. You can play these files on your computer by using QuickTime Player; in fact, this format is part of the QuickTime media layer. That's why almost every digicam under the sun provides you with QuickTime on its bundled CD.

Playing the files is one thing, but hooking them together and editing them is another altogether. If you can't use the DV-editing software that came with your computer, then what do you do?

In this hack, the answer is QuickTime—the Pro version (*http://www.apple.com/quicktime/buy*; $29.99), that is. Sure, the free QuickTime Player you download from Apple lets your *watch* the movies. But the Pro version that you buy online lets you *edit* them too, plus do a lot more. QuickTime Pro is an extremely versatile digital-media application. But for the moment, we'll focus on editing, stitching, and trimming.

The first thing you need to do is transfer the movie files from your camera to your computer. If you don't know how to do this already, you'll have to crack open your owner's manual. Every camera/computer combination is a little different. If you're lucky, your camera has what we call *Mass Storage Device connectivity*. That means it appears on your computer like a regular hard drive. Nikon, Olympus, and Kyocera cameras usually have this feature. In those cases, you simply open the "hard drive," drag the movie files out, and save them on your computer. In case you don't know which ones are the movie files, they will usually have an *.avi* extension.

Now, you need to purchase the Pro version of QuickTime. Go to *http://www.apple.com/quicktime/* and click Upgrade to QuickTime Pro. You'll receive a software key that unlocks the Pro features in the Player version of QuickTime.

Okay, you're in business. You need to learn only three commands in QuickTime Pro to edit your movies:

Trim
: This command enables your to snip off the yucky stuff on either end of your movie. Simply move the bottom triangles on the scrubber bar to the endpoints of the content you want to keep, as shown in Figure 4-1.

When you then go to the Edit drop-down menu and select Trim, the gray area will be kept and the white area on the scrubber bar will be trimmed away. You've successfully snipped off the content you didn't want to keep.

Figure 4-1. Editing in QuickTime Pro

Add

Most digital cameras allow you to shoot only a few minutes of video at a time. So, to construct your movie, you have to combine these short clips into a longer presentation. Use the Add command for this purpose.

The procedure is similar to copying and pasting in a text document. First, move the bottom triangles to select the section of the video you want to copy, as shown in Figure 4-1. Then, go to the Edit drop-down menu and select Copy. That will put the snippet on the clipboard.

Now, open the snippet to which you want to add the copied content. Move the top triangle on the scrubber bar to the end of that clip; then, go to the Edit menu and select Add. The video that you copied to the clipboard is now added to the second snippet, including the sound that was recorded with it.

Make Movie Self-Contained
After you build your movie from the snippets you captured with your camera, choose the Save As command from the File drop-down menu. Give your movie a name, and be sure to click on the Make Movie Self-Contained button. This will *flatten* all the layers you've added into one movie, without any dependencies.

Congratulations! You just made your first featurette from short clips captured with your digital still camera. Regardless of the platform on which you created the movie, you can play the movie on both Mac and Windows. If you want, you can even burn it to CD and share with friends.

To help you make the best minimovies possible, here are a few shooting tips:

- Capture all your video at the same frame rate, preferably 15 fps (frames per second) if possible.
- Hold the camera steady and don't pan. You can't lock in the exposure when shooting video with most digicams, so you don't want lighting changes in your shots. The camera will try to adjust, creating annoying lighting shifts. Instead, focus on your subject, shoot 15 to 60 seconds of video, and then pause. Set up your next shot and go from there.
- Don't use Automatic White Balance. You could get color shifts in the middle of your clip, which you will find most annoying. Instead, choose one of the presets, such as Cloudy, to lock in the white balance setting.
- Frame your subjects tightly, as shown in Figure 4-1. Digicam movies are usually only 320×240 pixels in dimension, so you can't afford to stand back too far; if your frame is too wide, your viewers won't recognize the subject.
- Beware of the microphone port. It's usually close to your fingers. Don't cover it up or run your hand across it while recording.

Now it's time to go shoot something and combine those clips into a movie. Look out, Martin Scorsese!

—*Derrick Story*

Make Videos with Windows Movie Maker

You can make better home movies and other videos with XP's built-in video maker.

XP is Microsoft's most media-aware operating system, and it comes with built-in software for making and editing videos and home movies: Windows Movie Maker. (To run it, choose Start → Programs → Windows Movie Maker.) But making videos properly with it can be tricky, so check out these tips on how to make better home movies and videos.

For many more multimedia hacks specific to Windows XP, check out *Windows XP Hacks* by Preston Gralla (O'Reilly), from which this hack is excerpted.

Capturing the Video Properly

Windows Movie Maker lets you edit movies, and add special effects and titles, but it all starts with capturing the video properly. So, first make sure you bring the video into your PC in the best way.

Capturing analog video. If you have an analog video camera or videotape, you need some way of turning those analog signals into digital data [Hack #37]. You can do this via a video capture board, or by using a device you can attach to your FireWire or USB port. If you're going the route of a video capture board, make sure the board has XP-certified drivers; otherwise, you might run into trouble. To find out whether a board has XP-certified drivers, search the Windows Compatibility List at *www.microsoft.com/windows/catalog*.

If you have a USB port, you can import analog video with DVD Express, Instant DVD 2.0, or Instant DVD+DV, all available from *http://www.adstech.com*. They're hardware/software combinations; to get the video into your PC, connect the analog video device to the USB Instant Video or USB InstantDVD device, and then connect a USB cable from the device to the USB port on your PC. (A similar product, called the Dazzle Digital Video Creator, will do the same thing. For details, go to *http://www.dazzle.com*.)

Check your system documentation to see what type of USB port you have. If you have a USB 1.1 port, you won't be able to import high-quality video, and you'd be better off installing a video capture card. USB 2.0 will work fine, though.

If you have a FireWire-enabled PC, you're also in luck, because its high-speed capacity is also suitable for importing video. You'll have to buy extra hardware, called SCM Microsystems Dazzle Hollywood DV-Bridge. Plug your RCA cable or S-Video cable into Hollywood DV-Bridge, and then plug a FireWire cable from Hollywood DV-Bridge into your FireWire port, and you'll be able to send video to your PC. It's available at online stores such as *http://www.buy.com*.

Once you've set up the hardware and your camera, recording the video is easy. Open Windows Movie Maker, turn on your camera or video converter, and choose File → Capture Video.

Capturing digital video. If you have a digital video camera or webcam, you shouldn't need any extra hardware to capture video from it, as long as you have a FireWire port on your PC. These devices generally include built-in FireWire ports (the cameras might call the port an IEEE-1394 port or an i. LINK port). If you don't have a FireWire port on your PC, you can install a FireWire port card. These generally cost much less than $100. Make sure the card is Open Host Controller Interface (OHCI)-compliant.

When you plug your digital camera into a FireWire port and turn it on, Windows will ask you what you want to do with the camera. Tell it you want to Record in Movie Maker, and it will launch Movie Maker to the Record dialog box.

Best Settings for Recording Video

Before you start recording, you'll need to name your video file, as well as select where to save it. Windows should default to your Videos folder, which is a good place. After assigning your video a name, along with a place on your hard drive, you need to select how you want Movie Maker to record your video.

You will then be able to select how Movie Maker should digitize your footage, as shown in Figure 4-2. This is your chance to change your video settings, and choosing the proper setting is perhaps the most important step in creating your video.

Look at the Other Settings drop-down box in Figure 4-2. By clicking on the radio button, the drop-down box becomes active and you choose the quality of the video you're creating, which is the most important setting. What you choose for this setting will depend on the input source; digital video cameras, for example, let you record at a higher quality than analog video cameras, so they will give you a wider range of options. Movie Maker comes with a number of preset profiles, including three basic ones: "DV-AVI (NTSC)," "High quality video (NTSC)," and "Video for broadband (512 Kbps)." When you choose your profile, Movie Maker tells you how many

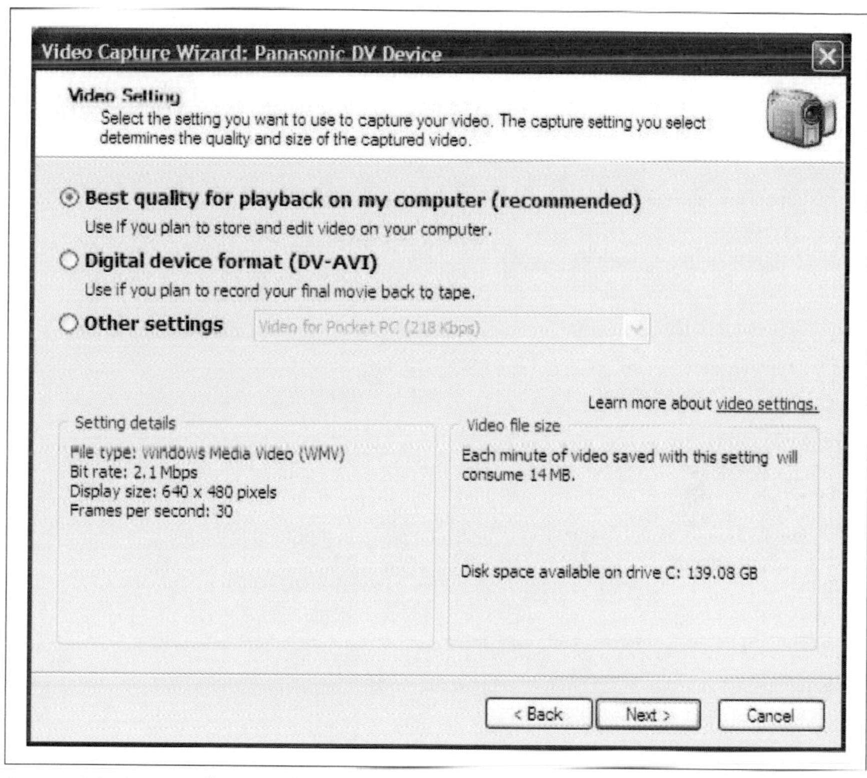

Figure 4-2. Options for recording video in Windows Movie Maker

hours and minutes of recording time you have, based on your disk space and the disk requirements of the profile. For example, you might have 193 hours of recording time based on the DV-AVI setting, but 1,630 hours based on the "Video for broadband" setting.

Those three profiles aren't your only choices, though. You can choose from a much wider variety of profiles (as a general rule, I suggest doing that), based on what you plan to do with the eventual video. Do you plan to post the video on the Web? Just play it back at home? Run it on a personal digital assistant? These other profiles are designed for specific purposes like those.

To select a profile, choose one from the Other Settings drop-down list, shown in Figure 4-3, with a range of profiles from which you can choose. They're prebuilt for specific uses. For example, if you're recording video to post on the Web, select one of the "Video for broadband" profiles. If you're recording for color PDA devices, select one of the "Video for Pocket PC" profiles. And, if you're recording to display your project on a standard TV, choose the DV-AVI profile.

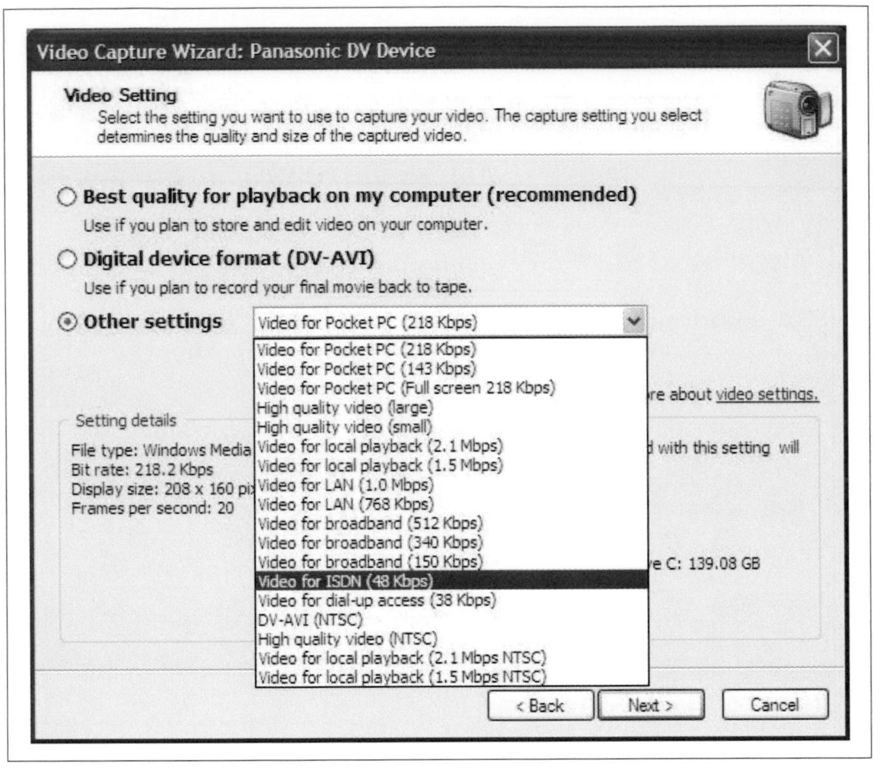

Figure 4-3. Choosing from additional preset profiles

Whenever you choose a profile, you'll see underneath it the frame size of the video and the frames per second. Here's what the settings mean:

File type
: The format of the file that will be created. Some formats can play using only certain media players, like Windows Media Player.

Bit rate
: The bit rate of the recorded video—the higher the bit rate, the greater the quality.

Display size
: The size of the video, in pixels—for example, 740×480, or 320×240.

Frames per second
: The number of frames captured per second. For smooth video, you need 30 frames per second, which is the "high-quality" setting. The medium- and low-quality settings record at 15 frames per second.

To help you make the best choice among profiles, Table 4-1 shows the settings for all the Movie Maker profiles.

Table 4-1. Settings for Movie Maker profiles

Profile name	Display size	Bit rate	Frames per second	Disk space per minute
Video for Pocket PC (218Kbps)	208 × 160 pixels	218.2 kilobits per second (Kbps)	20	1MB
Video for Pocket PC (143Kbps)	208 × 160 pixels	143.3Kbps	8	1MB
Video for Pocket PC (Full screen 218Kbps)	320 × 240 pixels	218Kbps	15	1MB
High quality video (large)	640 × 480 pixels	variable	30	variable
High quality video (small)	320 × 240 pixels	variable	30	variable
Video for local playback (2.1Mbps)	640 × 480 pixels	2.1Mbps	30	14MB
Video for local playback (1.5 Mbps)	640 × 480 pixels	1.5Mbps	30	10MB
Video for LAN (1.0Mbps)	640 × 480 pixels	1.0Mbps	30	7MB
Video for LAN (768Kbps)	640 × 480 pixels	768Kbps	30	5MB
Video for broadband (512Kbps)	320 × 240 pixels	512Kbps	30	3MB
Video for broadband (340Kbps)	320 × 240 pixels	340Kbps	30	2MB
Video for broadband (150Kbps)	320 × 240 pixels	150Kbps	15	1MB
Video for ISDN (48Kbps)	160 × 120 pixels	48Kbps	15	351KB
Video for dial-up access (38Kbps)	160 × 120 pixels	38Kbps	15	278KB
DV-AVI	720 × 400 pixels	25.0Mbps	30	178MB
High quality video (NTSC)	720 × 480 pixels	variable	30	variable
Video for local playback (2.1Mbps NTSC)	720 × 480 pixels	2.1Mbps	30	14MB
Video for local playback (1.5Mbps NTSC)	720 × 480 pixels	1.5Mbps	30	10MB

Capturing Methods

After selecting your profile, you need to choose whether to capture the entire tape or just part of it. If you choose to capture the entire tape, Movie Maker will rewind your tape for you (if you're using DV), and capture your footage automatically. If you choose to capture only part of it, you will need to cue your tape to the section you would like to capture. You can capture more than one section manually, without having to go through the entire process of naming your video and selecting a profile again. Figure 4-4 shows the Capture Wizard dialog box, set to capture the entire tape currently in the camera.

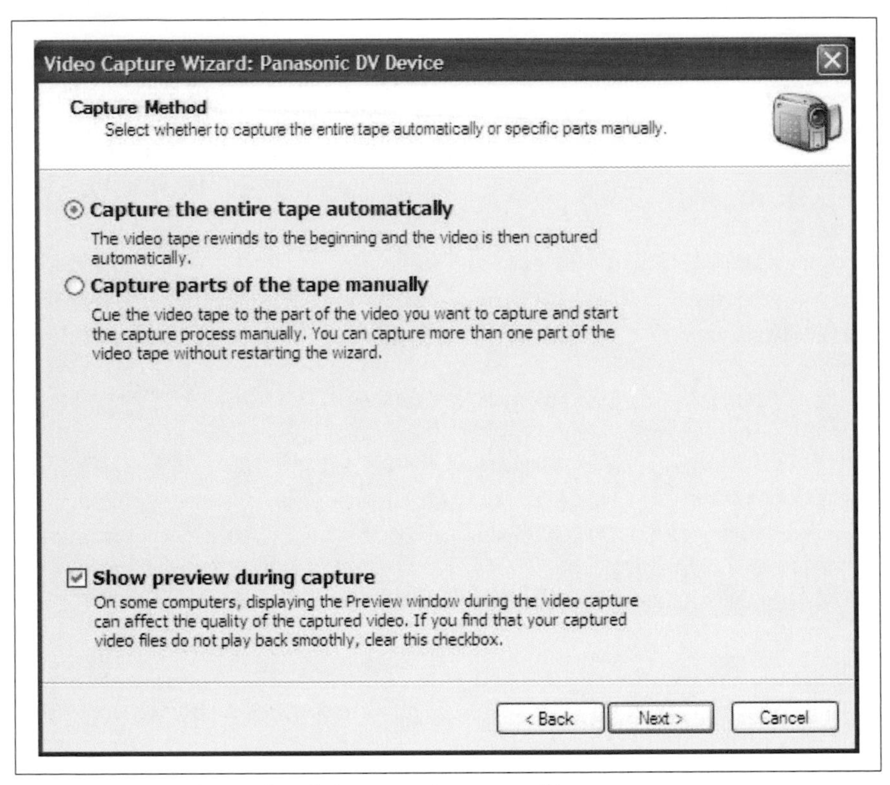

Figure 4-4. Your choice: the whole tape or just parts of it

You can also select to view your footage while Movie Maker is performing the capture sequence. This is a nice feature because you can make sure you are capturing the footage you expect to use…no surprises later on. Once you've made your selection, click the Next button and Movie Maker will proceed to capture your footage.

Making Your Own DVDs

If you use Movie Maker to make or copy your own videos and burn them to DVDs [Hack #79], consider these tips:

- The USB 1.0 standard is not fast enough to connect a camera or other video input to your PC. Its throughput of 11Mbps isn't fast enough for capturing high-quality video, which is 30 frames per second with 24-bit color at a resolution of 640×480, and requires speeds of at least 210Mbps. USB 2.0, which has a speed of 480Mbps, and FireWire, which has a speed of 400Mbps, will work, however.

- Make sure you have a substantial amount of free hard-disk space if you're going to burn your videos to DVDs. The video will be cached onto your hard disk before it's burned to DVDs, so you'll typically need several free gigabytes.
- Defragment your hard drive before creating and burning DVDs for best performance. If you have a second hard drive, use that for DVD creation rather than your primary hard drive. Regardless of the speed of your CPU, turn off any background applications that are running when you import video and create your DVD.
- If you're burning high-quality video to a DVD, figure that you'll be able to fit about an hour's worth on a single DVD. At a lower quality (lower bit rate), you can fit up to about two hours on a DVD. Keep in mind, though, that if you write at the lower bit rate, the DVD might not be able to be played on a set-top DVD player, though it will work on your PC's DVD player.
- There's no single accepted standard for DVD burning, so not all DVD disks that you burn will work on all set-top DVD players. Generally, most set-top DVD players will play DVD-R discs, but all of them might not play DVD-RW or DVD+RW disks. Manufacturer information can't always be trusted, but check the companies' web sites for the latest details.
- After you've created your video and you're ready to burn it to a DVD, set aside plenty of time. It can take up to two hours to burn a one-hour DVD, depending on your CPU and drive speed.

See Also

- If you want features beyond those offered by Windows Movie Maker, try a variety of software from Ulead Software, including Ulead Video-Studio, Ulead DVD Movie Factory, and Ulead DVD Workshop. They go far beyond basic video-editing tools, and they let you use transitions and add special effects and menus. The software also includes backgrounds, preset layouts, and music you can add to your videos. In addition, they will burn to DVD, VCD, and SVCD and can save files in a variety of video formats. They're shareware and available to try for free, but you're expected to pay if you keep using them. For details, go to *http://www.ulead.com*.

—Preston Gralla

Make a Movie Using iMovie

Apple includes its iMovie video-editing application with its computers. You can use it to create movies of almost any length.

If you've purchased a new Macintosh computer, the iMovie HD application can be found in the *Applications* folder. If you have an older Mac, or an older version of iMovie, you can purchase it as a part of the iLife application suite for $79. The iLife suite includes iDVD (for creating DVDs), GarageBand (for creating music), iTunes (for organizing music), and iPhoto (for organizing digital pictures), in addition to iMovie HD.

Using iMovie HD, you can edit your digital video and then distribute it on videotape, on the Web **[Hack #82]**, via email, or even on a cell phone **[Hack #89]**. But to do so, you first have to transfer your video to your computer. You can import video into iMovie using USB, IEEE-1394 (a.k.a. FireWire, i.LINK), or directly from an iSight camera.

Transferring Video

In order to get video from your camera to your computer, you need to connect the two using the appropriate cable.

If you have connected your camera using a USB cable, you can import your video by dragging the files off your camera (which appears as a hard disk on your Desktop) and into the iMovie window. You can also drag the files onto your hard disk and then import them into iMovie at a later date.

If you are using a DV, HDV, or iSight camera, you should import your video using a FireWire cable and iMovie's built-in process. Once your camera is connected, launch iMovie HD, name your project, and then switch to Camera mode. You can switch to Camera mode by clicking on the switch located just below the large video view. Figure 4-5 shows the iMovie interface and highlights the Camera mode switch.

If you want to transfer your entire tape, make sure your videotape is rewound to the beginning before proceeding to transfer your footage. You can rewind your tape by manually doing so on your camera or by pressing the Rewind button in the iMovie window.

To start the video transfer, click the Import button. The process will begin by playing your video, which will appear in the large monitor view just above the Import button. While importing, iMovie will perform Automatic Scene Detection and separate your footage into individual clips based on breaks in your video. These separate clips will appear on the right side of the iMovie window, in the Clips view, as shown in Figure 4-6.

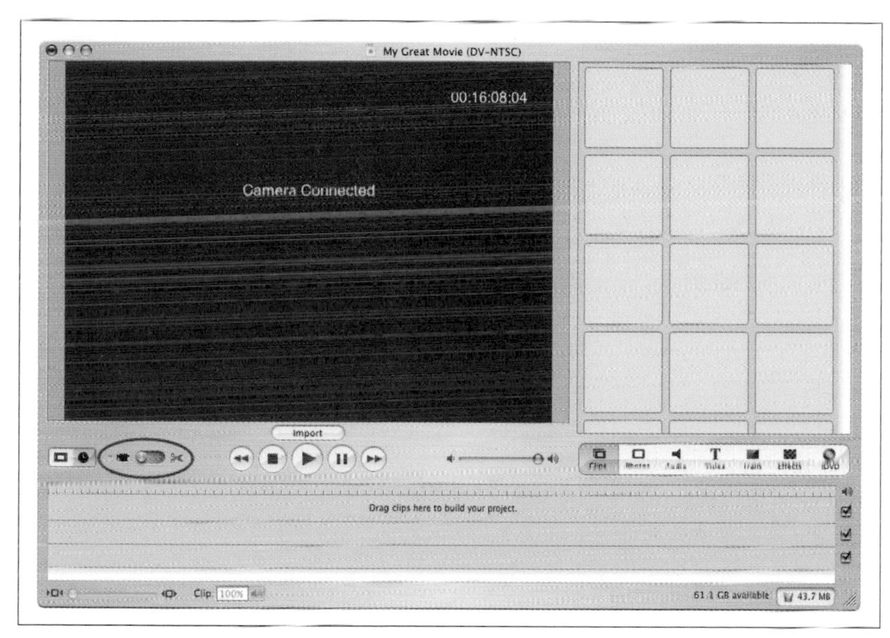

Figure 4-5. Camera mode in iMovie

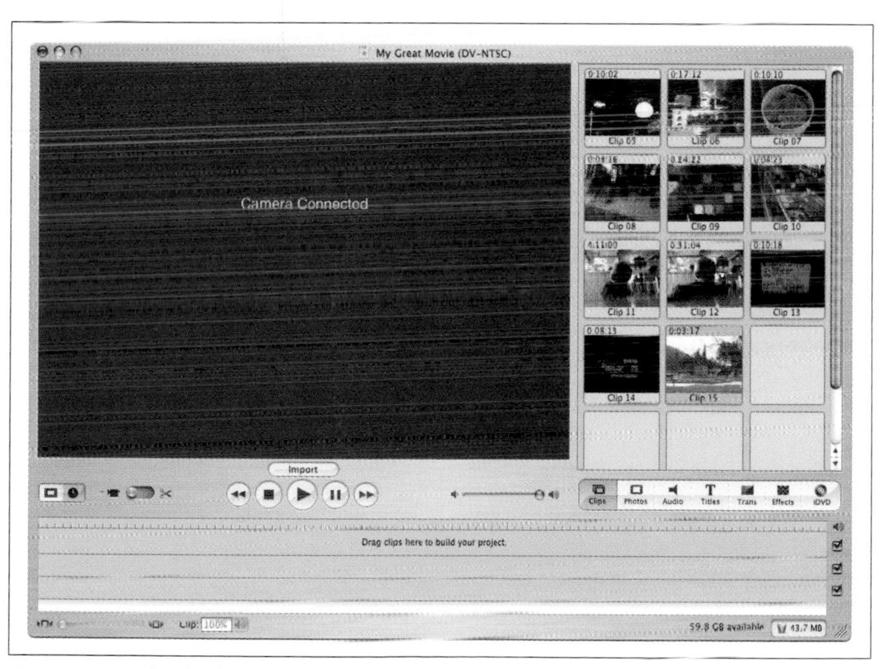

Figure 4-6. Individual clips, ready to be edited into a movie

Making a Movie Automatically

iMovie HD also has two features that will place your clips in your timeline automatically. The first simply places your clips on your timeline while you transfer your footage. To enable the feature, choose iMovie HD → Preferences → Import → Movie Timeline.

The other feature, called Magic iMovie, will not only place your clips on your timeline, but will also place transitions (such as dissolves) and add a sound track. If you want, iMovie will also create a DVD for you, or you can refine the timeline that has been created for you.

Both of these features can give you a jump-start on your project.

Editing Your Movie

To start editing your movie, switch from Camera mode to Cutting mode (this is the other setting on the Camera mode switch). You will then be able to work with your timeline. To add a clip to your timeline, simply drag it from the clips view onto your timeline. You can also alter the length of a clip by trimming, cropping, or splitting it.

While editing, you'll be able to add transitions and sound effects, create titles, and even add special effects. All of these features are available through buttons located under the Clips view.

Clips
: Provides access to the video clips iMovie has imported.

Photos
: Enables you to access and use pictures from your iPhoto library. You can also apply the Ken Burns effect when placing pictures on your timeline. This effect performs a pan and zoom over your picture to add motion to an otherwise static image.

Audio
: Lists audio built-in to iMovie, such as sound effects, as well as music and audio files in your iTunes library.

Titles
: Allows you to add a variety of titles to your movie, including titles as simple as those used for music videos to ones that spin in 3D and bounce around the screen.

Trans
: Makes a list of transitions available, to be used between clips on your timeline.

Effects
: If you want to alter the look of your video, you can select from a variety of effects, including Aged Film, Fog, and Letterbox.

iDVD
: iMovie and iDVD are well integrated, so you can add Chapters to be used on a DVD directly from the iMovie interface. You can also create an iDVD project by simply clicking one button.

The Clips view will change based on the button you select. Figure 4-7 shows the Effects view after the Aged Film effect has been applied to a clip in the timeline.

Figure 4-7. The Effects view

Creating a DVD

As previously mentioned, iMovie and iDVD are well integrated. From the one-click DVD feature of Magic iMovie, to the easy creation of Chapters and iDVD project files, iMovie provides an easy transition to create a complete DVD for viewing on a DVD player. To create a DVD from your iMovie project, click the iDVD button and then press the Create iDVD Project button. Figure 4-8 shows iMovie in the process of creating an iDVD project.

Figure 4-8. Creating an iDVD project in iMovie

After creating your iDVD project, you will need to author your DVD **[Hack #79]**. Once you're satisfied with your DVD, click the Burn button on the iDVD window, insert a DVD, and wait for your disc to be created.

Graduating from iMovie

iMovie is an incredibly powerful and feature-rich editing application. However, it does have limitations. When you find those limitations too discouraging, you should evaluate other editing applications available. Two of the more powerful applications are Apple's Final Cut Express (*http://www.apple.com/finalcutexpress/*; $299) and Avid's Xpress DV (*http://www.avid.com/products/xpressdv/*; $695). Avid also offers a more limited, but free, version of its software called Avid FreeDV (*http://www.avid.com/freedv/*; free).

These applications are essentially "light" versions of applications that are used by television and film professionals on a daily basis.

Rotate Your Movie from Horizontal to Vertical

Who says you have to shoot all your movies horizontally? Just as with stills, sometimes it's fun to turn the camera on its side. But when you upload your movies to your computer, they're turned the wrong way! Here's how to fix that.

When making movies with your digicam, you don't want your compositions limited any more than you do when shooting stills. Imagine if someone told you that you could shoot only horizontal pictures for the rest of your life. You'd tell them where to go stick their memory card.

The problem with movie making is, you might shoot your video with a vertical orientation, but when you upload the snippets to your computer, everything is horizontal. And 9 out of 10 chiropractors will tell you not to crane your neck sideways to watch these movies.

Fortunately for the health of your entire viewing audience, there is a simple fix.

Rotating a Video

After you upload the video to your computer, open a snippet in QuickTime Pro, the versatile movie viewing/editing application.

If you've not yet upgraded to QuickTime Pro (*http://www.apple.com/quicktime/buy*; $29.99), I can't recommend the move more highly.

From the Movie drop-down menu, choose Get Movie Properties. You've just tapped into one of the most powerful areas of QuickTime Pro. There are two drop-down menus at the top of this dialog box. From the left one, choose Video Track, and from the right one, select Size, as shown in Figure 4-9.

You'll see that the dialog box changes content and options as you choose different items from the drop-down menus. In this case, two of the goodies you can access are found in those rotation arrows in the lower-right corner. Click on the one that rotates your movie in the desired direction, as shown in Figure 4-10. Like magic, your movie and its controls are now oriented the way you originally intended.

Either way you do it, you can now edit, trim, and stitch together movie clips to your heart's content. Of course, all of your movies have to be oriented the same way; otherwise, you'll get some strange-looking results.

Figure 4-9. Making selections in the QuickTime Movie Properties dialog box

Editing a Vertical Video

Your editing system is expecting "normal" video in a horizontal position, so you need to create a custom setting. Fortunately, this is an easy task to achieve; the only setting you'll have to change is the height and width of the video. In fact, all you have to do is switch the two numbers.

Setting up your timeline. To edit your vertical footage, you need to create a new project and change your timeline settings accordingly. For example, if you shot on DV, change your width and height from 720×480 (720×576 PAL) to 480×720 (576×720 PAL):

Final Cut
: Sequence → Settings → General

Premiere
: Custom Settings → General → Editing Mode → Video for Windows

Figure 4-10. Rotating a movie in the QuickTime Movie Properties dialog box

If your editing system does not allow you to alter the timeline settings, and you are stuck at a resolution such as 720 × 480, you can edit your video as is and then export it as a QuickTime movie. You can then perform a rotation on the complete project.

After setting up your project to handle a vertical video, you need to alter your footage to match.

Importing your footage. Even though you've shot your video vertically, neither your camera nor your editing system will be aware of that fact. Therefore, it is essential that you rotate your video 90 degrees, as outlined previously. Because your video has already been transferred to your computer, and rotated, all you need to do is import the footage into your editing system.

After importing your footage, you can proceed to edit your movie just as you normally would. You will know if your video imports correctly as soon as you place it on your timeline. Figure 4-11 shows a vertical video as imported by an editing system without rotation, and Figure 4-12 shows the same image after it has been rotated appropriately.

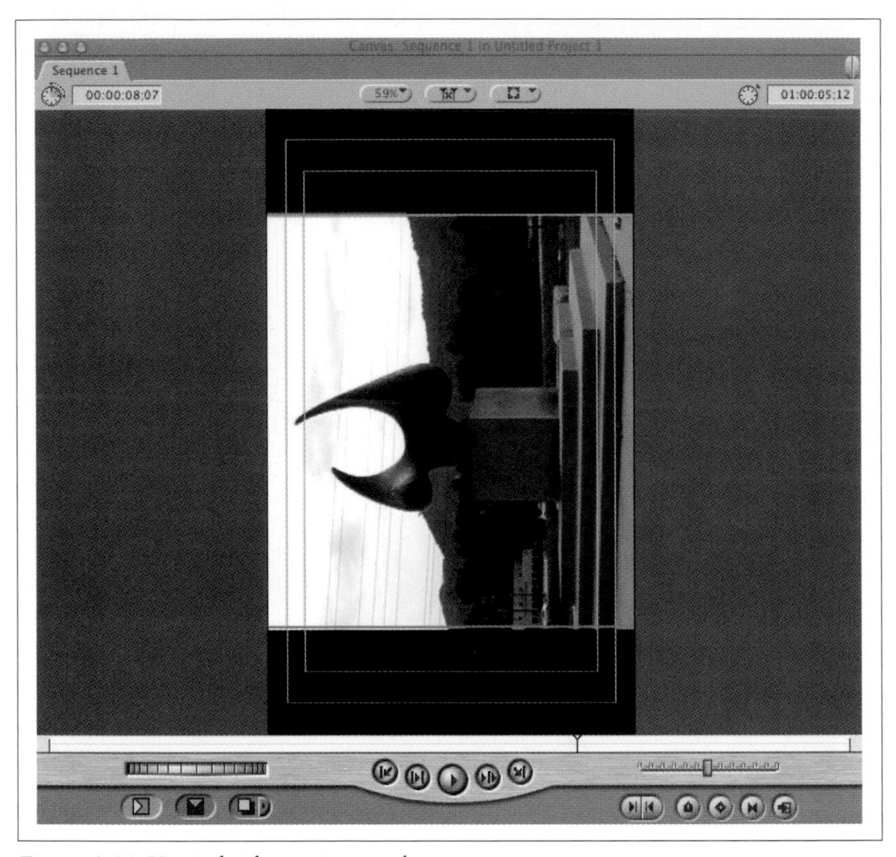

Figure 4-11. Vertical video, as imported

There's something inherently interesting about a vertically framed movie. And it's an option I encourage you to try as appropriate subjects present themselves.

See Also

- Simple Rotate (*http://www.imovieplugins.com/plugs/simplerot.html*) is a $1.50 shareware iMovie special effects plug-in.

—Derrick Story

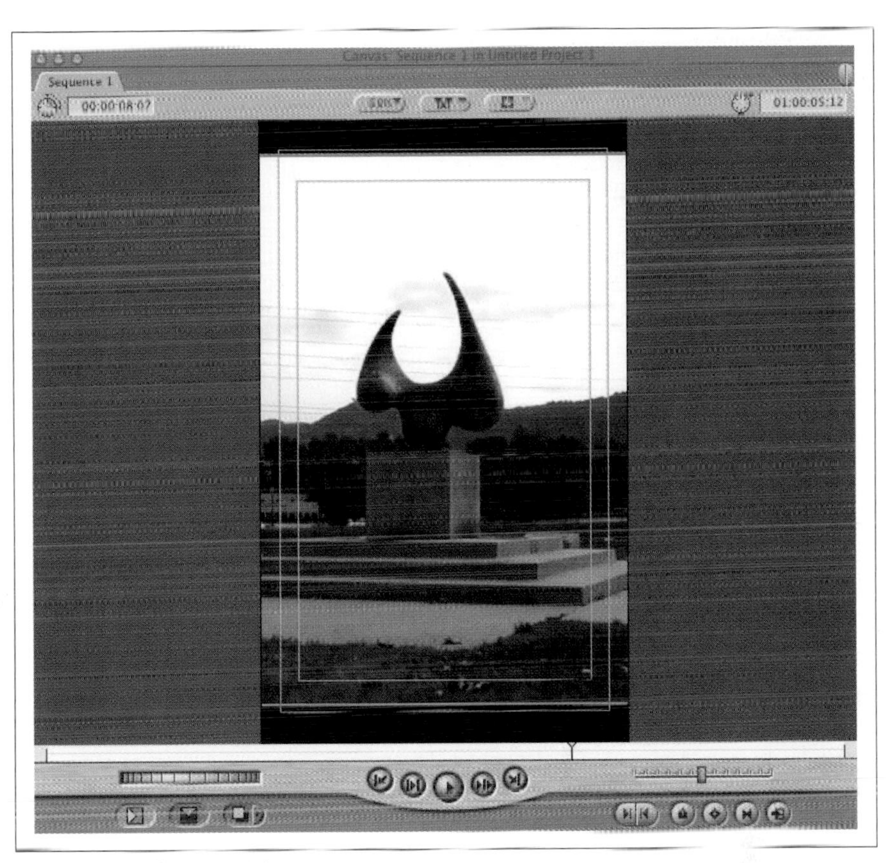

Figure 4-12. The rotated footage

Create a Submaster for Editing

When editing, sometimes you have too much footage or too many changes to easily manage a project. Submastering helps editors deal with these situations.

Whether you are dealing with a small project or a large one, changes are an inevitable result of the editing process. As you are editing, the more changes you make, the more your mind begins to swirl. This is especially true when your changes are being influenced by a third party, such as a director or producer. By creating a *submaster*, you will limit the number of tapes you need to handle over time.

The practice of submastering can be confusing. In order to avoid confusion, make sure you have a good tape numbering scheme **[Hack #3]** in place and you are exceptionally organized.

When you're dealing with a lot of footage, that's a perfect time to create a submaster. You can create a timeline and go through your footage to locate *selects* (i.e., footage that you know you want to use). You should not be editing this footage into anything really meaningful, but rather grabbing the footage you find you will use in your final timeline.

You can create a *select reel* by digitizing only the footage you plan to use or by sifting through all of your footage. If you are going to digitize only what you plan to use, you will probably log your footage ahead of time **[Hack #5]**.

When you have created a timeline that contains enough footage to fill a tape, you can output and create your submaster. After creating your submaster, you can then create a new timeline and continue the selection process. You should not delete your timeline, as it is a reference to where your footage originated.

Creating a Submaster

The process of creating a submaster is the same as the process of creating a master tape. Once you have assembled your timeline, simply output it to tape.

If you have digitized your footage using compression, such as OfflineRT for Final Cut or 15:1 for Avid, you cannot submaster your footage. In order for the process to work, your footage *must* be uncompressed.

Avid
: Clip → Digital Cut

Final Cut
: File → Print to Video...

Movie Maker
: File → Save Movie File...

Premiere
: File → Export → Export to Tape

iMovie
: File → Share → Videocamera → Share button

The result is a tape you can use to edit with in the future. Keeping a copy of your project that relates to the tape you have created is a good process. In other words, save a copy of your editing project immediately after outputting to tape.

If you want to continue to work on the project after outputting your submaster, you should work on a *copy* of the saved project. Otherwise, if you continue to work on the project, you will lose any references you may need at a later date. This is an essential step and a good habit to get into, even if you are not using the submaster process.

Using a Submaster

You use a submaster exactly as you use master tapes. A submaster effectively becomes a master tape within your project. The only difference is that it is a compilation of footage from various master tapes.

When using a submaster, what is most important is that you can distinguish it from your other tapes. You should make sure your tape-numbering scheme allows for you to make the required distinction. A simple number like PWCut1, representing Peter's Wedding Cut #1, works quite well. Figure 4-13 shows a timeline utilizing a submaster.

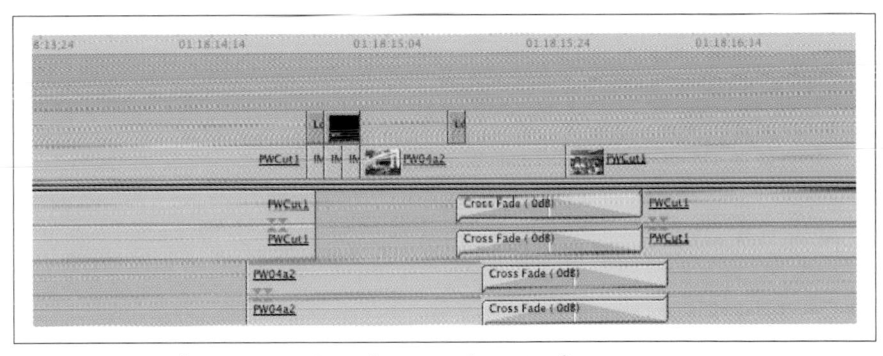

Figure 4-13. A submaster, PWCut1, being used in a timeline

If you have created a submaster from an edited project, you can create a new timeline, digitize the submaster, and then cut *into* the submaster. By doing this, you will have a much cleaner timeline and, if your system has been slowing down (i.e., dropping frames), possibly allow your system to perform better. The performance improvement will occur because your hard drive will not have to jump around so much to play a lot of different files.

Use a Television to Avoid Surprises

A computer monitor doesn't operate in the same way as a television. When editing, if you use only a monitor, you might get a few surprises when you watch your project on a TV.

Televisions display images using *interlacing* to create an image on screen. With interlacing, every frame of video is made up of two *fields*: an upper and a lower. These fields are *drawn* to the screen in alternating patterns, so that one field draws and then the other draws, to create a frame of video. So, if your television displays 29.97 frames per second, it displays 59.94 fields per second.

All DV format footage draws the lower field first.

Computer monitors display images using *progressive scanning*, and at variable rates. Progressive scanning draws the image to the screen in one pass. Because of the difference between a television and a computer monitor, what you view on your computer monitor when editing isn't necessarily what will appear on your television. By attaching a television to your editing system, you won't run into any surprise images when you're done. Figure 4-14 shows an image in which each field is from a different frame.

When attaching your camera to your editing system, you might discover that your video displays on your camera's LCD viewfinder (if it has one). The vast majority of LCD viewfinders do not interlace the image they display. If you want to use the LCD viewfinder, instead of attaching an actual television, make sure the LCD is interlacing the image.

To wire your system together, attach your camera to your editing system using an IEEE-1394 cable (a.k.a. FireWire or i.LINK). From your camera, run a video cable to your television, just as you would if you were going to play video directly from your camera. This might involve using RCA cables or a specific cable supplied by the manufacturer of your camera. For example, the Panasonic PV-GS120 uses a mini-AV cable to send its audio/video signal, whereas the Canon XL2 provides options of BNC, S-Video, and RCA connections.

Once you've wired your system, your audio and video signal should be traveling in the following way:

Figure 4-14. Two fields from different frames, making an odd-looking image

- From your editing system to your camera (where it converts from digital to analog)
- From your camera to your television

Figure 4-15 illustrates the connections required for viewing video on an external television.

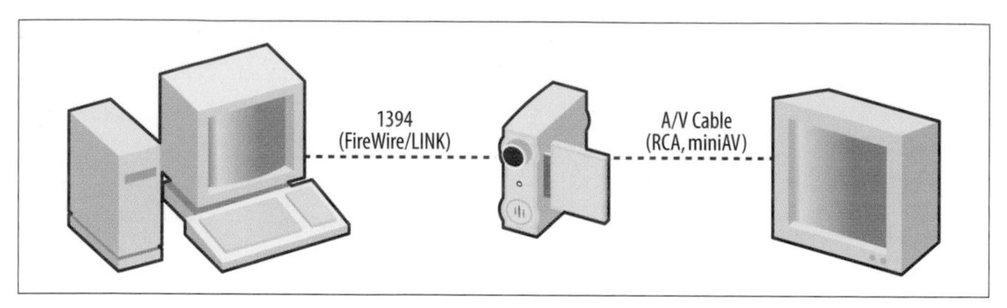

Figure 4-15. Cable connections from computer to camera to television

You can then turn on your editing system, your camera, and your television. (Yes, your electric company should send you a nice holiday gift to say, "thank you," but they won't.)

To *send* your video signal from your editing system to your television, you need to configure your system appropriately:

Avid
: Automatically configures for external sources

Final Cut
: View → External Video → All Frames

Premiere
: Project → Project Settings → General → Playback Settings... button → Video Playback → Play Video on DV Hardware

iMovie
: iMovie HD → Preferences... → Playback → Play DV project video through to DV camera

Once configured, the video you are working with on your editing system will appear on the television.

When using this setup, you might notice that the audio and video seem out of sync. More than likely, you are viewing the video on your television, while you are hearing the audio from your computer. Because the signal from your editing system needs to travel over the various cables and hardware (such as your camera) to your television, there is a slight delay from the time the signal is sent to the time it is displayed. Because the audio is not being heard from the same source (your television), it is out of sync with what you are viewing.

The solution to the problem is to monitor both your audio and your video from the same source. So, if you are using the television for video reference, you should make sure your audio is being heard through it as well. This might require you to configure your system to send the audio signal externally, in addition to your video signal.

Working with an external television provides a few more benefits, beyond being able to avoid incorrectly interlaced frames. For one, your television will display only the video image, so it will probably appear larger than it does on your monitor. Additionally, if you are working with other people, they won't have to peer over your shoulder while you work. Finally, being able to view your video on a television can be gratifying, even if it is in a narcissistic way.

Convert a Closed Caption File to a Script

If you've closed captioned your project, you can use a small amount of Perl code to extract a script for others to read.

In the business of video entertainment, if you sell a project to someone, she will probably request a script with timing information, also known as an *As Broadcast Script* in television. However, with a documentary or reality-style program, you might not have a script to give them.

Creating an As Broadcast Script is a time-consuming and tedious process. It involves listening to a section of audio and typing exactly what is heard. Often, a transcriber has to listen to a section of audio three or more times in order to accurately capture what is being said.

However, a closed caption file is regularly created in order to fulfill government requirements. With this hack, you can use the information in that closed caption file to create a script after the fact.

Getting Motivated

Even if you've never programmed a computer before, you might someday find yourself motivated enough to type a few lines of Perl, a computer programming language. If faced with transcribing for hours on end, tediously listening to audio over and over, or spending 5 minutes typing in a program and another 20 minutes formatting its results, I'll bet 9 times out of 10, you'll choose the program. In fact, once you've used the program once, you might become addicted to using it.

The following Perl program was written for a television production company that had 140 hours of video that needed As Broadcast Scripts within two weeks. A rough estimate is that it takes 5 hours for a transcriber to complete one hour of video. Therefore, there was roughly 560 hours of work to be completed.

Looking at a Closed Caption File

The process of closed captioning involves inserting codes onto line 21 of a video signal. If that doesn't make sense, think about how NTSC DV is 720× 480, which means the video signal is 720 lines wide by 480 lines high. Closed caption data would be placed on line 21 of 480.

NTSC broadcast video contains 525 lines of horizontal resolution.

Just like most computer documents, a closed caption file can be opened in a regular text editor. The results, however, aren't pretty. Here's an excerpt from an actual closed caption file (*.tds* file extension):

```
ù01000013
úFÛ14,Û14)Û13tÛ17"Û11.Coming up:Û14tÛ17"Whoa, wait.
ù01000112
úFÛ14 Û14.Û14tÛ17!Û11.That's got meÛ14,Û14/
ù01000209
úFÛ14 Û14rÛ17#a little nervous.Û14,Û14/
ù01000316
úFÛ14 Û14RThese guys are not likeÛ14rÛ17!my friends back home.Û14,Û14/
```

Looking at the contents, you can see where there are sentences being spoken. In this example, you can clearly see the words `Coming up`. The rest of it is messy.

But if you look closely, you might discover that there is a pattern: `ù(`*`some number`*`)`, then a sentence, then another `ù(`*`some number`*`)`, and so on. The number following the ù is eight characters long. Coincidentally, so is timecode.

So, you can read `ù01005108` as `01:00:51;08`.

The Code

Perl works well with text. It can be intimidating to look at and difficult to read, but it can perform wonderful tasks and save a lot of time when used. If you are using Mac OS X or Linux, Perl is most likely already installed on your computer. If you are using Windows, you can download Perl from Active State (*http://www.activestate.com/Products/ActivePerl/*; free).

The following is a Perl script to reformat *.tds* closed caption files:

```
#!/usr/bin/perl
# for Caption Center (.tds) files
while (<>) {
# remove everything before and including BeginData
```

```
$_ =~ s/.*BeginData//g;
# remove all of the >>
$_ =~ s/>> //g;
# remove all of the úF
$_ =~ s/\x9cF//g;
# remove all of the Û followed by 3 characters
$_ =~ s/\x9e(...)/ /g;
# reformat the timecodes… i.e. ù01005108 to 01:00:51;08
$_ =~ s/\x9d([0-9][0-9])([0-9][0-9])([0-9][0-9])([0-9][0-9])
/\r\1:\2:\3\;\4\t\t/g;
# replace all of the places where there is a tab-tab-return with a single
tab
$_ =~ s/\t\t\r/\t/g;
# print everything back out
print $_;
}
```

That's it. That's the entire application. Fin. Done. Out.

Save the file to your computer and name it *CCConverter.pl*—or whatever you want; it's your application after all.

Running the Hack

Once you've written your Perl application, all you have to do is run your closed caption files through it. To do so, enter the following on the command line:

```
perl CCConverter.pl <MyScriptCC.tds >MyScript.txt
```

To run a command line program on:

- Mac OS X: open the Terminal application, located in the Utilities folder.
- Windows: open the Run... item from the Start menu, then type `command`.

When running the Perl application, it will be easiest to do so from the same directory where the application resides.

You'll want to enter the name of your closed caption file for *MyScriptCC*.tds and then name the resulting file as *MyScript*.txt. The < indicates to Perl which file you want to read and the > indicates which file you want to write.

The Results

Here is the original excerpt, as converted by the Perl program:

```
01:00:00;13    Coming up:  Whoa, wait.
01:00:01;12    That's got me
01:00:02;09    a little nervous.
01:00:03;16    These guys are not like  my friends back home.
```

So, with the three or more hours, you save per script, you should find plenty of time to explore more hacks, spend time with your family, or maybe even get outdoors and breathe some fresh air.

Make a Tough Cut Easy on Your Viewers

Whether you're working from a single source or multiple sources of video, occasionally an edit "feels" wrong. Using a simple editing technique, you can help an edit become less jarring—or even unnoticeable.

Every edited piece of video attempts to tell a story, from the 30-second commercial to the three-hour documentary. To tell a story using video, it is necessary to cut from one section of video to another. Yet, the two sections of video you edit together don't always *flow* together. To overcome the jarring nature of some edits, you can use a technique called a *split edit* (sometimes referred to as an *L-cut* or *J-cut*) to help smooth the transition.

Creating a Split Edit

To create a split edit, you need to be able to edit the audio and video portions of your footage separately in your timeline. This means that for some editing systems, such as Final Cut Pro, you need to either *unlink* or lock your audio and video tracks. Other systems, such as Avid Xpress, allow you to create split edits with no additional steps.

You can create a split edit by continuing the audio section of your footage after the video section has been cut, or vice versa. In other words, the audio (or video) will continue playing after the video (or audio) has ended. To do this, you need to *roll* the appropriate track backward or forward in your timeline.

Creating an L-cut. To create an L-cut, you need to have the audio portion of your footage play longer than the video portion. You can either roll the video track backward in your timeline or roll the audio track forward in your timeline. You can use this type of edit to cut to someone listening during a conversation. Figure 4-16 shows an L-cut in an Avid timeline. Notice the audio trailing the video.

Creating a J-cut. As you might expect, a J-cut is the opposite of an L-cut. Therefore, you need to roll your video forward in your timeline or roll your audio backward. A J-cut is useful to cover jarring edits, as the audio prepares the viewer's mind for a change in imagery. Figure 4-17 shows a J-cut in an Avid timeline. Notice the video trailing the audio.

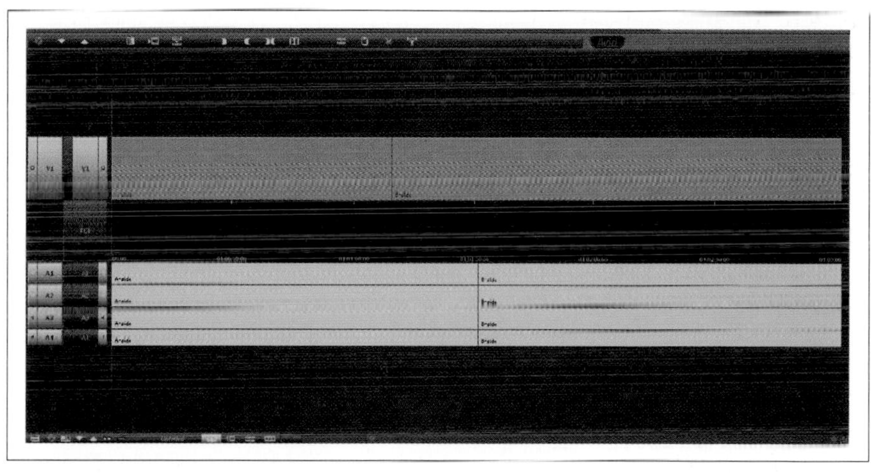

Figure 4-16. An L-cut in Avid

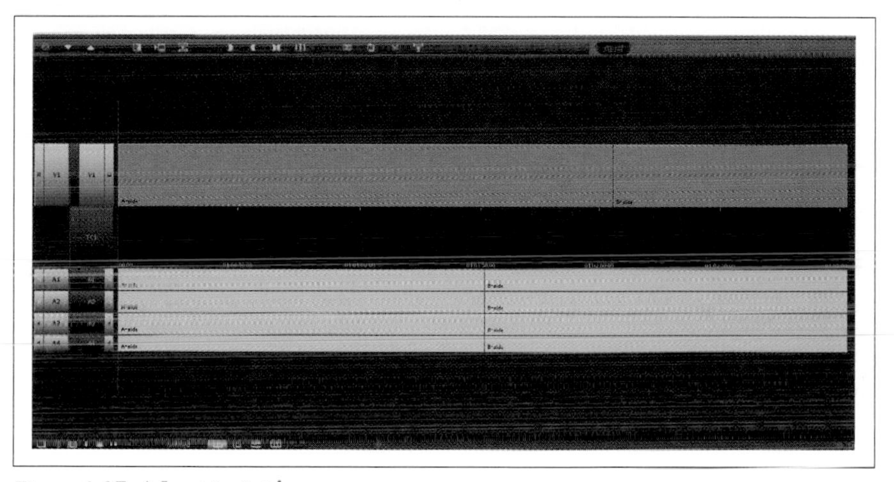

Figure 4-17. A J-cut in Avid

Using a Split Edit

Although audio comprises only a small amount of an audio-video signal, it is a vital part of the movie-making process. Using audio, you will be able to smooth over cuts that would otherwise be jarring and disruptive to your story. The composition of your footage, along with the type of transition you are attempting to accomplish, will determine which type of cut to use.

If you are editing a conversation, you can use a split-edit to transition from one speaker to another. The result is that the person *listening* is on screen for a portion of the conversation. Using a split edit in this situation creates a much more visually appealing scene, even though the reason for the appeal is the underlying audio.

You can use also split edits to overcome bad video or audio. For example, if your footage has an unwanted composition, such as someone talking underneath a shot, you can use a J-cut to introduce another angle of the shot, or perhaps a shot of a commentator, while still keeping the necessary audio.

After learning to use split edits effectively, you'll not only use them regularly, but you'll also become more aware of their use in the movies you watch.

Hacking the Hack

In *Rope*, Alfred Hitchcock attempted to produce a cut-free movie. Although he was limited by the size of the film magazines, resulting in only 10 minutes of useable footage at one time, he created a true piece of art. Of the nine cuts in the movie, only four are noticeable.

Hitchcock accomplished the feat by planning ahead and shooting action that would fill the frame at the points he needed to make a cut. For example, he would have someone walk in front of the camera when he was running out of film. When the person filled the frame, and the viewer was unable to see anything other than the fabric of actor's clothing, a cut would occur. The following shot would simply continue where the previous one ended: a full frame of fabric.

By combining Hitchcock's seamless style with the technique of split edits, you can create your own, modern-day, cut-free video.

Create an Interactive Video Catalog

We've all seen dramatizations of, or heard about, the day when you could be watching a television show and click on an actor's shirt to find out about it or even buy it. That day is today.

When viewing digital video on a computer, you gain some cool features not available on a DVD or videotape, such as linking to a URL **[Hack #84]** and allowing people to jump to bookmarks in a movie **[Hack #49]**. Another feature, called a *hotspot*, enables viewers to make parts of a video react when they click on corresponding sections of the image on screen. By combining hotspots with URLs and bookmarks, you can create an interactive video catalog.

Producing the Video

Producing a video is challenging. With the addition of interactivity, another dimension comes into play. If you know you will be producing an interactive catalog, keep a few (hopefully, obvious) thoughts in mind when shooting.

First, attempt to keep the item you are making interactive unobstructed by other objects. This will allow viewers to click on the item at any point during a scene. For example, if you would like a viewer to click on a television, don't have someone standing in front of it.

You should also keep the item stable within the scene. The best solution is to use a tripod, even if you are shooting with multiple cameras. Keeping the item steady within the frame allows viewers to easily click on it because they won't be chasing it around with their mouse.

Finally, keep in mind the size of the object in relation to the rest of the image. If your viewers are going to be able to buy a television, make sure the television isn't off in the distance or obscured somewhere in the background.

If your item is smaller than you would prefer, you can help it stand out in the frame when editing. One method is to briefly superimpose a circle around the item at the beginning of a scene. Another method is to make everything in the scene black and white, except the item **[Hack #74]**. Whatever your approach, you should have it in mind before your shoot your scene.

Creating a Hotspot

How you create your hotspots depends on the application you choose to use. There are three capable applications available: LiveStage Pro (*http://www.totallyhip.com/*; $449), cleaner (*http://www.discreet.com/cleaner/*; $549), and VideoClix (*http://www.videoclix.com/*; $349). Unfortunately, all of the applications are available only on Macintosh.

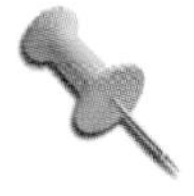

A previous version of cleaner for Windows (Version 5) did include the ability to create hotspots, as well as other interactive elements, using a feature called EventStream. To create hotspots using Windows, you need to either obtain a copy of cleaner 5 or pressure Discreet to incorporate EventStream into an update.

Because tutorials for creating hotspots are available online for both LiveStage Pro and VideoClix, this hack will cover cleaner only.

After launching cleaner, you can import video by selecting Batch → Add Files... and choosing your video file from your hard drive. To create a hotspot, you need to edit the EventStream data. To do so, either double-click on the video's icon or select Windows → Project. This brings the Project window forward.

Within the Project window, choose EventStream → Edit (the Edit button is just below and to the right of the tab) to bring up the EventStream window. To add a hotspot, you need to locate the section of video where you would like to place the hotspot and click the Add button to create a new event.

In the Event Type pop-up menu, select Hotspot. You can then click and drag over the section of video you would like to allow your viewers to click. Additionally, you can choose whether the shape of the hotspot, as well as assign it a Label for reference. Figure 4-18 shows a hotspot being created with an Oval shape.

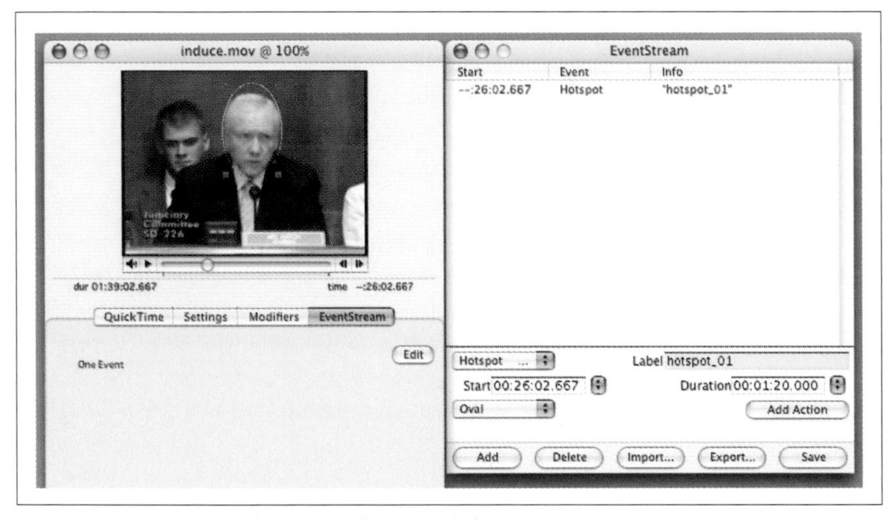

Figure 4-18. Creating a hotspot with an Oval shape

Even though you've now created a hotspot, you still have a few other tasks to complete. First, unless you want the hotspot to last for the length of your video, you need to decide its duration. Fortunately, cleaner makes it fairly simple to determine how long to enable a hotspot: just click on the small pop-up menu to the right of the Duration text field. The menu offers three choices: To In Point, To Out Point, or To Current Time. Selecting any of these will enter the respective duration into the Duration text field.

Finally, create an *action* for your hotspot by clicking on the Add Action button. This creates a related event, which executes when the viewer clicks on the hotspot. The two types of actions you will be interested in, at least when creating a catalog, are the Open URL action and the Go to Time action. Figure 4-19 shows an Open URL action being entered for the referenced hotspot. This action causes the specified URL to open in a viewer's web browser.

Figure 4-19. Creating an Open URL action

The way in which you implement your catalog will determine which action you choose. If you are planning to integrate a web site into the process—to allow viewers to make a purchase online, for example—use the Open URL action and enter the URL where you would like your viewers to be sent. If you plan on having your video display within a web page, use the Target entry to have the URL opened in a separate frame (HTML knowledge is helpful in this instance).

If, on the other hand, you would like to provide more information on the item, use the Go to Time action. This action makes your video jump to another section within itself, such as a video of the item close up, along with dialogue explaining the item in detail. The Go to Time action can be jarring, yet exciting, when experienced for the first time.

When used together in the same video, the Open URL and Go to Time actions can create quite compelling content. To create more hotspots in your video, just rinse and repeat. When you're finished, don't forget to click on the Save button in the EventStream window! Otherwise, all of your events will be missing when you do finally encode your video for delivery.

See Also

- LiveStage Pro (*http://www.totallyhip.com*; $449.95)
- Video Clix (*http://www.videoclix.com*; $349–499)

Hacking the Hack

Just because you can use hotspots to create a catalog, that doesn't mean they're limited to that use. One alternative is to create a "choose your own adventure" type of movie. By using a hotspot and an action to jump to another section of your movie, you can allow viewers to click on areas of the video to influence the story.

For example, you could create a scene in which a character needs to choose between two doors to possibly win a prize (think *Let's Make a Deal*). Upon reaching the decision stage, you could allow the viewer to click on a door and, based on her selection, jump to the appropriate section of your video. Quite obviously, you need to shoot two scenes, one for each door.

However you use hotspots, one thing is certain: your project will take a step beyond traditional "passive" video.

Fix Timecode Problems on an Existing Tape

Broken timecode will cause you headaches in the long run. Fix it early or invest in Excedrin.

If you have timecode that jumps from one time to an *earlier* time, you will encounter problems when digitizing your footage. For example, if your tape's timecode is 00:34:23;00 and jumps to 00:00:00;00, you have a problem. It is best to fix problems in your process as early as possible. One way to *avoid* this problem is to black and code your tapes [Hack #4] ahead of time.

Some people attempt to overcome timecode problems while digitizing by using tape numbers to indicate which breakpoint to locate. For example, tape *TCP001* would be the section of tape before the jump and tape *TCP001.1* would be the section of tape after the break. They continue to increase the last number for each break. Not only does this approach *not* solve the problem, but it might cause additional problems in the future with tape numbering and EDLs [Hack #3].

If you are going to fix timecode on a tape, you have to transfer it to another tape or onto your computer. The process of transferring your footage can cause you to lose quality, which you do not want to do. You only want to fix the timecode.

Avoiding a Loss in Quality

In the past, with analog formats such as VHS and Beta, video footage would lose quality when it was copied from one tape to another. Many people refer to this loss of quality as *generation loss*. Fortunately, in the digital world, you can create a copy of your footage without losing any quality.

In order to avoid generation loss, you need to make a *digital* copy of your footage. To create a digital copy, you must first create a digital connection between two capable devices. Most digital video cameras and decks have an IEEE 1394 connector to create a digital connection.

IEEE-1394 is commonly referred to as FireWire® or i.LINK®.

IEEE-1394 connectors come in three types:

- 4-pin
- 6-pin
- 9-pin

For the most part, you will not have to deal with 9-pin connections. The fun part, though, is figuring out what type of cable you need to connect your two machines. You have three options:

- 4-pin to 4-pin
- 4-pin to 6-pin
- 6-pin to 6-pin

Many sales people I have encountered do not know the difference between the connections, so you should take the time to figure out what you need before shopping for your connection cable. Figure 4-20 illustrates the differences between these different cable types.

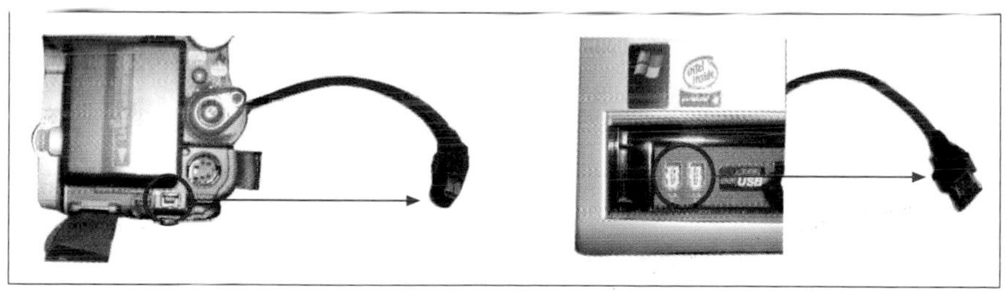

Figure 4-20. Choosing the right cable connection for your machines

Machine to Machine

If you are fortunate enough to have more than one digital video machine—either a camera or a deck—you can create a digital duplicate of your tape.

To create a digital copy, you cannot use RCA connections, because those connections are analog. If you use RCA cables for your connection, you will lose a generation of quality.

To duplicate your tape, connect your two machines using the IEEE-1394 cable. Place your master in one machine and a blank tape in the other. Press Play on the machine with your master tape and Record on the machine with the blank tape. While your master tape has broken timecode, only the footage (both audio and video) will be recorded to your new copy. The new copy will have an exact digital copy of you footage, but with new, continuous timecode. If you have a black and coded tape **[Hack #4]**, you can use it to ensure your timecode is continuous.

Digitize and Output

If you have only your digital video camera, all is not lost. You can still fix your timecode problem by using your editing system to digitize your footage and then performing a digital output. To create a digital copy of your footage, you have to digitize your tape using the same codec it was acquired with. For example, if you are using DVCPRO, you should capture your footage using the DVCPRO setting.

There are different types of digital video. You should be aware of which type you are using. Most consumer digital video cameras record DV, but there are also DVCPRO, DVCAM, and HDV formats.

You should also make sure you have enough hard drive space to hold a complete tape. For DV, you will need approximately 13GB of space per hour of footage.

To digitize your footage, connect your camera or deck to your editing system. Once connected, you need to capture the footage from your tape to your system. Each editing system is different, but the end result is the same. Here's how to get started in a variety of popular editing systems:

Avid
: Tools → Capture

Final Cut
: File → Log and Capture...

Movie Maker
: File → Capture Video

Premiere
File → Capture

iMovie
Switch to Camera mode and then click the Record button

If your editing system allows you to Ignore Timecode Breaks, enable that feature.

Once you have digitized your footage, place the entirety of it on a timeline and then export your footage to a new tape:

Avid
Clip → Digital Cut

Final Cut
File → Print to Video...

Movie Maker
File → Save Movie File...

Premiere
File → Export → Export to Tape

iMovie
File → Share → Videocamera → Share button

You could record back out to your original tape, but if something goes horribly wrong, such as an ugly power surge, you could lose your footage.

Just like using two cameras or decks, when you output your footage to tape, you will have an exact digital copy of the footage on your master tape . . . except for the timecode problems.

Finish Up

When the process is complete, make sure you label the duplicate tape as your *new* master and keep track of which tape is which. As always, proper labeling and organization are key to successfully completing your project **[Hack #1]**.

Add Bookmarks to Your Movie

Both QuickTime and Windows Media have a feature to allow viewers to jump to specific points in a video. Using this feature, you can create a casting reel for your movie.

When casting for a movie, you have to audition a lot of people. Even after narrowing down your list of candidates, you still want to view their auditions easily. You might even want to distribute video to your colleagues for their input. You can either pass along one videotape or DVD per candidate, or you can create a single digital video file and allow people to jump from one candidate to another easily.

To enable people to move from one section of video to another, you need to create locators in the video. In QuickTime these locators are called *chapters*, and in Windows Media they are called *markers*. For the sake of consistency, I will refer to them as *bookmarks* when referring to their use in both systems.

Although both QuickTime and Windows Media provide functionality for bookmarks, the way you create them is completely different in each platform. For QuickTime, you use QuickTime Pro (*http://www.apple.com/quicktime/upgrade/*; $29.99) and a text editor. For Windows Media, you use the Windows Media File Editor, which is a part of the Windows Media Encoder 9 Series of tools available from Microsoft (*http://www.microsoft.com/windows/windowsmedia/9series/encoder/default.aspx*; free). If you plan on distributing both types of video, you can use Discreet's cleaner (*http://www.discreet.com/products/cleaner*; $549) to create bookmarks and have both types of video created from it.

If you are using Final Cut Pro or Premiere to edit, you can easily create chapters using *markers* within your editing system. After adding the appropriate markers, your bookmarks will automatically be included when exporting your sequence to a compatible media format.

Within Final Cut, you can to edit a marker by choosing Mark → Markers → Edit... Then, you can type `<CHAPTER>` in the Comment text box or click the Add Chapter Marker button.

Within Premiere, you can edit a marker by selecting Marker → Edit Sequence Marker... Then, you simply enter the chapter's title in the Chapter text field.

Creating Chapters Using QuickTime Pro

If you have QuickTime and a text editor installed on your computer, you have the tools to create a chapter track. However, you will need to upgrade QuickTime Pro to unlock the features necessary to create chapter tracks. Upgrading also unlocks the ability to perform minor editing, apply filters and effects [Hack #39], export movies for use on a cell phone [Hack #89], and a slew of other features.

Creating a chapter list. To create the chapter track, you need to create a list of chapters you would like people to be able to access. Using a text editor, such as Text Edit, simply create a list of chapters, each separated by a line break. For example, on a casting reel, you would create a chapter for each actor, most likely by name:

Oxford Wells
Tony Clark
Rich Gable
Joey Harper

After creating your list, save it as a text file.

After you have created your initial chapter list, launch QuickTime Player, choose File → Import..., and then select the text file containing your chapters. This action will import the text file and create a QuickTime movie from it. Figure 4-21 shows a text file, as imported by QuickTime Player.

Figure 4-21. An imported text file for use as a chapter track

Oddly enough, immediately after importing the chapter list, you need to export it. Selecting File → Export opens an export window with a few options. The configuration of these options is important, because they will affect the chapter track.

Select Export → Text to Text, and then choose Use → Text with Descriptors from the available pop-up menus to create an appropriate text file. Additionally, you should click the Options... button and select to Show Text,

Descriptors, and Time. Also select "Show time relative to start of:" → Movie and "Show fractions of seconds as:" → 1/30 from the Options window, as shown in Figure 4-22.

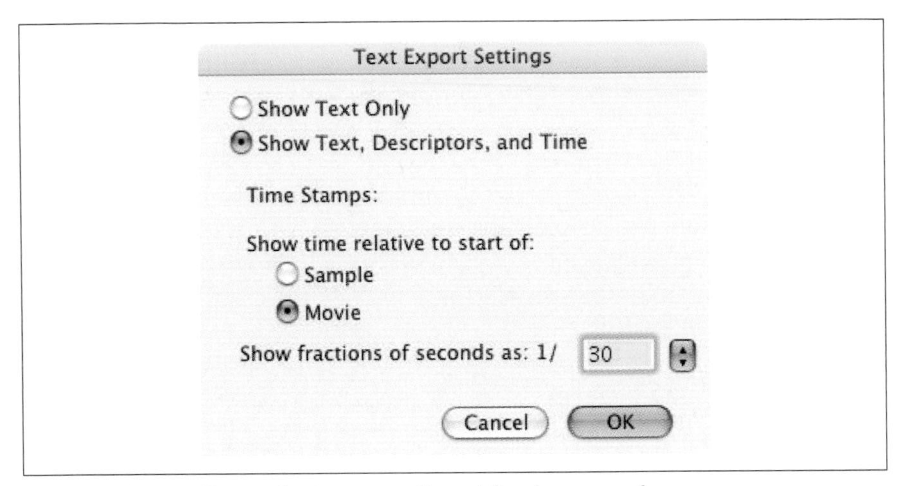

Figure 4-22. Text Export Settings as configured for chapter track use

Editing a text descriptor file. After exporting the QuickTime descriptors file, you need to open the file in a text editor. You also need to open the movie you want to add the chapter track to in QuickTime Player. Once the movie is opened in QuickTime Player, choose Movie → Get Movie Properties and then select Movie from the first pop-up menu and Time from the second pop-up menu.

When you open the text file, as exported from QuickTime Player, you will notice a bunch of words and numbers enclosed in braces and brackets, as well as a particular format to the text. *Do not* alter any of the information *except* for the timecode inside the brackets ([ ]) and the respective chapter title, unless you feel comfortable experimenting. For example, the following is for a chapter at 00:11:10.08 and titled Tony Clark:

```
[00:11:10.08]
{textBox: 0, 0, 50, 160}Tony Clark
```

Take note that the timecode is formatted as `[HH:MM:SS.FF]` and the title of the chapter appears on the line below the timecode, just after some QuickTime required formatting.

The Properties window displays the Current Time (where you are in the movie's timeline), the Duration of the movie, and the Selection Start and

Duration times, which is basically representative of In and Out points. As you change your location in the timeline of the movie, the Current Time will change accordingly.

When you have found a position where you would like a viewer to jump to, such as the beginning of a person's audition, type the Current Time into the appropriate location in your text file. You should continue the process of locating the time and typing it, until you have entered information for all of the chapters in your movie. Figure 4-23 shows a movie, its Properties window, and a text file being edited for use as a chapter track.

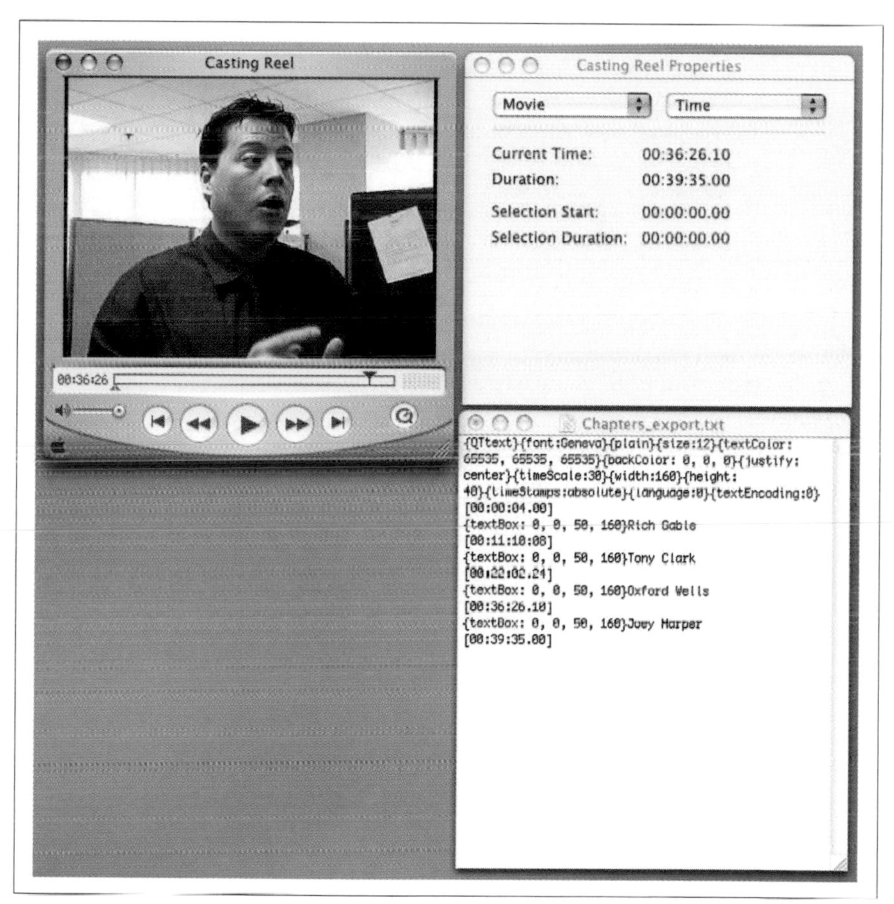

Figure 4-23. Creating a chapter track as a text file

You cannot have times in your chapter track run out of order. Therefore, if your initial list is not in the same order as your video, you need to rearrange the list as necessary. For example, if you initially had Joey Harper listed first, but discover he appears fourth in the video, you need to move his name to fourth in the list.

The last timecode in your text file should be the duration of the movie. So, if your movie's Duration is 00:39:35.00, enter your final timecode as `[00:39:35.00]`. When you're finished, save your text file and then import it into QuickTime Player by choosing File → Import... and selecting that file from the dialog box.

Creating a chapter track from a descriptors file. After importing your descriptor file into QuickTime Player, choose Edit → Select All and then Edit → Copy. Then, paste the copied movie into the movie you would like to add the chapter track. Selecting the movie and then choosing Edit → Add Scaled enables your movie to use the imported text as a chapter track.

In the Properties window of your movie, choose your newly added Text Track from the first pop-up menu and Make Chapter from the second pop-up menu. Then, click the Set Chapter Owner Track button and select either the audio or video track as the Owner. Figure 4-24 shows the Properties window with a newly created chapter track.

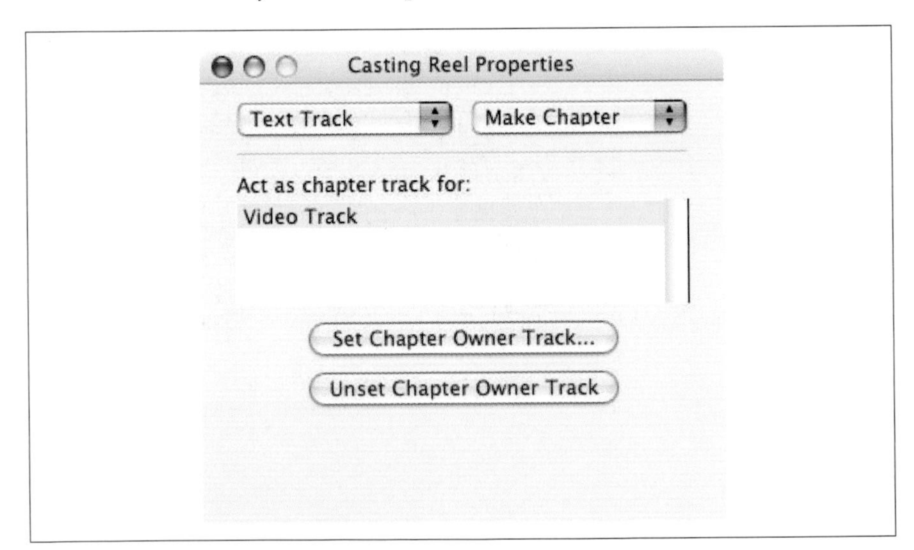

Figure 4-24. A newly created chapter track with a Video Track owner

A feature called Preloading makes sure your chapter track loads before your movie does. Although QuickTime Player might automatically enable Preloading on your chapter track, you should make sure it is enabled. From the second pop-up menu in the Properties window, select Preload and make sure the Preload checkbox is checked.

Finalizing a movie with a chapter track. After creating the chapter track, you will notice a small pop-up menu inside your movie's window. As shown in Figure 4-25, this pop-up contains the chapter titles and will allow a viewer to easily jump from one section of your movie to another. You will also notice that there is a small black box in the corner of your movie with the title of each chapter.

Figure 4-25. A chapter track with an annoying black box in the corner

You most likely don't want the side effect of the black box, so you can turn it off by choosing Edit → Enable Tracks and turning off the Text Track. Then, use the chapter track pop-up to jump from one place in your movie to another, to make sure it works as you expect. If it does not, you will need to repeat the process until it does and pay careful attention to the timecodes you enter as well as their respective chapter titles.

When you are satisfied with your movie, choose File → Save As... and select the "Make movie self-contained" radio button. Whew! That was fun.

Creating Markers Using the Windows Media File Editor

The Windows Media File Editor is installed, along with a suite of other applications, when you download and install the Windows Media Encoder 9 Series of tools. Using the File Editor is straightforward, but it does require your video to be in the correct format, either WMV or ASF. Within the application, you can also add commands to display text, open applications, open URLs, or perform other scripting events.

Opening a video. After launching Windows Media File Editor, you need to open a video file to edit. To do so, choose File → Open... and locate your movie. The application opens Windows Media Files (*.wmv* and *.wma* extensions) or ASF files only. If your video is in another format, you'll need to transcode it **[Hack #29]**. You can also transcode between various Windows Media formats using Microsoft's Windows Media Encoder 9 Series. Figure 4-26 shows a Windows Media Video named Audition being opened using Windows Media File Editor.

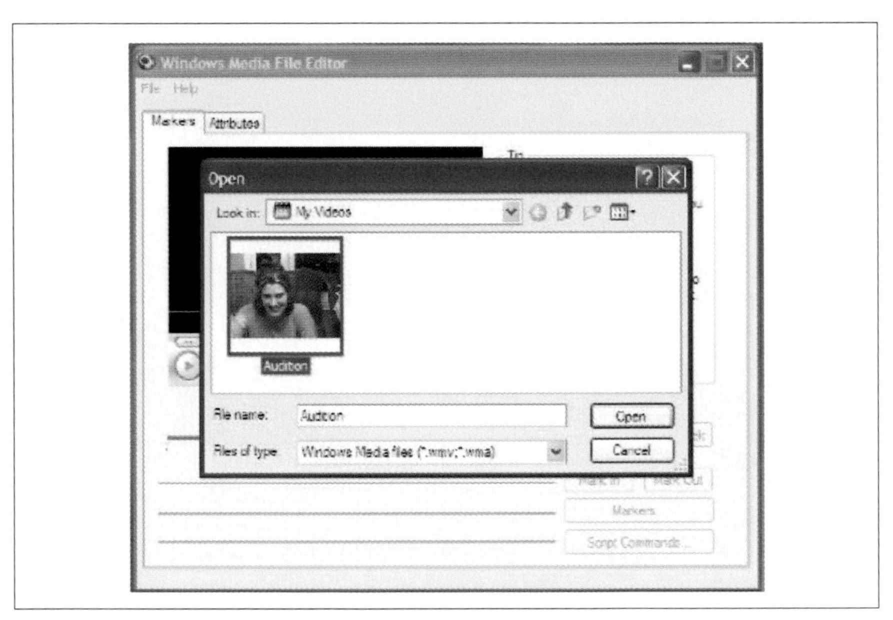

Figure 4-26. Opening a video using Windows Media File Editor

After opening your video, you can play it just as you would in a media player. However, this is more than a media player, so you will notice that you can add markers, add script commands, and enter information for various attributes like the Title, Author, and Copyright information.

Adding the markers. To add markers, scan the video and stop at a point where you would like to add a marker. Then, click the Markers... button to open the Markers window. If you know the times for each marker you would like to add, you will be able to add your markers a lot faster.

The Markers window provides a list of all the markers associated with the video file. Clicking on the Add button brings up a dialog box to enter the Marker Properties for both the Name and Time. The current time (where you are in your movie) will be entered in the Time field, so the only thing you need to do is enter the Name for the specific marker, such as an actor's name. Figure 4-27 shows a marker with the Name of Joey Harper and the Time of 00:36:26.1 being entered.

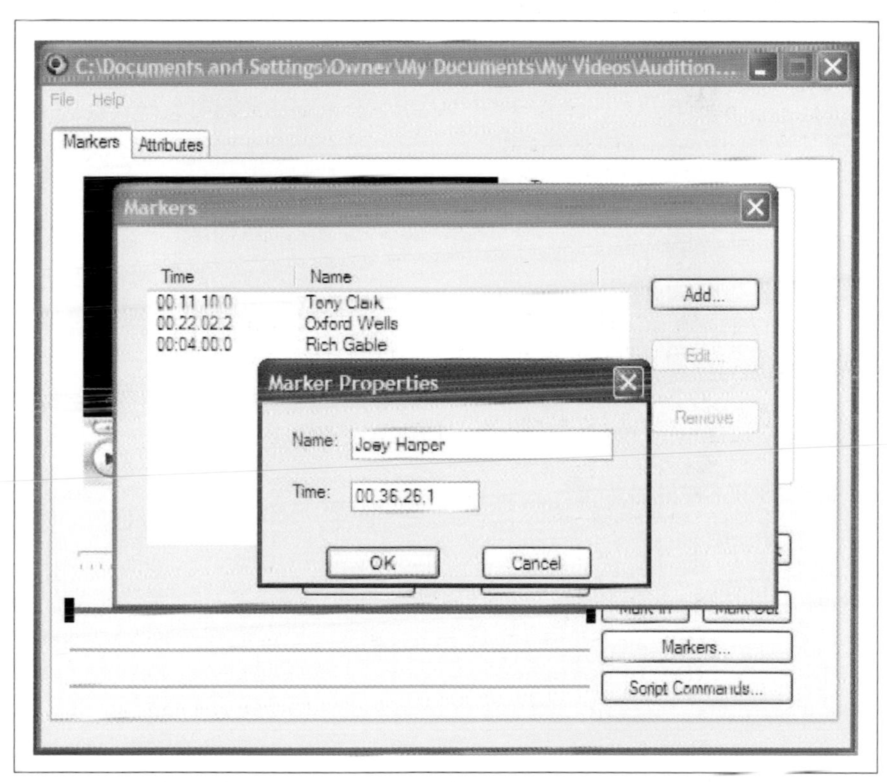

Figure 4-27. Editing the properties for a marker

When you have entered the Name and Time, click the OK button to close the Marker Properties window. If you do not know the time for any additional markers, click the OK button to close the Markers window and locate the next point in your movie. You will need to repeat the process for each marker you would like to add.

Saving a marked movie. After entering all of your markers, you need to save your video file. To save the markers inside the current file, choose File → Save and Index. This will save your markers to the video file you are currently working on.

After you've saved and indexed your movie, you will be able to jump from one marker to another by choosing View → File Markers and choosing a marker. Figure 4-28 shows the markers for the actors in the Audition video.

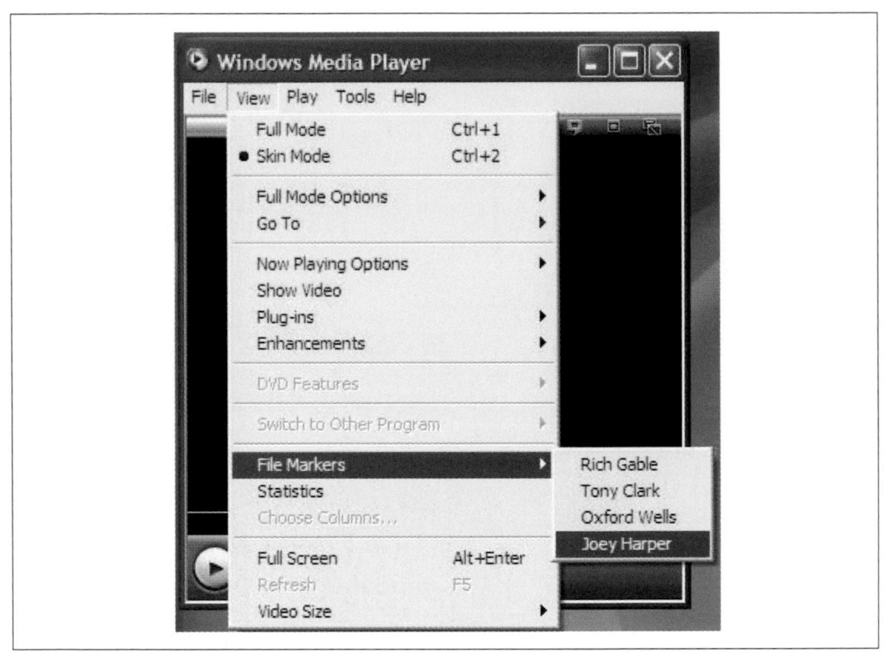

Figure 4-28. File markers available through Windows Media Player

If you would like to save your movie in a different format, choose File → Save As and Index. Performing the Save As function enables you to export your video as either a Windows Media file or an ASF file. Although not a big deal, and limited, it is a nice feature to know about when it's needed.

Using cleaner to Create Bookmarks

If you own Discreet's cleaner, and you want an easier-to-understand solution, or you plan on distributing both QuickTime and Windows Media files, using cleaner to create bookmarks is the perfect solution. cleaner refers to the feature as a *chapter event*.

To begin, launch cleaner and simply drag the movie you would like to work with to the Batch window. Then, choose Windows → Project or double click on the movie's icon to activate the Project window.

From the Project window, click the EventStream tab and then click the Edit button, which brings up the EventStream window, allowing you to add the Chapter information. Figure 4-29 shows chapters being added to a movie using EventStream.

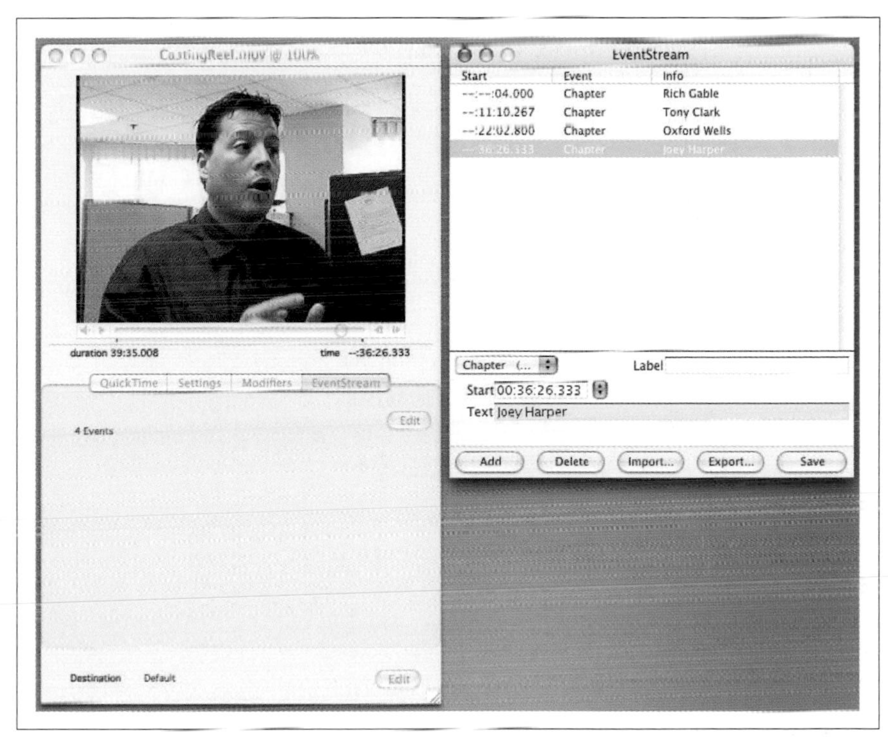

Figure 4-29. Adding a chapter using cleaner

To create a new chapter, locate the position on the movie's timeline (in the Project window) where you would like the Chapter to start. Then, in the EventStream window, click the Add button and select Chapter (QT, WM) from the pop-up menu. cleaner will automatically enter the Start time for you, so you just need to type the Text entry as you would like it to appear in your movie's chapter list.

Merely continue the process of locating the start time, clicking the Add button, and adding Text until you have added all of the chapters you want. When you have finished, click the Save button and close the EventStream window. To create your final movie, choose Batch → Encode or click the Start button in the Batch window.

That's all you need to do with cleaner—straightforward and easy.

Hacking the Hack

Just because I've presented this for use as a casting-reel solution doesn't mean it's limited to such use. You can create a video catalog, where people can jump to various products. Or, you can create a how-to video, where viewers can jump from one step to the next.

If you want to get creative, you can combine the bookmark feature along with the ability to open a URL **[Hack #84]** and open a web page related to the scene in the movie. For example, if you have created a video catalog, when a user is presented with an item on screen, a web page could open with details about the item, such as the current price or ordering information. You can also take your approach one step further by creating an interactive video catalog **[Hack #47]**.

Create Faster DVD Navigation

Navigating DVD menus doesn't have to be slow and tedious. You can use a simple setting in your authoring software to speed up the next/previous page actions.

We've all seen them: pages and pages of menus for jumping to a particular chapter in a movie. Many of the extra features, such as cast bios and original scripts, also contain multiple pages of information. DVD menus require the user to interact with on-screen buttons.

You use directional controls (up, down, left, right) to navigate to a particular button and then use a Select or Enter control to activate the button. The process is fairly straightforward, but it can be a bit tedious…especially if you're trying to get to page 5 of 7 quickly.

Challenging the Norm

What if you could skip the activate step? Most DVD authoring tools support an option known as auto-activation. When the user navigates to the button using a remote control and a consumer player, the button is automatically activated. On a computer using a software DVD player, this feature is usually disabled; otherwise, simply passing your mouse over a button would send you flinging off to some distance section of the DVD.

Well, it's not that bad, but it would be annoying if you had to be careful where you moved your mouse when trying to use the DVD menu on your computer.

Consider Figure 4-30. When the user lands on this menu, the currently selected button is the Main button. You can move up to select an actual chapter, or you can move left or right to go to the previous or next page of chapters, respectively. Rather than making the user press the Right button on the remote and then the Select button, you can trigger the Next Page button automatically when the user navigates to it.

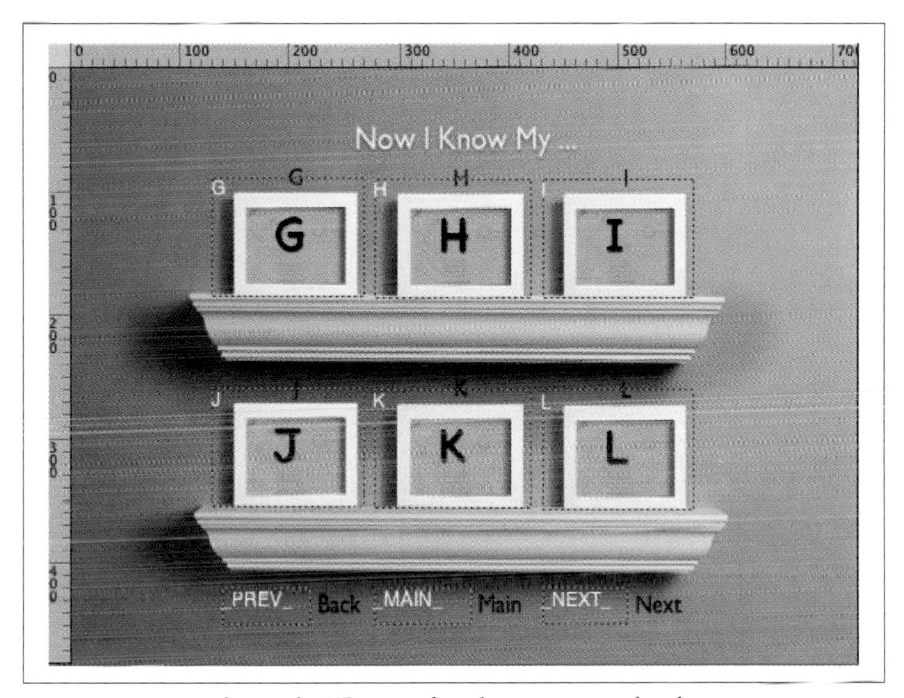

Figure 4-30. A sample paged DVD menu for selecting a particular chapter

Figure 3-31 shows the property inspector for the Next button in Apple's DVD Studio Pro. The Auto Action option has been highlighted. Check that box and you're done! When the menu comes up, pressing the Right button on the remote control immediately navigates to the next page of the menu.

Do the same for the Previous button, but don't set that action on any other buttons. You want to use the Auto Action sparingly. If, for example, you set the Auto Action on all of the chapter buttons, you could never get to all of the chapters. Pressing the Up button on your remote control would take you

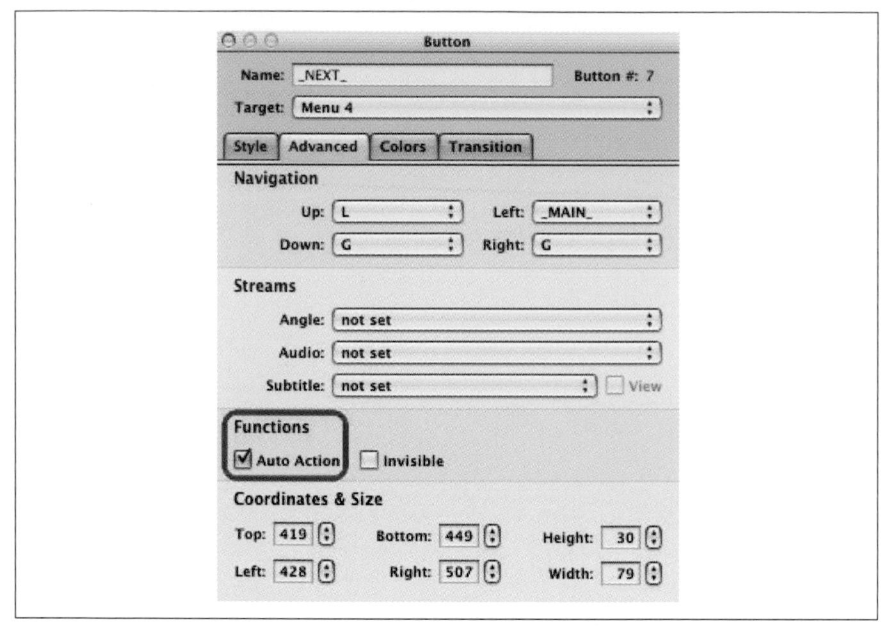

Figure 4-31. Configuring the Next button to activate automatically

to the K chapter index and you would immediately jump to the K section of the video clip. You would never have the chance to navigate to any of the other chapters.

Hacking the Hack

Any interesting variation on this hack can be used to produce reasonably functional slide shows with your DVD. Consider the slide shown in Figure 4-32. You can make the slide the background image of your menu and place the same Previous, Main, and Next buttons on it. Default to the Main button as before, and now you can navigate your entire slide show using the Left and Right arrows on your remote control.

Notice how big the proposed buttons are. If you play this DVD on your laptop, you will want a large target area, to make it easy to navigate with your mouse.

Of course, you'll also want to use the Invisible option in your authoring software, to make sure the buttons don't obscure the slides. Figure 4-33 highlights this option in DVD Studio Pro.

Even if your authoring tool doesn't have an explicit Invisible option, you can usually trick it by supplying a completely transparent GIF or PNG as the image to display for the button.

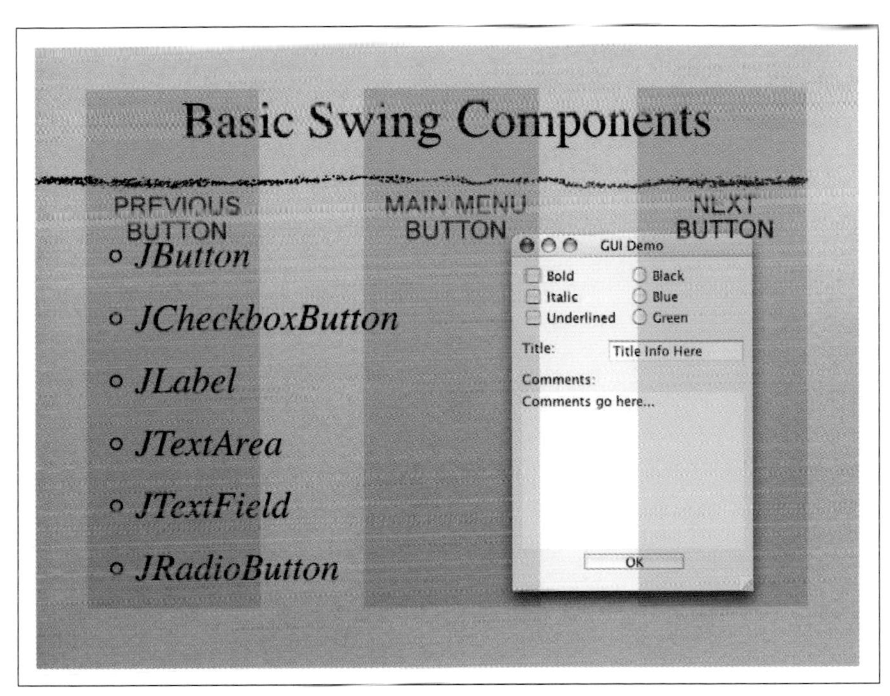

Figure 4-32. A sample slide ripe for auto-action navigation

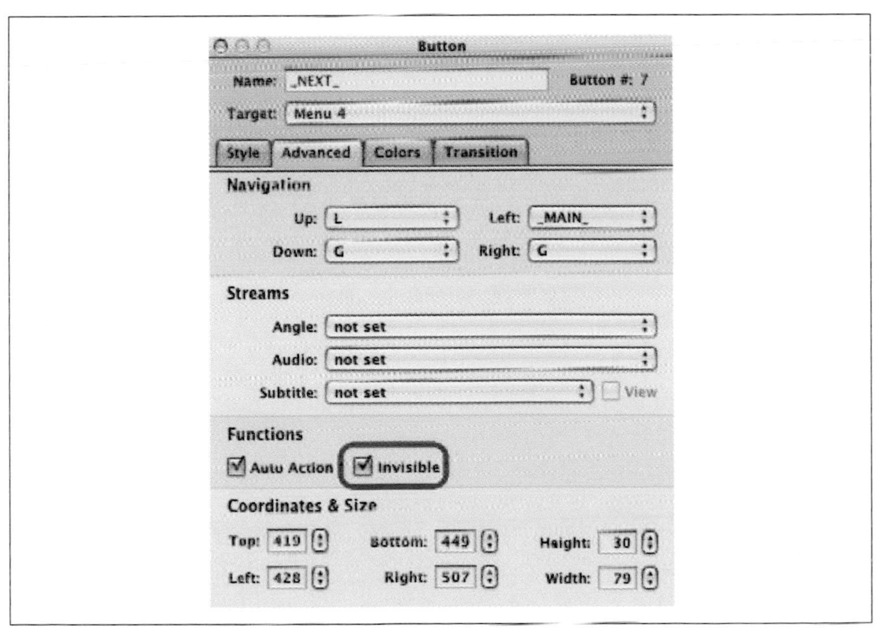

Figure 4-33. Selecting the Invisible option in DVD Studio Pro

—Marc Loy

CHAPTER FIVE

Audio

Hacks 51–60

Audio is often an afterthought in video production, even though it's a vital part of it. There is something very special about audio and its effect on human emotion. Music, sound effects, and even the clarity of dialogue, will all have an effect on your audience. Whether you're shooting your child's first birthday or the next epic movie, you should always pay attention to your audio's characteristics. No one wants to watch a video and constantly have to ask, "What'd she say?"

Mobile Audio Booth

You've probably got a perfectly good sound booth at your house, and you don't even know it.

Recording clear dialogue can often be a significant challenge. Outdoor shoots battle unwanted noise from wind **[Hack #52]**, aircraft, traffic, and more, while indoor shots often have to contend with interference from room echoes, central air systems, computer fans, and electrical hum **[Hack #56]**. When these factors are combined with a limited audio toolset, capturing clear, understandable on-set dialogue that will hold up over a background music score becomes a nearly impossible task.

Even the most expensive professional productions have to contend with some of these challenges. Often the solution is to hold an Additional Dialogue Recording (ADR) session in a recording studio after a shoot, where the actor or actress records their lines in a noise-free, echo-free sound booth. Another advantage to ADR is that the performer can focus solely on their delivery of the dialogue, while having the script in front of them, free from concerns about body position, hitting marks, and other on-set distractions. The dialogue recorded in the studio is then used to replace the on-set audio, resulting in crystal clear sound in the finished movie.

Finding a Sound Booth

On a limited budget, however, renting a sound studio for an ADR session could easily be out of reach. There is however a readily accessible no-budget alternative: record your ADR inside a parked car.

Most modern car interiors are specifically designed to be quiet, deadening much of the sound from outside and limiting echoes within the cabin itself. More expensive luxury cars tend to have the best audio characteristics, but even many economy cars have better audio isolation characteristics than typical rooms that lack deliberate soundproofing. Another advantage to using a car is that...well...it's quite mobile. So, even if your studio or "home base" is in a noisy area, you can simply drive your "sound booth" to any accessible, quiet, location when you're ready to record.

Preparing on Set

Even when planning to use ADR, it's generally helpful to take a best effort approach to recording dialogue on set. This raw audio can be useful as reference for editing prior to ADR, and can also be played back during an ADR session to help a performer recall the particular timing and rhythm they used when delivering the lines for the camera.

Before leaving the set, it is also a good idea to record several minutes of the natural, ambient sounds **[Hack #55]** (yes, all that noise you're actually hoping to get rid of!) at the location, with no dialogue or other movie-making related sounds, for use in the final mix. While these sounds can be distracting to the story if they are too loud, they can still be important to establishing the character of the location and solidifying the scene's environment in the viewer's perception.

Finding a Location

As mentioned before, it's best to drive your automotive ADR studio to a quiet location. It can also be helpful to have a rough edit of the scenes ready to playback (on a camcorder or any audio player) as a guide for the dialogue performance in the ADR session.

For recording, a digital recorder with a good microphone is best. Virtually all DV cameras record CD-quality digital sound when set to 16-bit mode. If you use a DV camera, make sure that it is *not* set to 12-bit audio mode when recording, because this setting results in a lower recording quality.

If possible, avoid using a camera's built-in microphone because they can sometimes pick up extra noise from the camera's tape drive. An external condenser (or *shotgun*) microphone is best, because that will cancel out sound that is not coming from directly in front of the pickup element, rather than a dynamic mic, which picks up sound from all directions.

Be sure to have one person on hand (other than the performer) who is responsible for running the recorder and listening to the audio while it records. This will allow the performer to focus on the dialog, and help ensure that you do, in fact, get a good recording. This person should also take careful notes regarding the timecode locations of each recording **[Hack #11]**, so finding the good takes will be easier when capturing and editing the final scenes.

Finally, when delivering the dialogue, the performers should strive to use the same timing and emotion as was performed on set for the camera.

Editing Your Audio

In closing, here are a few tips that can help with synchronizing the ADR takes with the camera shots from the original set:

- Use your editing application's audio waveform display to show the audio wave on the timeline. This will help make it possible to align the audio and video visually.
- Use cue marks to *tag* important syllables seen in the video and the matching syllables in the ADR track.
- Use a short fade-in and fade-out on the ends of ADR tracks to avoid noticeable pops and clicks in the final audio.
- If an ADR track is too far out-of-sync with a character's on-set actions, consider cutting to a different angle to show another character listening to what is being said, or something else related to the dialog so that viewer never sees the discrepancy.
- Remember to mix in just a little of the ambient, background noise recorded on-set, to help establish the location's environment without obscuring the important dialog.

As with any creative process or technique, using and editing ADR will take practice to perfect, but hopefully, the ideas and tips presented here will help you on your way to creating crystal-clear dialogue for your projects.

—Nick Jushchyhsyn

Reduce a Microphone's Wind Noise

Microphones are designed to pick up sound, but some sounds, such as wind, are unwelcome guests. A windsock can help reduce the sound of wind as it blows past a microphone.

Many sports fanatics have probably seen cameramen flanked by a group of personnel, one holding a long pole with a funny-looking, furry object hanging off the end. Well, both the pole and the fuzzy object have an important role in capturing quality audio. The pole, better known as a boom **[Hack #53]**, helps position the fuzzy object, better known as a *microphone* (covered in a windsock), to capture the necessary audio.

Designing a Windsock

The first step in creating a windsock is to purchase the necessary fabric. Many fabric stores carry fake fur, which provides the best wind noise reduction. In fact, the longer the hair, the better. And while you're at it, get some of the gaudiest stuff you can find, because it's not going to show up on camera anyhow. (Okay, that last part isn't necessary, but it's fun!)

The amount of fabric you need to purchase depends on the type and size of microphone you are going to cover. If you purchase one-quarter to one-half yard of fabric, you should have plenty of material to work with. You can always explain to a salesperson what you are trying to do, if you have questions. (I know I'm a little out of my element inside a fabric store.)

Once you've got your fabric, lay it out, fur-side down, and place your microphone on top of it. Then trace two outlines of your microphone, allowing a little extra room for what will become the seam. This is usually about 5/8 of an inch. After tracing your microphone, cut out the two sections, as shown in Figure 5-1.

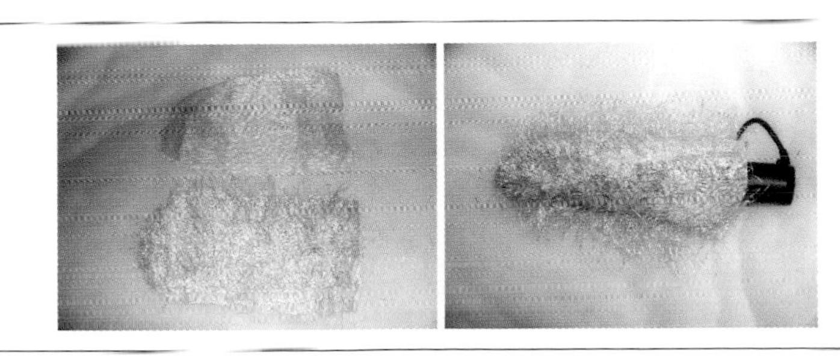

Figure 5-1. Two halves (left) and a completed windsock (right)

Place the two sections fur-side to fur-side and sew around the edge of the pattern. After sewing the two sides together, turn your new windsock right-side out and slide it onto your microphone. If the sock is too loose, you will probably have to rip the stitching and sew the sock again, this time with a larger seam.

The best time to use your windsock is, well, in the wind. Seriously, though, you can use a windsock in most situations to simply reduce the amount of extraneous audio your microphone can pick up.

Resist the temptation to *double up* your windsock (by adding a second one over the first), because doing so will reduce the overall quality of your audio.

Creating a Windsock in a Pinch

You might find there are times where you don't have a windsock with you, either because you haven't made one yet, you lost it, or you simply forgot it. Fortunately, not all is lost in such a situation. If you, or someone around you, are wearing athletic socks, you're in luck.

The *in-a-pinch* windsock is easy to create. Turn your athletic sock inside-out. Then, place the sock over your microphone and proceed as if you've created a perfectly normal windsock.

HACK #53 Make Your Own Boom

To capture better audio than that offered by your camera, you'll need a long pole and a decent microphone.

A *boom*, sometimes called a *boompole* or *fishpole*, is essentially a long pole used to hold and maneuver a microphone, which is at one end of the pole. Using a boom allows you to get the microphone closer to your subject, while still keeping it out of the frame's composition. Although you can purchase professional-quality booms online (search for `boompole`), it is easy to build your own.

Finding a Pole

First, and foremost, a boom is a pole, so that's what you'll need. The length of the pole will determine how far away from the subject you can be, while still recording audio. You can use something as simple as a broomstick or as elaborate as a telescoping aluminum pole. The most flexible option is to locate a telescoping pole, as it will allow you to adjust the length of your boom to fit whatever situation you might find yourself.

Purchasing a Microphone

Audio is an important aspect of digital video, yet it is often an afterthought. Many microphones are on the market, and each one has its pros and cons. For use with a boom, you should purchase a *shotgun* microphone, which looks like a long, thin cylinder, as shown in Figure 5-2. I recommend evaluating the microphones available from Audio-Technica (*http://www.audio-technica.com/*) and Sennheiser (*http://www.sennheiserusa.com/*), among any others you might come across.

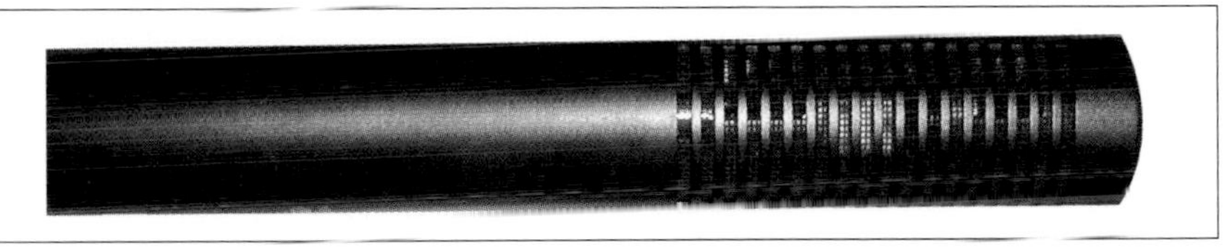

Figure 5-2. A shotgun microphone

You might also be able to rent a microphone, depending on where you live and the businesses in your area.

Unless you already have a microphone, purchasing a separate microphone is the most important aspect of creating your own boom. With the dollars you save by building your own boom, you can put your money toward purchasing a better microphone. After all, if you audience can't hear what someone is saying on camera, you might as well make a silent movie.

Measuring for Cable

The next piece to the boom puzzle is to determine the type of connection that is available on your camera and what type is required by your microphone. Most professional-quality cameras have XLR connectors, while most consumer-quality cameras have either RCA, 1/4", or a mini-jack for connecting an external audio source. Figure 5-3 shows the different connection types available: XLR, RCA, 1/4", and mini.

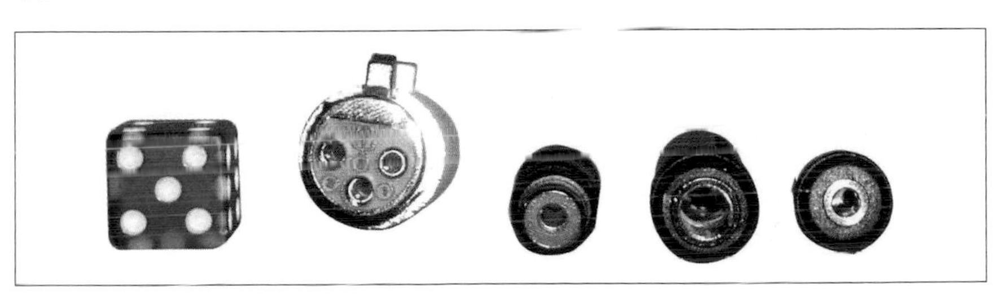

Figure 5-3. XLR, RCA, 1/4", and mini-jack connectors

You will need to purchase an audio cable that is longer than the length of your boom, as well as having connectors that will fit into your camera. Ideally, your audio cable will be much longer than your pole—at least six feet longer—because you will need to run the cable from the top of the pole (where the microphone will be) to the bottom of the pole, and then off to wherever the camera may be. For example, if you have an eight-foot pole, you will want about 14 feet of audio cabling. So, when purchasing your audio cable, remember that it is better to have too much cable than not enough.

Putting It All Together

After gathering all of the necessary items, you need to put them all to use. If you have a telescoping pole, you should place the microphone on the smaller end of the pole. How you attach your microphone to your pole depends on whether your pole is wood or metal, as well as the design of your microphone.

When attaching your microphone, you should keep in mind that the microphone is *designed to pick up sound*. Depending on how you attach your camera, it might also pick up the sound of the boom operator's hands moving against the pole. Therefore, you should create a solid mount point where the microphone won't rub or bang against anything, as well as separate it from the pole. After all, you're trying to create a system for recording better audio than without the boom. Figure 5-4 shows a boom microphone.

Once you've attached your microphone, you need to attach the cable to the microphone and run it down the length of the pole. To keep the cable out of your way, you can use either zip-ties or Velcro to hold the cable against the pole. If you don't have a foam cover for your microphone, or if you are interested in cutting potential wind noise, you will probably want a windsock [Hack #52].

Building your own boom will help you to record better audio, and better audio will lead to better video and a better experience for your viewers.

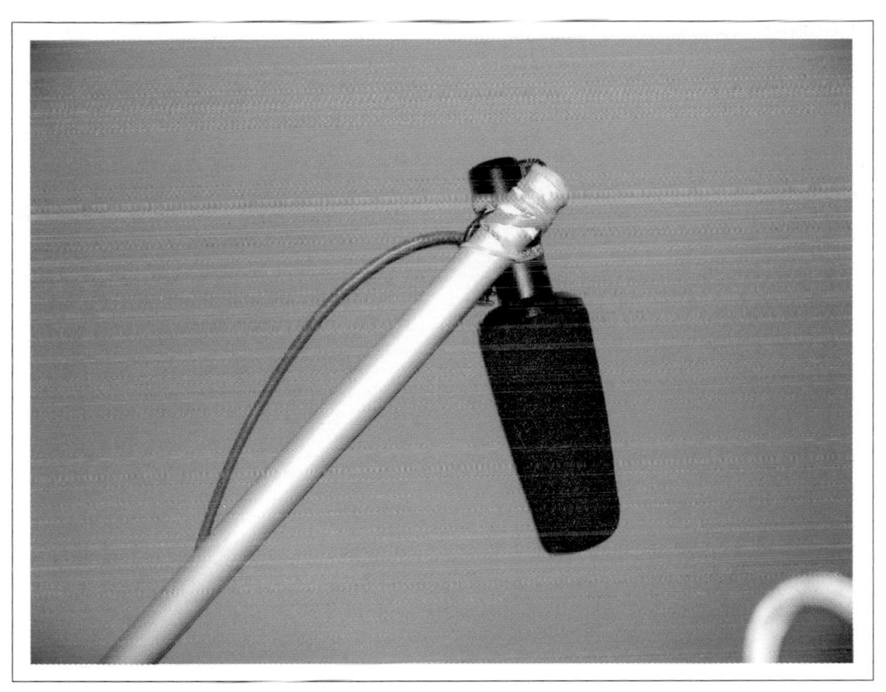

Figure 5-4. Mic on a stick

Export Your Audio for Mixing

Using an application designed specifically for audio, such as Pro Tools, can make a tremendous difference in how your audio sounds. But first, you need to get your audio to your mixing system.

Video-editing applications have come a long way in their audio-editing abilities. But at the end of the day, they are focused on video editing and fall short of the capabilities you need to make you audio really sound great. For this reason, applications such as Audacity (*http://audacity.sourceforge.net*; free, open source), Adobe's Audition (*http://www.adobe.com*; $299), Apple's Logic (*http://www.apple.com*; $299 Express, $699 Pro), and DigiDesign's Pro Tools (*http://www.digidesign.com*; $495–$2,895) exist and excel.

So, you want to mix your audio to make it sound better. How do you go from your editing system to your mixing system? It's easy.

Finding Out What You Need

You will need to find out which audio formats your audio application supports, in addition to finding out which audio formats your video-editing application can export. Each audio system has different capabilities, but they all serve the same purpose: to work with audio. Because of this, you can rely on the fact that they'll be able to import and export standard audio formats, in addition to whatever proprietary formats they use.

These are some of the standard formats every audio system should support:

- Audio Interchange File Format (AIFF)
- Audio (AU)
- Musical Instrument Digital Interface (MIDI)
- Waveform Audio (WAV)

If you are editing using Final Cut Pro or Avid Xpress, and plan on mixing using Pro Tools, you might be able to use the Open Media Framework (OMF) format to exchange your audio. Be aware that not all Pro Tools systems will be able to translate an OMF and that the process can be trying.

After you determine which format you can both import into your audio application and export from your video-editing application, you are on your way to audio bliss.

Preparing to Mix

Before you actually export your audio, you should create a short (preferably one frame long) audible cue at the beginning of your video. This cue is referred to as a *two pop*, and usually includes a video cue, as well. For safety, you should also add a cue to the end of your video. By adding the audible cues at the beginning and end of your video, you will be able to realign them when you import the audio back into your system and know with confidence that your audio is in sync with your video.

Also try to make the process of mixing your audio as easy as possible. One step you can take is to make sure your audio is cleanly separated **[Hack #2]**, so that it is easy to discern which track of audio contains which type of audio. For example, if you need to replace a piece of music, you will know it's not on your dialogue tracks, because your dialogue and music are on separate tracks. If you edited your project without taking this into concern, it will prove extremely worthwhile to spend the extra time and move your tracks as necessary before proceeding.

Exporting Your Audio

After preparing your audio for mix, it's time to send it off. Here's how to export your audio:

Avid
: File → Export → select setting

Final Cut
: File → Export → Audio to AIFF(s)... (or Audio to OMF...)

Premiere
: File → Export → Audio

After the export, you should have at least one audio file for every two audio tracks. For example, if you have four tracks of audio in your timeline, you should have at least two audio files (the first would comprise tracks one and two, and the second would comprise tracks three and four). These audio files will then be imported into your audio mixing application.

Importing Your Audio After Mixing

After your audio has been mixed, you will have one or more audio files, depending on how it has been exported from the audio-mixing application. If there is only one file, it is a complete, mixed version of your audio. If there is more than one file, you most likely have mixed audio, in the same configuration as you exported. Either way, you need to import the files into your video editing system:

Avid
: File → Import

Final Cut
: File → Import → Files...

Premiere
: File → Import

You are creating a new version of your video (the mixed version), so you should create a duplicate timeline and label it accordingly. Then, place your mixed audio onto your duplicate timeline and realign your two-pop markers. If everything has been done correctly, your audio and video will be in sync. Once your mixed audio has been placed, you can safely remove your old, unmixed audio.

Cover Missing Audio with Room Tone

Sometimes, such as when a crewmember sneezes, you need to eliminate audio completely, but missing audio is immediately noticeable, even to those who have an untrained ear. Capturing "room tone" helps remedy this problem.

When you're editing, sooner or later you will discover an edit that must eliminate audio in your timeline. Trying to avoid a certain word that someone said, using footage that was recorded after a scene ended, or hearing a car horn honking in the background are just a few examples of how this can occur. When you remove audio from a scene in these circumstances, you wind up removing audio from the scene completely.

A complete absence of audio is both noticeable and disturbing to your audience. In order to fix the problem, you need to *fill the hole* with nondescript audio. This nondescript audio is often called *room tone*, because it is simply audio that has been recorded in a silent room. Even though the room might be "silent," there is audio to be captured, no matter how quiet it seems to be.

Ed Golya, a multiple Emmy award–winning audio editor, likes to joke that he is going to use "the sound of clouds passing by" when he adds room tone to a scene.

Recording Room Tone at the Right Time

To anticipate potential problems while editing, the best approach to recording room tone is to record it while shooting a scene. The most common technique is to have everyone on the set stand silently for 30 to 60 seconds after completing a scene. During this time, record your surroundings to tape.

While doing this, you should be clear to everyone that you are recording room tone, possibly by holding up a slate or sign in front of the camera. If you do not make it easy for yourself or your editor to locate sections of room tone, it will be very frustrating. To use the section of room tone, simply locate it and use the necessary amount of audio to fill in where there is silence.

Stealing Room Tone from the Scene

If you do not have an intentional recording of room tone available, you might be able to *steal* some from elsewhere in your scene. To do so, play the raw, unedited footage for your scene and listen for sections where there is no movement, talking, or distinguishing characteristics (such as music lyrics).

Once you've located a section that is usable, copy the audio to your timeline. This solution can work well if you need to cover only a brief section of silence. However, the longer the period of silence, the more difficult it becomes to locate an equal length of room tone.

If you need to, you might be able to place the same short section back-to-back several times on your timeline. Doing so essentially provides a longer *fill*. You will probably notice, however, that you can repeat the process only so many times before you notice the room tone repeating. It might sound odd, but you can hear room tone repeating, and when you hear it, you know it.

Recording Room Tone After the Fact

If you don't have room tone that was recorded while on set and you can't locate a decent section of audio to substitute, you can try recording room tone when you require it.

The best option in this circumstance is to return to the location where the scene was shot and record audio while there. Your hope is that the audio you record will be close enough to the original that it will be nearly indistinguishable from what was originally recorded. If you don't have access to the original location, you can try recording in an empty room and hope for the best outcome.

Covering Your Tracks

In order to not draw attention to the fact you've altered the audio, use small audio dissolves between the room tone you edit into the sequence and natural audio that already exists. These dissolves should be short; usually a few frames will suffice. It might take a little bit more work to smooth out your audio, but if you can keep your audience engaged in the scene, and not distracted by technical difficulties, you'll have a much better chance at keeping their interest.

Make Your Audio Sound Better

Audio might take up only a small section of the audiovisual signal, but it has substantial impact on how your audience perceives your project.

Have you ever watched a low-budget, independent movie and been distracted by a subtle audio problem? It could have been something as simple as the buzz of fluorescent lights, but the sound drew your attention away from the scene and suddenly you were thinking about the buzz. If there is a slight hum—caused by an air conditioner, for example—your audience can easily become distracted. Removing hum and other extraneous noise makes an appreciable difference in your audio . . . and it can be removed easily with the right tool.

If you're producing a movie and you discover your audio has areas that need to be *cleaned*, don't fret. You have a few options at your disposal. The first, and easiest, option is to purchase Bias's SoundSoap (*http://www.bias-inc.com/products/soundsoap/*; $99), which is available for both Macintosh and Windows. The second option is to use the audio filters available within your editing program. Finally, you can take your audio to a professional audio mixer and have her do the cleanup work for you. This last option might yield the best results, but your wallet will be a lot lighter afterward.

Whatever approach you take to cleaning your audio, always keep a backup of your original. Besides providing a safety net should something go wrong, it's always nice to go back to the original audio to hear whether or not there has been an improvement.

Cleaning Audio Using SoundSoap

When it comes to removing audio noise, such as hums, buzzes, rumbles, and the like, there is no easier tool to use than Bias's SoundSoap. It works either as a standalone application or as a plug-in to a compatible DirectX or Virtual Studio Technology (VST) application, most which are audio editing programs, such as Bias's own Peak (*http://www.bias-inc.com/products/peak/*; $499) or Steinberg's Cubase (*http://www.steinberg.com*; $129–$799). Using the SoundSoap is similar in each case.

If you are using SoundSoap in standalone mode, you need to open a media file to work with. Selecting File → Open Media File allows you to select a QuickTime-compatible file (which includes DV, AVI, WAV, MP3, and MP4) from your hard disk. SoundSoap then imports the file for you to use.

The default settings work perfectly fine, right out of the box. However, you can fine-tune which sounds are removed by using the Noise Tuner, Noise Reduction, Preserve Voice, Remove Hum, and Remove Rumble features. The Wash window in the top center of the interface indicates visually how much noise is being removed from the audio signal; blue represents your desired audio and red represents noise.

The Noise Tuner function works like an old radio dial, where you can "tune into" the noise you would like to remove. You can play your media file by clicking the Play button on the bottom of the interface. While your file is playing, you can use the Noise Tuner to locate the noise you would like to remove.

The amount of noise reduction can be altered using the Noise Reduction dial. It works in the same way as the Noise Tuner dial. When doing noise reduction, you will notice that it removes the "highs" of your audio, so it can dull your audio if you are not careful.

If you are working with people talking—and, well, you probably are—you should turn on the Preserve Voice feature. Doing so removes extraneous audio that is outside of the human tonal range. This is especially helpful for on-camera interviews.

Remove Hum and Remove Rumble are specific in what they remove from your audio. Using the Remove Hum feature removes the typical hum associated with electrical interference, such as poorly wired lighting. If you are using 120 VAC (North America), you should remove 60Hz hum; if you're using 220 VAC, you should remove 50Hz hum. Here's a quick-and-dirty guide: if you are using NTSC, remove 60Hz, and if you are using PAL, remove 50Hz.

The Learn Noise button works best on small sections of audio, where only noise is present. If you have a scene in which there are two to three seconds without dialogue, you can play across that section and click the Learn Noise button. SoundSoap analyzes the audio and attempts to remove similar sounds from the rest of your scene.

Figure 5-5 shows the SoundSoap interface with the Noise Tuner and Noise Reduction dials tuned and the Preserve Voice function enabled.

You can also mark specific areas by setting In and Out points, much as you would in a video-editing application. To set an In point, choose Edit → Set In Point (or type i) and to set an Out point, choose Edit → Set Out Point (or type o). Setting In and Out points allows you to remove noise from a specific section of your audio only.

When you are satisfied with the job SoundSoap has done, click the Apply button to save the changes to your media file. One approach to using SoundSoap is to find a section of your audio you would like to improve by marking In and Out points, find a small two-second section of noise-only audio, have SoundSoap learn the noise, make any changes you deem necessary (such as removing a hum), and finally, apply the changes to your file.

Removing Hums and Buzzes Using Your Editing System

Most video-editing systems now include some level of support for audio editing. The audio-editing support often comes in the form of audio effects or filters. In order to remove distractions such as hums, hisses, and buzzes, you want to be able to suppress certain frequency ranges.

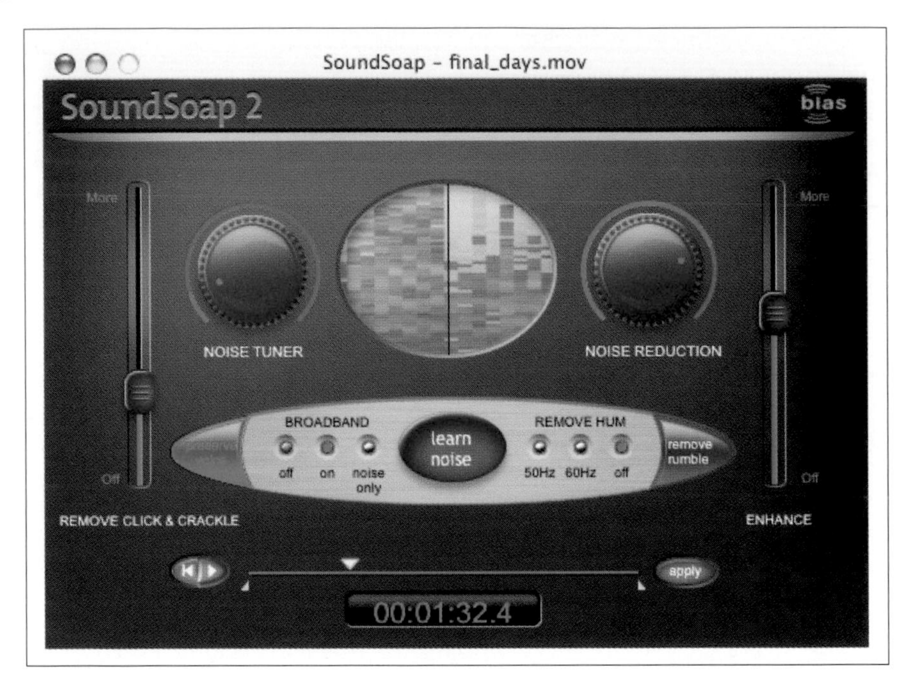

Figure 5-5. Scrubbing audio clean using SoundSoap

A good approach to removing unwanted noise is to use *high-pass* or *low-pass* filters. These filters *pass through* either high or low frequencies, respectively. You might want to use more than a few of them to accomplish a cleaner sound. For example, if there is a rumble in your audio, you can apply a high-pass filter to help suppress the lower frequency. Or, if you discover there is a hiss in your audio, you can apply a low-pass filter to help block-out the hiss from the audio signal. Figure 5-6 shows Final Cut Pro's Low Pass Filter in use.

✓ Low Pass Filter		
Frequency		2600
Q		0.71
Mono 2 Filters		
✓ Low Pass Filter		
Frequency		2600
Q		0.71

Figure 5-6. Using a low-pass filter in Final Cut Pro

A *notch* filter is similar to high-pass and low-pass filters, but it blocks a specific frequency range instead of passing it through. The notch filter can also work well at removing audio noise. If it's available, you can also try using an equalizer (EQ) filter.

Some editing systems, such as Premiere Pro, have a DeNoiser filter. Others, such as Final Cut Pro, have a Hum Remover filter. Ultimately, you should evaluate what filters are available on your system and experiment to see if they can accomplish what you want to do.

Hiring an Audio Mixer

There are professional audio mixers in just about every metropolitan area around the world. Additionally, some very talented hobbyists are available. If you plan on having a professional mix and edit your audio, you should not attempt to alter the audio in any way. Almost every audio professional wants the purest audio they can use in order to work their magic.

Before your session with the person who is going to mix your audio, you should find out which type of audio file he will need. Also ask if there are certain specifications he will need, such as whether he would like 48KHz or 44.1KHz audio. Depending on the system he is using, you might have to provide an OMF, SDII, or AIFF file. You should make sure your editing system can provide the correct file format, or find an alternate solution, such as providing the audio on videotape.

Fool Your Audience's Perception

What you hear isn't always what you see.

Producing and editing a video project involves working with both audio and video. Most people know they can use audio to smooth over cuts [Hack #46] and cover other problems [Hacks #55 and #56]. But you can also use the combination of audio and video to trick your viewers into seeing one thing, but hearing another.

The McGurk effect is an excellent demonstration of this fact.

For more information on the McGurk effect, and other visual tricks, check out *Mind Hacks* (O'Reilly) by Tom Stafford and Matt Webb. You can also search online for "McGurk effect" to find out more of the scientific reasons behind it, as well as view some demonstration videos.

Creating the Effect

The McGurk effect occurs when you see something and it influences what you hear, and vice-versa. Harry McGurk and John MacDonald explained the effect in a 1976 article titled "Hearing lips and seeing voices" (*Nature* 246, 746–748). The effect can be seen in video where a person is mouthing "ga," but the audio is "ba," and what you think you hear is "da." To view a sample online, visit *http://www.media.uio.no/personer/arntm/McGurk_english.html*. To create the effect, you will need to acquire audio and video from two different sources.

Acquiring the video. As mentioned, to create the effect requires video of someone mouthing "ga." The best way to record the footage is to have someone simply say "ga," because you'll remove out the audio portion when you edit the project together. To make the editing process easier, you will want the person on camera (you?) to speak the word at least a handful of times. Something like "ga...ga...ga... ga...ga..." will work well.

Acquiring the audio. Your audio portion can be spoken by the same person in the video, or by someone else. You will want the audio to consist of the sound "ba," and it should be paced at the same rate as the video portion. If you need a visual cue, you can always play back your video segment, with the volume turned off, while recording your audio segment. So, for your audio segment you should have "ba...ba...ba...ba...ba..."

Editing the footage. Once you've acquired your footage, import it to your editing system. Then, add the video (ga ga) to your timeline. After adding the video to your timeline, you need to remove the audio portion of the signal:

Avid
: Use the Lift/Overwrite Tool to select the audio and lift it from the timeline.

Final Cut
: Press F4 (function 4) and the number of the track you would like to lock (e.g., 1). Then, select the audio tracks and delete them from the timeline.

Movie Maker
: Right-click on the audio and select Mute from the menu.

Premiere
: Lock the video track, using the lock button on the timeline. Then, delete the associated audio.

iMovie
: Click on the clip in the timeline, and then choose Advanced → Extract Audio. Click on the new audio clip and choose Edit → Clear.

Some systems have a Lock feature available within the timeline. If your system offers the feature, it is just another way for you to lock your video track and then remove the associated audio track. Figure 5-7 shows the interface for Adobe's Premiere Pro and the Lock feature available in the timeline.

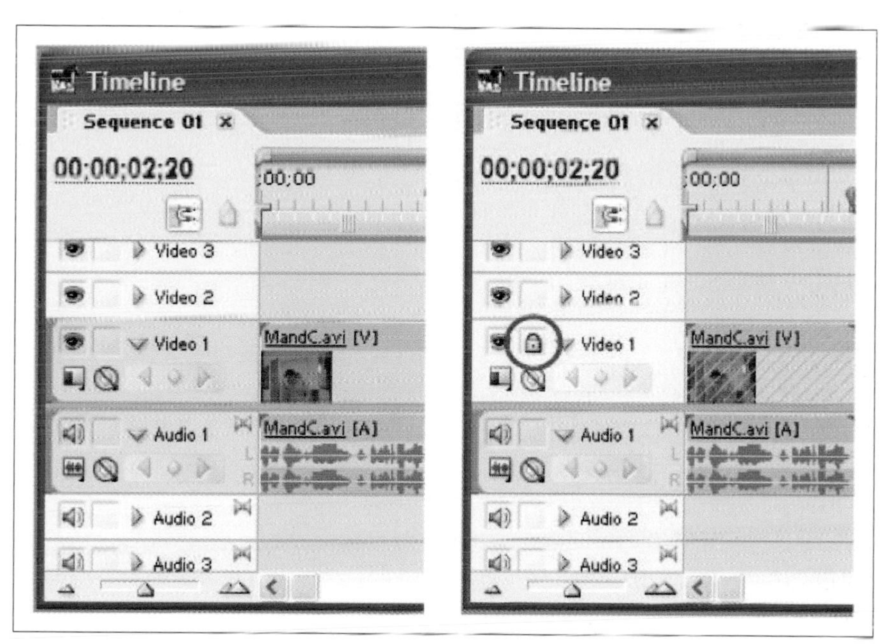

Figure 5-7. Locking a track in Premiere Pro

After removing the audio from your "ga ga" footage, you need to add your "ba ba" audio to your timeline. To do this, follow a process similar process to the one used for the video:

Avid
: Use the Lift/Overwrite Tool to select the video and lift it from the timeline.

Final Cut
: Press F5 (function 5) and the number of the track you would like to lock (e.g., 1). Repeat the process for your second track, if you've used stereo audio. Then, select the video track and delete it from the timeline.

Movie Maker
: Drag the "ba ba" footage to the Audio/Music track. Only the audio will be added to your timeline.

Premiere
: Lock the audio tracks, using the Lock button on the timeline. Then, delete the associated video.

iMovie
: Click on the clip in the timeline, and then choose Advanced → Extract Audio. Click on the video clip and choose Edit → Clear.

Once you have your independent video and audio clips, you need to line them up. Drag your audio until it lies below your video in the timeline. From here, you need to nudge your audio forward or backward in order to have it match your video. You might also need to cut your audio footage into sections, so that you can line up each "ba" with each "ga." Once you've lined up your audio and video segments, play your timeline.

The effect should be apparent (at least to 98% of the population). Even though you know that the audio is "ba," your mind will translate it to "da" to match what you are seeing. You can close your eyes and play the video to make sure nothing "funny" has happened to your audio.

Using the Effect in Real Projects

Now that you've experimented with the McGurk effect, you might be wondering how you can use it in a *real-world* scenario. Well, the effect obviously works well with speech. As a visual-only experiment, have a friend speak the words "olive juice." Then, ask her to only mouth the words. What is it that you think she's saying?

Taking the visual aspect of what your mind interprets, now have someone stand behind your friend and have him speak "I love you" while she mouths "olive juice." Beside the strange fact that you now have a woman telling you she loves you in a man's voice, what is it that you're *hearing*? What is it that you're *seeing*?

This type of editing occurs frequently when major motion pictures are shown on television. For example, some four-letter words cannot be broadcast on television. Therefore, select words are edited over the original censored words to make the movie acceptable for broadcast. You will notice this when someone says, "Forget you!" but you know they said something else in the original version.

But you don't have to be limited to trying to match words to people's mouths. Editing audio can prove very useful, especially when you need to re-craft a story after it's been shot. A good example of this is when you have a wide-shot of a scene and would like to change what's being said. You can use audio from another scene and place it underneath your wide-shot, and then try to lineup the audio with someone's lips moving. In a wide-shot it will be difficult for your audience to lip-read, and they will rely on (and trust) what they hear.

> The most impressive use of this effect I have personally experienced was on *The Larry Sanders Show*. We had a scene where one character, Artie, had his back to the camera and another, Larry, was speaking a line. We needed to drop one of Larry's lines, but we couldn't cut away. So, the editor borrowed one of Artie's lines from a take we weren't going to use and placed it where Larry was speaking. When watching the scene, even though Larry's lips were moving, your focus turned to Artie and your mind ignored the fact that Larry was supposed to be talking.

So, don't forget your audio tracks. Not only is audio half of the medium (even though it takes up a small percentage of the signal), but it also can help you rewrite scenes, cover mistakes, and move your story forward when your video falls short.

HACK #58 Get Music for Free

Music is one of the most important aspects of a movie. If you know what you are doing, you can get great music for free. Sometimes you can even get someone else to pay the artists who provide it!

You'd be amazed at how much the music that you don't even consciously notice can make a huge difference in how you are affected by the visuals of a movie. We subconsciously rely on the cues music gives us to know how to react. Different music can make the same video clip seem alternately uplifting, sad, calm, or heart-poundingly scary. Most horror movies even over-rely on music...you know when someone's gonna get butchered when the music gets really spooky.

This is one reason *The Blair Witch Project* was so groundbreaking. Oh sure, everyone talked about the low budget and record-high box office return, but not many people talked about the fact that there is no music in this very scary horror movie! Okay, there was one song, but it's what they're listening to in the car on the way out the woods.

It is a testament to the power of marketing that the producers actually did manage to make and sell a soundtrack for this movie though. It's called *Josh's Blair Witch Mix* and it's supposed to be the tape that police found in Josh Leonard's car after his "death." Josh told me he never listened to any of the bands featured, and three of the songs on it were recorded two years after the tape was allegedly found.

But yeah, music makes or breaks a scene.

Using Music in a Movie

In mainstream moviemaking, Hollywood pays a lot of money for classic songs from the past. Or puts out the same ridiculous amounts of cash to license use of the latest hits by the latest (often disposable) stars du jour. Some of this even amounts to product placement in reverse, where the moviemakers pay the product to place it. Or more likely, the movie studio owns the record company too, or at least has the same corporate parent. All these machinations are just more reasons to make and support independent movies.

You probably can't afford the rights to use any of this popular music in your movie. And you can't afford to "just use it anyway." You'll never be able to broadcast or sell your movie if you have uncleared music in it.

Tarnation, Jonathan Caouette's autobiographical documentary that recently became a surprise hit at Cannes Film Festival, cost the director just $218 to make—that is, before the director paid for the music and video clips playing in the background. Purchasing those rights bumped the film's budget to about $400,000.

But you can get great music for your independent efforts, without paying anything.

Finding Music for Free

What you want to do is find quality music by unknown artists. There are a lot of bands who would be honored to have their tunes in a cool movie. And it helps them too. Again, it's win/win. As it should be.

Great places to post for music are, of course, craigslist (*http://www.craigslist.org*) and MySpace.com (*http://www.myspace.com*). You can also go to MP3.com (*http://www.mp3.com*), or other MP3 sites, and listen to songs and contact the artists.

I usually post or email something that quickly describes the project, and names the terms, like this:

> SUBJECT: Need music for movie, specifically some more background music to use under interviews. Any style except hard rock might work. MUST BE ALL ORIGINAL COMPOSITIONS. No uncleared samples or nicked melodies will be accepted.
>
> Please go to *http://www.kittyfeet.com* to check out the project.
>
> Email me a small MP3 or better yet, a link, and I'll let you know if it's something we can use. We won't use it without talking to you. This does not pay, but we will be submitting cue sheets to BMI and ASCAP, so residuals are possible from them if the movie is broadcast, and there's a really good chance it will be.
>
> You retain copyright.
>
> Much respect,
>
> Michael W. Dean

Basically, what this all amounts to is being the music director for your movie. Hollywood movies hire dedicated music directors, who make six-figure incomes for basically having the right Rolodex. Again, you probably can't afford to hire them, but you can do their job yourself.

Being a Music Director

A large portion of being a good music director consists of being able to process huge amounts of data quickly and accurately. As soon as you hear an MP3 you like, you should ask the artist to send you a CD. I usually send them a cut-and-pasted document that basically says, "Do we have permission to consider using this, under the terms we described in our post?"

If the artist emails back affirmative, save the email and go ahead with the process of checking out more tunes by them. But don't say "We're gonna use your tune for sure" until you're sure you're going to use it. You don't want to get people's hopes up. And you have to make people understand that just because you don't use something, it's not a reflection on their music. The best music in the world might not be suitable for a certain scene. It's all very subjective, and you'll get a better ear for it the more you do it.

When producing my latest movie, I got about 70 CDs of music submitted. That's about 70 hours of music, from which I had to pick about 70 minutes of music. So I learned to listen like a music director.

When you get a CD, pop it into your computer and listen to about 20 seconds each of the beginning, middle and end of a tune. It's easier to tell which songs won't work than to tell which ones will work. You need to do a process of elimination, and generally cross off about six or seven cuts on a 10-cut disc.

After eliminating most of the cuts, listen to about a minute of the remaining ones. (I like to make the marks right on the CD cover with a Sharpie.) Then, from the two or three cuts you finally consider, rip uncompressed audio copies of them using Windows Media Player. You can then use those audio files while editing.

Keeping organized. Give the files descriptive names like *BobBartosik Quartet-slow jazz_Sax_Gasam.wav* rather than names like *song14.wav*. Having the type of music in the title (slow jazz) makes it easier when you're editing to find something that might work. And having the song title (Sax Gasm) and the artist (Bob Bartosik Quartet) makes it a lot easier when you finalize your rough cut and have to gather written release forms and create the end credits.

Never put a dot in the first part of your filenames. They are an accepted character, but they can cause stuff to behave oddly in the editing program. Put them between the filename and the extension only.

So, you go through a couple CDs and end up the night with a few possible songs out of the many. Then start the next day with your video editor by playing him the tunes you've culled the previous night. If you trust his ear, he can make the next round of decisions. If you are your own editor, you get to make the decision!

You should winnow about five or six tunes down to one or two. Then that one or two should get transferred to your editing system and go in the music bin. Then whenever you need music, you can just check out stuff in that bin and drag potential tunes into the timeline. Not all of the songs will necessarily get used.

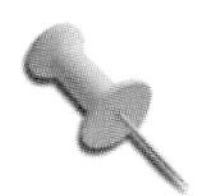

If you're editing in a quick process, and you like someone's tune, you might not even want to wait until you get his CD in the mail. Occasionally, I'll use an artist's MP3 in the timeline to check stuff out, and end up keeping it in there for a rough cut. But you really do want to get the uncompressed (*.wav* or *.aiff* from taken from a CD) version of their contribution if you end up using it. Replace it later if you use an MP3 as a dummy for mocking stuff up, because *.wav* files sound better than even the best MP3.

Obtaining a release. For the songs that you do use, you should email a release form to the artist, and ask them to sign it and send it back.

Here's a summary of the release form I wrote and used for my movie:

> Music Release Form
>
> I, ______________ (herein also called "The Artist") hereby agree to have my song ______________________ used in the movie currently called *Movie Title Goes Here*, directed by Michael W. Dean (herein also called "The Filmmaker".)
>
> ...
>
> No payment is expected or implied.
>
> ...
>
> The artist will still own the copyright on the song. The artist will always still be permitted to perform the song live at any time, and to record it.
>
> ...
>
> Michael W. Dean, the Filmmaker, wants me to know, in writing, that he loves me and will use my music for good, and never for evil.
>
> Date:_______________
>
> Signature:__________________________________
>
> Printed Name:______________________________
>
> BMI or ASCAP affiliation:_____________________________

If you'd like a copy of the complete release form, it is downloadable at the bottom of this web page: *http://www.kittyfeet.com/30bucks/*.

The line at the end of the release about ASCAP and BMI is the cool part. If the artist registers with one of those organizations (a person can only belong to one or the other, not both) then you, as the moviemaker, can submit cue sheets for the movie to each organization, listing all the songs used, and the names and affiliations of each artist. Then, allegedly, if the movie is broadcast, the musician might get a small check at some point, from money paid out from the broadcasting network or cable channel. I say allegedly, because I do not understand all the workings of huge corporations that mainly cater to big media, but I have a friend who swears she did it on an indie movie and it worked.

Joining BMI or ASCAP. You can join these associations at *http://www.ascap.com* and *http://www.bmi.com*, respectively. It's free to join and there are no dues. Cue sheet information and downloads are at *http://www.ascap.com/about/payment/identifying.html* and *http://www.bmi.com/library/brochures/cuesheet.asp*, as well.

When I was on a major label with my band, Bomb, I was a member of one of these groups and got a few checks. But that's long expired. I may re-up my membership (since I created a little of the music used in my movies), but I'm still stalling on it.

There is a contract you have to sign in order to join. I looked at the ASCAP one and it had a clause that seemed to say that by joining, I was agreeing to back any anti-piracy laws they lobby to pass. I didn't like that one bit, as it could incriminate some, um, friends of mine. I haven't read the BMI contract yet (they're very long), but I'd be willing to conjecture it's similar. There isn't much difference in these two organizations, as far as my casual query of several members can ascertain.

Being Fair to Those Who Help You

Please notice that the contract I wrote is very fair to the artist. It has to be: you're not paying them! But most artists are so hungry, they'd give away a lot more rights than those outlined just to be in a movie. But you shouldn't do it. I love art (and artists) too much...hopefully you feel the same.

I did make one concession that restricted the contract on the artist a little more for one tune only: the song "Friends" by Aaron Jones. That's the song that's rolling during the final credits of my most recent movie, and as such defines the movie a lot more than the stuff in the background under people talking.

I just didn't want it to end up as the theme song for another movie, so I restricted the artist to not use the song in another movie. I wanted it to stand out in our movie. It has a very climactic effect, redemptive even, after the very end of the movie. You'll see when you see the movie. I picked that song very carefully; basically picked it as the one, very specially out of 70 hours of music.

Again, don't forget to make sure you credit everyone who helps out, both on your web site (give them a link too) and in the movie. And make sure that you mail them a copy of the completed DVD (as soon as it's done, without them having to bug you for it). And write everyone a thank you note.

Give a good movie!

—Michael W. Dean

Wrangle Your Music and Sound Library

Using Apple's iTunes, you can create a complete digital library of your music and sound effects.

With hundreds of millions songs sold through the iTunes Music Store and countless iPods in the hands of consumers, Apple's iTunes is a tremendous success. But the iTunes application (*http://www.apple.com/itunes/download/*; free), which runs on Macintosh and Windows systems, can be used for more than ripping and buying music. You can use it to organize your music and sound effects library, as well as keep track of what tracks you've used in your projects.

Organizing Your Audio

To organize your audio, import it into iTunes. Just load a CD into your computer and launch the application, if it doesn't launch automatically. More often than not, you will discover that someone has already entered the CD's information into CDDB (*http://www.gracenote.com*), which iTunes uses to automatically enter track names. Figure 5-8 shows a CD of household sound effects, with track information retrieved from CDDB.

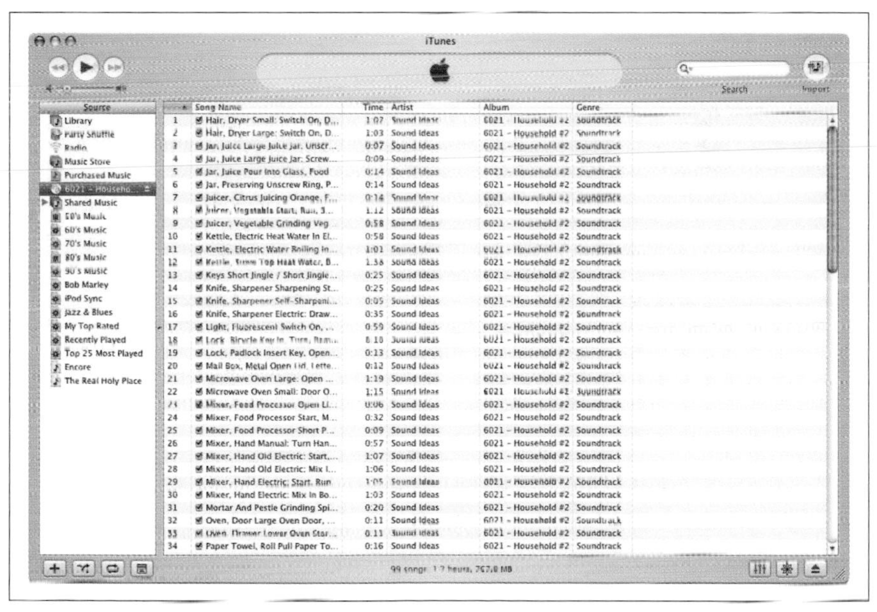

Figure 5-8. Listing sound effects using iTunes

Although importing all of your music and sound effects in this manner can be time consuming, the amount of time you save on the *back end* is well worth it. This is because the columns in the iTunes window allow you to sort by Song, Genre, Time, and more. Additionally, your entire library can be searched.

Cataloging Your Audio

iTunes provides you the ability to add information, known as *metadata*, to your songs. Adding metadata to your songs allows you to catalog your audio in a more productive way. For example, there is no Sound Effects genre in iTunes, but you can add it as a Custom selection. Figure 5-9 shows the metadata, and some additions, for a group of tracks.

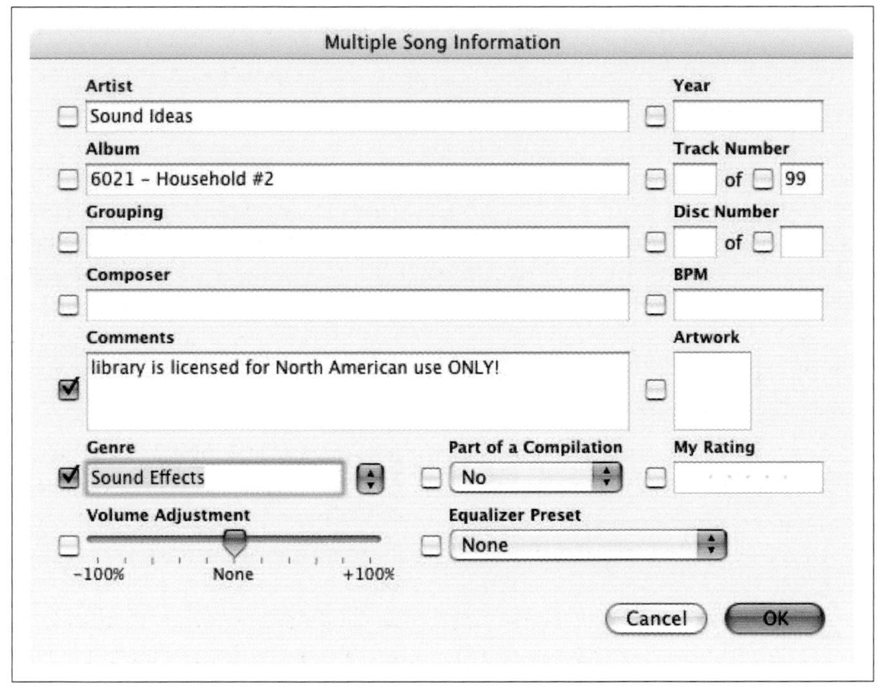

Figure 5-9. Adding information for a CD

To change information for a track, or group of tracks, select the file and choose File → Get Info. If you have selected more than one track, iTunes will present a confirmation box to make sure you are not doing something unintended. You will then be able to change and add information for the selected tracks.

The Comments column can be of particular use, especially if you have tracks with license restrictions. You might find it useful to display the Comments metadata in the main window by default (you can do so in the Edit → View Options... window).

Searching Your Audio

The advantage of using iTunes for your audio library is its search function. Some music and effects libraries contain thousands of tracks, so finding a particular track without iTunes can be time consuming and frustrating. Even if the library is organized well, and you have a good paper catalog, thumbing through it doesn't always prove to be the most effective use of time, because you still need to locate the CD and listen to the particular track to see if it matches what you are searching for.

Locating tracks. Simply enter what you are looking for in the search box in the upper-right corner of the application window to search across all of the metadata for all of your tracks in your iTunes library. This functionality allows you to find songs or effects quickly. Additionally, if you've entered additional metadata, such as the cost of a song, you can find that as well by using the built-in search function. If you would like to restrict your search to Artists, Albums, Composers, or Songs, you can do so by clicking on the small magnifying glass in the search box and selecting the appropriate item.

The search function is especially useful when you're looking for a sound effect. For example, if you are looking for the sound of juice pouring into a glass, you can try searching for `juice` or pour. Figure 5-10 shows a search for the word `juice` and the result from a library of over 6,500 tracks.

Importing tracks. When you have located a song or effect you would like to use in your project, you have a few options. If your editing system is aware of iTunes, you will be able to import songs directly. If your editing system is not aware of iTunes, the most basic and reliable solution is to locate the original CD or audio file where you obtained the track and then import it into your editing system.

Another option to import the audio track is to find where iTunes maintains its library: choose Preferences → Advanced and look at the entry for the *iTunes Music* folder location. When importing from the iTunes library, you might need to navigate a series of folders, most likely Artist and Album. For example, if you are attempting to import *Jar, Juice Pour Into Glass, Food* from your library, you have to navigate through the *Sound Ideas* folder (the

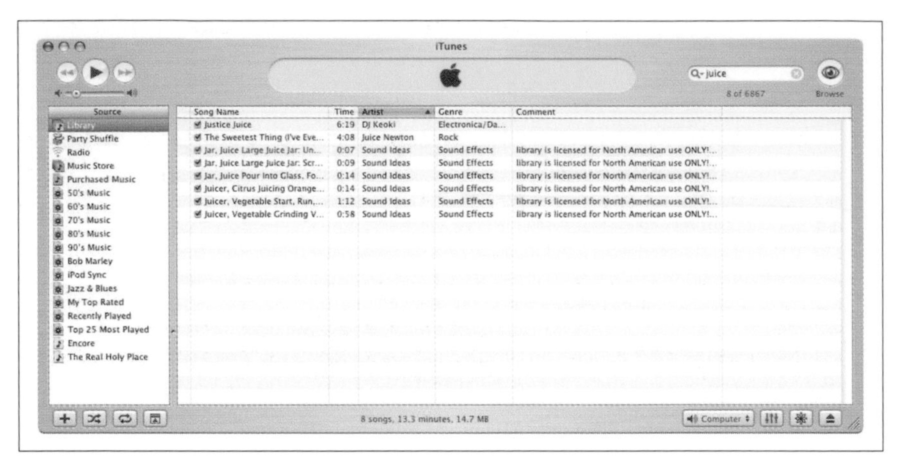

Figure 5-10. Finding sound effects using iTunes

Artist of the CD), then *6021 - Household #2* (the Album of the CD), and then select the appropriate track. When using this method, it is helpful to have both the Artist and Album columns visible in the iTunes window.

Using Playlists

iTunes does a great job in organizing music tracks, and you can possibly get away with just having one big library. However, you might want to create *playlists* to help keep your library more organized, particularly if you are wrangling a large sound library. Additionally, the iTunes Smart Playlists automatically update themselves when you add tracks that fit within certain criteria. As an additional benefit, when you select a playlist and then search, the search is limited to the tracks contained within the playlist.

Knowing the cost of a track. When in post-production, questioning the cost of audio is not uncommon. In fact, I'd say it's uncommon that this question *doesn't* arise. If you enter the licensing cost of a track in the Comments field, you can easily find out how much it costs to license.

You can also create Smart Playlists to contain tracks that cost a certain amount. For example, you could have playlists for royalty-free music or $99 music. For your royalty-free music, just enter `royalty-free` in the Comments field. Then, create a Smart Playlist (File → New Smart Playlist...) called Royalty Free to contain only tracks in which Comments → contains → `royalty-free`. Then, do the same process for tracks that cost $99.

Staying on top of your usage. Another common question about music regards exactly what you have used. In other words, someone might ask what music you used in a particular project—often a year or more *after* you've completed it. The unfortunate answer is often, "Um, I don't know."

Well, again, iTunes can help you out. By creating a basic playlist and naming it the same as your project, you can add the songs and effects you use in your project to it. To create a Playlist, choose File → New Playlist. Then name your Playlist and simply drag audio tracks from your Library to the Playlist in order to add them.

If at any time, you need to know what music or effects you used, simply click on your project's playlist and find the information you need. A nice side effect of the project playlist is that, if you've also entered the associated costs, you can scan down the column and get a quick approximation of how much you've spent!

Hacking the Hack

Don't be afraid to be creative! If you're willing to step outside the bounds iTunes has set up, you can bend the application to fit your needs.

Hacking costs. If you want to enter the cost information for your tracks in the Year field, go right ahead. By doing so, you can create *smarter playlists*, in which the collections are of tracks costing between $5,000 and $9,999, for example. If you take this approach, however, you might discover that the year is limited to four digits, so your licensing fees couldn't exceed $9,999, which is probably fine for everyone except those looking to license a good Rolling Stones song.

Bottlenecking a group. If you are working with more than one editor, you can create a central library of your approved music and sound effects. The setup requires one system to be the *server*, which is where the library will be shared. Because iTunes uses Rendezvous to find other iTunes players, all of your systems will be able to find the server and will be able to search, listen, and browse the tracks.

In order for the editors to be able to import the music into their project, the folder where the music is stored on the server has to be shared across the network. This is different than sharing music using iTunes.

There are a number advantages of using a central system. First, editors are able to use only *authorized* tracks, because they need to obtain their music from the shared playlist. This should curtail rogue editors from sneaking in unlicensed music and effects. You can ensure that no unlicensed tracks wind up in your final project by relinking all of your audio tracks with only the drive in which the iTunes library resides. So long as no one has been able to write files to the drive, any unlicensed tracks will be *offline*.

Cut to the Beat of Your Music

Whether editing a music video, your friend's wedding, or the next great action movie, music is a tremendous addition to help set the mood for a scene. If you can cut to the beat of the music, your scene will benefit even more.

Digital video is an audiovisual medium, yet the audio portion of the signal tends to be neglected until late in production. If you are attempting to enhance a scene using music, you should choose music that is appropriate to the mood you are trying to create. Even after selecting the *right* music, if your editing doesn't take the music into account, your scene will not accomplish what you want and your audience will actually become distracted by your selection.

Selecting Your Music

Your choice of music is very important, because the wrong selection can completely change the mood of your scene. For example, choosing up-beat techno music is not the best choice for a scene involving a heartfelt apology. If you are planning on distributing your movie, you should also make sure you have the legal right to use the music **[Hack #58]** you have chosen.

There are a lot of royalty-free music libraries available, specifically for use in motion picture production. Some are reasonably priced, depending on your planned method of distribution. A simple search on the Web for `"royalty free music"` should yield plenty of results. Many companies even allow you to search and listen to their music on the Web before you order or sign an agreement.

Importing Your Music

Once you have selected the music you would like to use, import it into your editing system. In order to obtain the best result, import the best quality audio you can acquire. If you have the physical source of the music, such as a compact disc, import the song directly from that.

I am not a person who debates audio quality and compression. My belief is you should try to get the best quality audio file you can obtain. If it sounds good to you, that's all that matters. Should someone ever challenge your choice, you end the conversation very quickly by telling him, "It's a creative decision."

If you are downloading the song, try to get at least *CD-quality* audio at 44.1KHz. Oftentimes, when downloading, you will not be given too many choices, so you can use one hard-and-fast rule: go for the larger file. If you are downloading a three-minute song and are given a choice between two files, choose the larger of the two, because it will be of better quality.

Marking Your Beats

After importing your music and, of course, your video footage, you can get to the fun task of editing your scene. You might find that as you are editing, you are *missing* the beats of the music. This can be especially frustrating if you are cutting a scene that is particularly music driven.

Instead of fighting with the music, you might find that adding markers to the music (or your timeline) can help you finesse your scene with more confidence. Setting the marks *to the beat* is easy; all you need to do is play the section of your timeline and press the keyboard shortcut for setting a mark on each beat. The end result will get you close, if not exact (depending on how fast your fingers are), marks for each beat of your music.

I have found that looking away from the computer monitor helps me concentrate on the music.

Figure 5-11 shows a section of music with enhanced beats.

To set your marks:

Avid
: Assign a keyboard shortcut to the Locator button.

Final Cut
: Mark → Markers → Add

Premiere
: Marker → Set Clip Marker → Unnumbered

iMovie
: Bookmarks → Add Bookmark

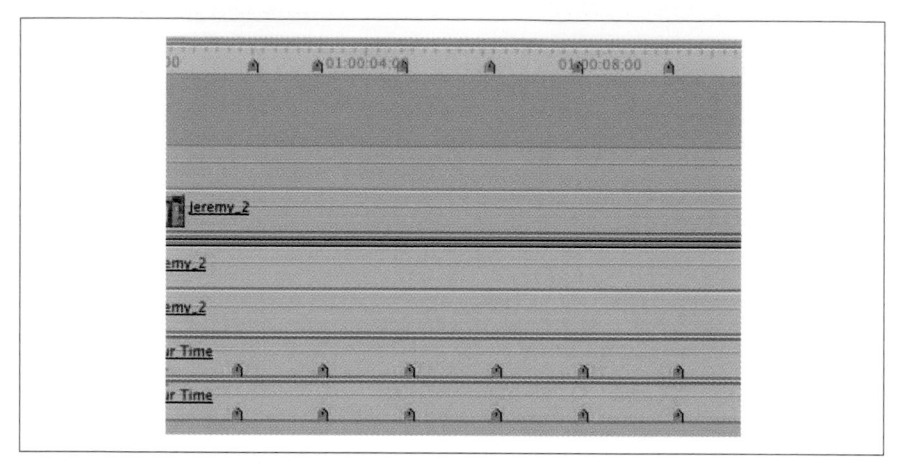

Figure 5-11. A section of music with enhanced beats

To make the process easier, learn what your particular system uses as a keyboard shortcut. By doing so, you will be able to just press a key (or combination of keys) on your keyboard, instead of using the mouse.

As soon as you have finished placing your marks, you can begin to make a few cuts using the markers as a guide. If you find that your cuts still don't hit on the beat, you might want to try finessing them one frame at a time in either direction. After a short while, you should discover your *personal offset*, which will be a reflection of your reflexes.

Using a Waveform for Visual Cues

You can also use the audio waveform as a guide to determine where the beats of your music occur. Most often, you will see a pattern of spikes in the waveform where the beats take place. Figure 5-12 shows a waveform with markers placed.

If you combine the process of marking the beats with the visual reference of the waveform, you will wind up with a reliable reference to the music's beat. When cutting to music, you can then use your markers as reference points where you can cut. By cutting in the same spots where your markers are, you'll be able to have your cuts match the beat of your music.

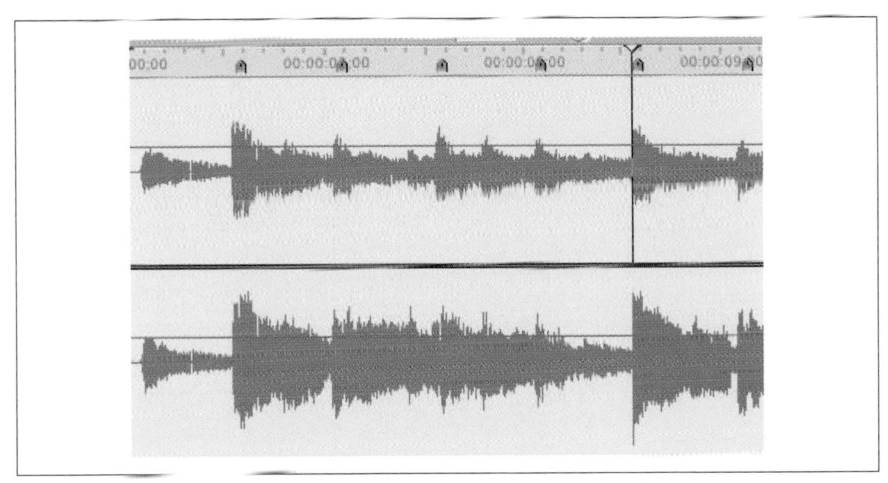

Figure 5-12. A waveform with markers placed

CHAPTER SIX

Effects
Hacks 61–74

Effects are where the editing process gets to be really fun. Even though you've already captured your footage, you can still manipulate it, change its look, and give it your own style. Special effects can be subtle, such as those in *Forrest Gump*, or blatant, such as those in *The Matrix*. Using effects, you can also fix problems that occurred while you were shooting.

Give Your Video a "Film Look"

You can get that expensive film look, even when using digital video.

Many people prefer the look of film to the look of video. Unfortunately, film is expensive to shoot with. But, if you are willing to plan ahead, shoot under certain conditions, and take a little time in post-production, then you can come close to giving your video a "film look." Figure 6-1 shows an unaffected frame of video, and the same frame with a film-look applied.

Figure 6-1. Video as shot (left) and with a "film look" applied (right)

Shooting Like Film

If you know you are going to process your video to look like film, you should plan accordingly before you shoot any footage. Part of the look of film is the way it is shot. Simple things, such as lighting a scene instead of relying on natural light, can make an enormous difference in the look of your video.

Also be aware of how you move your camera. For example, use a crane or dolly as opposed to panning or tilting, because the latter can cause a *stuttering* in your image. Additionally, get to know the features of your camera and the extremes to which they can be pushed. If there are features that allow you to reduce the sharpness or detail, alter the chrominance (a.k.a. *chroma*—includes hue and saturation), or change the exposure, use them.

Some cameras have features that enhance production value, including Frame Mode (where frames of video are progressively scanned, as opposed to interlaced), 24p (where video is shot at 24 frames per second), and the ability to change the shutter speed. New cameras and technologies are released throughout the year. So, if you are shopping for a camera and plan on achieving a film look, find out which features are currently available and which fulfill your requirements. Additionally, you should evaluate HDV cameras [Hack #34].

Using the Magic Bullet Plug-in

Magic Bullet for Editors (*http://www.redgiantsoftware.com*; $299) is a plug-in that offers a set of tools called Look Suite and Misfire. The plug-in does a fantastic job of accomplishing the look of film and comes with 55 preset *Looks*. It is available for Avid Xpress, Final Cut Pro, and Premiere Pro. A demonstration version is available.

Magic Bullet was designed by Stu Maschwitz, who is a former Industrial Light & Magic (ILM) visual effects artist. He has worked on a large number of major motion pictures, including *Casper*, *Twister*, *Men in Black*, *Deep Impact*, and *Star Wars: Episode I*.

Getting started. The easiest way to get started using Magic Bullet is to use one of the 55 preset Looks. Simply select the footage you would like to affect and apply the Look to it, by either dragging and dropping the preset onto the footage or selecting it from your Effects palette. Figure 6-2 provides a sampling of the prebuilt Looks available.

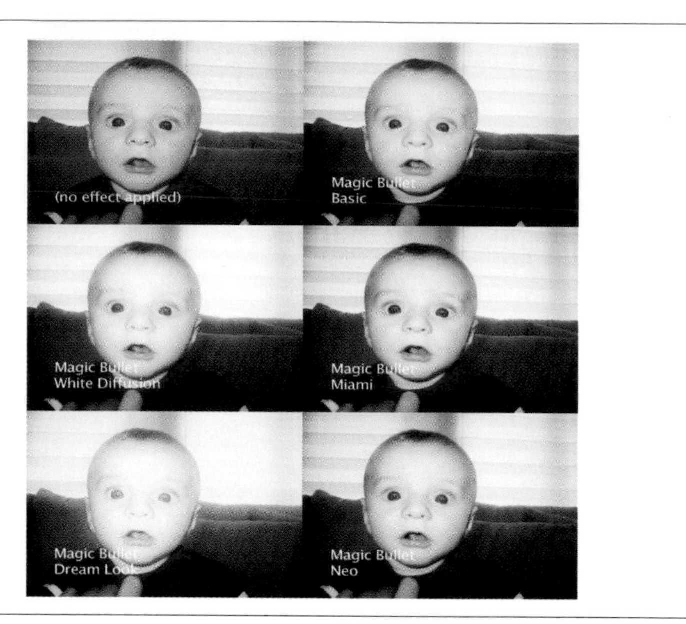

Figure 6-2. A small sample of the 55 available prebuilt Looks

Once you have applied one of the preset effects, open your affected clip in your Monitor/Viewer window. Then, look at the settings for that clip. This is probably the best way to understand how the various controls work together to achieve a given effect.

Using the Look Suite

The Look Suite consists of 26 different controls, in four different categories: Subject, Lens Filter, Camera, and Post. To demonstrate the controls, we'll use the Neo and Miami Looks.

If you are unfamiliar with any of the terms used by Magic Bullet, you can use either a glossary or dictionary of video terms to help you wade through the process of creating a film look. Joe Kane Productions provides a comprehensive glossary of video terms at *http://www.videoessentials.com/glossary.php*. Brad Hansen provides a good dictionary at *http://www.hansenb.pdx.edu/DMKB/dict/index.php*, through Portland State University.

Working with the Subject settings. The Subject category gives you the ability to alter the Pre-Saturation, Pre-Gamma, and Pre-Contrast levels. Make your image as generic as possible using these controls. Figure 6-3 shows the initial state of the image, with just the Subject levels set.

Figure 6-3. Using Subject settings to alter the Saturation, Gamma, and Contrast

Neo
- Pre Saturation: -30
- Pre Gamma: 0
- Pre Contrast: -30

Miami
- Pre Saturation: -10
- Pre Gamma: 0
- Pre Contrast: -10

You will probably notice there is a very slight difference between the three images. The Neo Look, having both the Saturation and Contrast set to -30, is more noticeable.

Working with the Lens Filter settings. The Lens Filter controls enable you to apply Black Diffusion, White Diffusion, and Gradient filters to your image. If you didn't shoot with a filter on your lens [Hack #30], you can accomplish a similar effect using these controls. Figure 6-4 shows the image as it becomes more unique after applying the following Lens Filter settings:

Neo

Black Diffusion
- Grade: 0
- Size: 10
- Highlight Bias: 0

White Diffusion
- Grade: 3
- Size: 3
- Highlight Bias: 70

Gradient

Grade: 0

Size: 85

Color: (black)

Highlight Squelch: 15

Fade: 50

Miami

Black Diffusion

Grade: 0

Size: 10

Highlight Bias: 0

White Diffusion

Grade: 3

Size: 3

Highlight Bias: 70

Gradient

Grade: 3

Size: 85

Color: (orange)

Highlight Squelch: 15

Fade: 50

Figure 6-4. Using Lens Filter settings to change the diffusion and color filters

With the Gradient Color set, along with the Diffusions, the Looks start to become unique. The Miami Look, with its Color set to a hue of orange, begins to have a very rich tone.

Working with the Camera settings. The Camera group has controls for 3-Strip Process and five controls for Tint (Tint, Tint Color, Tint Black, Tint Black Color, and Tint Black Threshold). Of most technical interest is the control for the 3-Strip Process, which attempts to mimic the Technicolor process of the 1930s. Figure 6-5 shows the following Camera effects on the image:

Neo

3-Strip Process: 0

Tint: 10

Tint Color: (olive green)

Tint Black: 50

Tint Black Color: (sky blue)

Tint Black Threshold: 25

Miami

3-Strip Process: 0

Tint: 0

Tint Color: (clover green)

Tint Black: 0

Tint Black Color: (navy blue)

Tint Black Threshold: 30

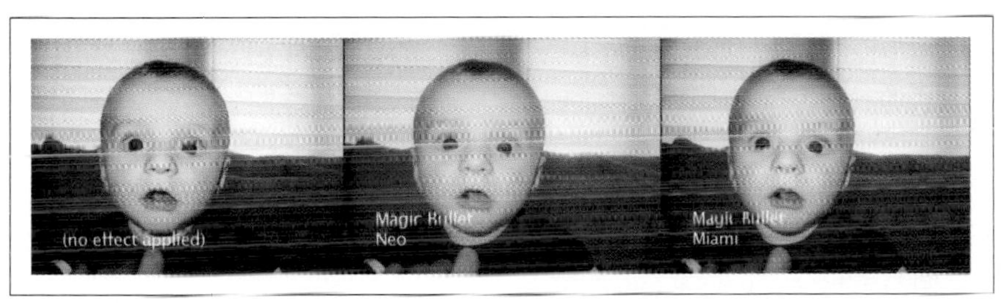

Figure 6-5. Changing the Camera settings, concentrating on Tint

Only slight changes are noticeable when applying the Camera filter. The Neo Look has started to take on its greenish hue because of the Tints being applied.

Working with the Post settings. Finally, the Post settings allow you to alter the tone and temperature of your image. The settings you can change are Warm/Cool, Warm/Cool Hue, Post Gamma, Post Contrast, and Post Saturation. You may have noticed that the Subject category allows you to set values on Saturation, Gamma, and Contrast. Figure 6-6 shows the final image, once the following Post settings have been applied.

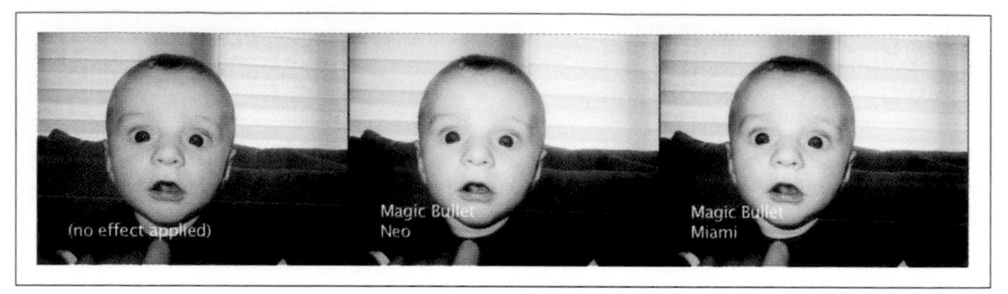

Figure 6-6. Concluding the process with the Post settings

Neo

Warm/Cool: -3

Warm/Cool Hue: -66.11

Post Gamma: 0

Post Contrast: 40

Post Saturation: -20

Miami

Warm/Cool: 0

Warm/Cool Hue: 0

Post Gamma: -3

Post Contrast: 30

Post Saturation: -10

The settings in Post work directly with the settings in the Subject category, so if you set your Pre-Saturation to +5 and your Post Saturation to -5, then your image's saturation level will become unaltered

Finally, the Looks are tweaked to bring out some extra details. The Warm/Cool settings on the Neo Look are brought down to *cool* the image, while they are set to zero for the Miami Look.

Using the Misfire Set of Tools

Misfire offers a variety of film-like artifacts, including scratches, dust, flicker, and grain. Every setting in Misfire can be turned on or off, so you can mix and match until you discover the look you prefer. Or, you can apply just one of the settings by using the individual effect:

Fading
: Lightens the image

Funk
: Varies the tone of the image

Splotches
: Discolors the image

Dust
: Places small black and white particles on the image

Flicker
: Makes the image's brightness change from frame to frame

Vignette
: Darkens the edges of the image

Displacement
: Slightly distorts the image

Micro Scratches
: Places small black lines vertically down the image

Grain
: Offers 10 variables to mimic film grain

Deep Scratches
: Creates a colored line vertically down the image

Basic Scratches
: Produces thin, wiggling lines

Gate Weave
: Causes the image to move slightly from side to side

Post Contrast
: Allows you (again) to work with the contrast

Each of the effects has to be seen in running video in order to convey the resulting look. Using the Misfire set of tools along with the prebuilt Looks provides an exceptional means of creating impressions not traditionally available when using video.

Rendering Your Project

Sadly, all the cool effects you get when using Magic Bullet come at a cost: time. The amount of time it takes to render your particular project will depend on how many effects you have applied to your video and how much computing horsepower you have at your disposal. It is not uncommon for an editor who has applied a lot of effects to set her system to render overnight, only to discover the next morning that the machine is still chugging away.

Be aware that Magic Bullet for Editors does *not* convert your footage to 24 frames per second. Film runs at 24 frames per second, so in order to give video a film "feel," it is necessary to remove frames and then apply a 3:2 pull-down. This process is quite involved and is beyond the capability of Magic Bullet for Editors.

If you would like to have your video take the one additional step toward looking and feeling like film, by appearing to run at 24 frames per second, you can either acquire your footage using a 24p camera or use a plug-in to apply the effect. Magic Bullet Suite (*http://www.redgiantsoftware.com*; $795) does perform the additional steps to create the film feel, but it is only available for Adobe After Effects.

In order to not be disappointed with the end result, always render small sections of your affected video. This will enable you to make sure you like the final "look" you are constructing. The process of affecting the look and feel of your video can be very time-consuming and should be left as the last thing you do before finalizing your audio.

HACK #62 Make Your Own Weather Report

As the famous song says, "I don't care what the weatherman says/When the weatherman says it's raining/You'll never hear me complaining," because it's easy to create your own weather reports.

No disaster movie is complete without a weather report on how the planet is being wracked by storms, hail, and frogs falling from the sky, along with the pointless request for everyone to stay calm as the world plunges into chaos. Or perhaps you want to add a weather report to your own personal newscast. Whatever the reason, creating your own fake weather report is not difficult.

To create the weather report, we are going to use weather maps downloaded from the Internet, video of you (or a friendly actor) presenting the weather report in front of a green screen **[Hack #22]**, and green screen compositing techniques **[Hack #70]**.

Grabbing the Weather

The first thing you'll need is a source for the weather images that you see on TV. The Internet provides a great way to obtain access to the same data and images that the professionals use. Government web sites such as the National Oceanographic and Atmospheric Administration (NOAA, *http://www.noaa.gov*) provide access to satellite images and weather maps produced by their weather experts.

The Weather Underground (*http://www.wunderground.com*) is another great source for weather maps. This site provides access to radar rain maps and personalized weather maps for the entire U.S. and many international locations. For a modest fee ($5 per year), the Weather Underground also allows you to get access to long-term data—a useful source if you want to show how storms move over time. Figure 6-7 shows a radar map of the San Francisco Bay Area, grabbed from the Weather Underground site. Whoever told you that it never rains in California is a liar, and here's the proof.

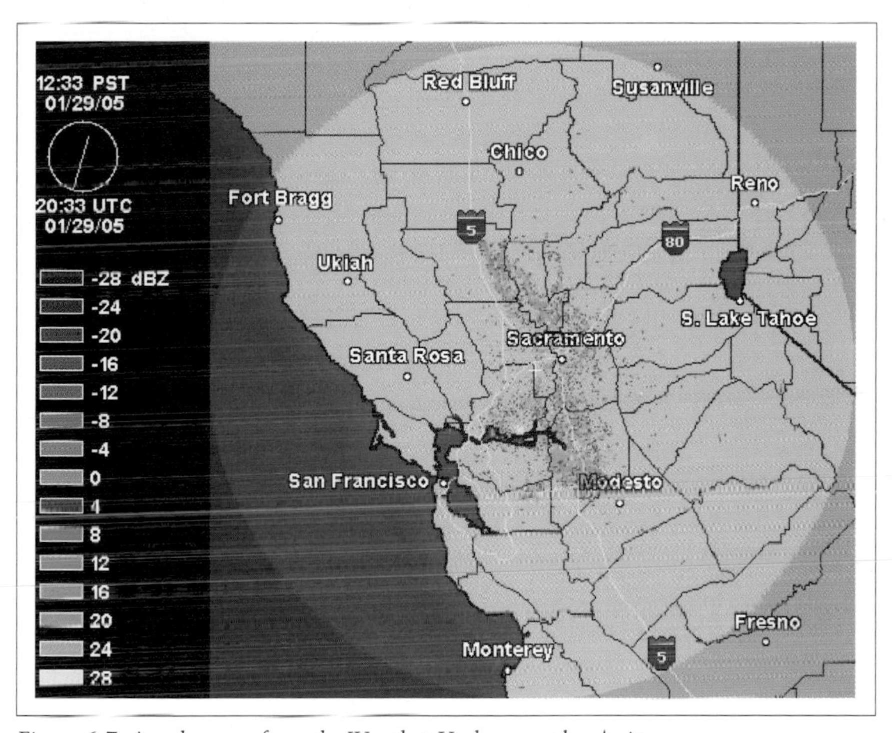

Figure 6-7. A radar map from the Weather Underground web site

To grab one of the images, just right-click (on a PC) or Control-click (on a Mac) and choose Save Picture As... from the menu. Most of these images are GIF files, which you can load into any image editing application.

If the file is animated (such as a radar map that shows the progress of a storm), it will probably be either an animated GIF or a Flash animation. If it is the former (you can tell by the *.gif* suffix in the filename), you can save them as described previously, but you'll need to use an application that understands animated GIFs. Photoshop can't, but ImageReady (which is bundled with most versions of Photoshop) can. If you don't have access to an application that can work with animated GIFs, you could try the demo version of Paint Shop Pro 9 from JASC (*http://ww.jasc.com*).

If the animation is in another format, such as a Flash animation, you'll need to use a screen-grabbing application, such as SnagIt (*http://www.techsmith.com*; $40) if you're using Windows. If you're on a Mac, use Grab's Timed Screen feature (Applications → Utilities → Grab → Capture → Timed Screen).

Editing the Images

You'll then need to tweak the images to look like the ones you see on TV. Watch the TV weather and see how they use these images: they usually show them full screen and add personalized touches based on the graphics that the station uses. You can do the same thing with any image-editing application. You'll find a selection of weather-related icons at *http://www.awesomeclipartforkids.com/cat.cfm?cat=Weather&sec=General*.

You can use these real images as a basis for your fake report, or you can build your own images. The U.S. Census bureau has a web site (*http://tiger.census.gov/cgi-bin/mapbrowse-tbl*) that produces custom maps that are in the public domain (meaning that you can pretty much do what you want with them without having to pay anyone). The site allows you to customize what the maps show.

ToolFarm has a tutorial (*http://www.toolfarm.com/tutorials/weather.html*) that describes how to create fake weather images using After Effects. The technique can also be used in other applications that can create fractal noise. The pros use expensive software, such as Curious maps (*http://www.curious-software.com*; $2,500–$7,500), but you can produce good results with most image-editing applications and a bit of patience.

Creating the Report

By taking the green screen image of your weatherman and compositing it over your weather map, you'll have the starting point of your weather report. Hey! You're a TV weatherman! The first time you try it, you will feel like a total idiot: you're standing in front of a green screen pointing at things that aren't there.

For that genuine disaster-movie feel, use stock footage of disasters in your weather report. The slightly inappropriately named FreeStock Footage (*http://www.freestockfootage.com)* offers a selection of moderately priced bad weather footage, or you could search the Moving Images section of the Internet Archive (*http://www.archive.org*) for public domain footage of chaos and mayhem. It's a lot cheaper than filming your own.

It is easier to point things out on the map if you can get an idea of what the final image looks like. Some compositing applications, such as Visual Communicator from Serious Magic (*http://www.seriousmagic.com*; $199–$695), provide a live preview of how the composited video will look, which makes it a lot easier. For Figure 6-8, I used Visual Communicator to create the composite and then used Premiere to add a fake news crawl at the bottom of the screen.

Figure 6-8. A fake TV weather report created using radar images from the Weather Underground, Visual Communicator, and Premiere

If it isn't possible to do a live composite, do a few test runs to get an idea of where features on the map appear so that you can point to them.

You can also use Post-it Notes to indicate where you should point while you record your report. You will need to hook up your television to your computer **[Hack #44]**, so that you can see your weather map. Then, prior to recording, stick the Post-it Notes to the television where you need to point. Finally, you'll need to be able to see yourself while you're doing your report **[Hack #11]** on the same television that you've marked. This will ultimately allow you to see where you're pointing. It may take some practice, but in the end, it's very fulfilling.

—*Richard Baguley*

Zoom In from a Satellite

Use a series of satellite images to establish a location.

Establishing the location of a scene can sometimes be critical to your story. Thanks to the United States Geological Survey (USGS), you can download satellite images of almost any location. You can use a series of these images to *zoom in* on your location.

There are a number of great web sites that allow you to both locate and view satellite images. I prefer to use TerraServer USA (*http://www.terraserver-usa.com*). The instructions in this hack pertain to that site, but the process can be applied to any other sites, as well.

Google Maps (*http://maps.google.com*) now also offers satellite images of locations, searchable by street address. Like the images provided by TerraServer, you can zoom in on a location (using a handy scroll bar and an altogether nifty interface). Though Google's images are in full color, many locations don't offer the same level of detail provided by TerraServer's images.

Find an Image of Your Location

On the TerraServer USA Home page, click the Advanced Find link, and then select the Address menu item. From the Address Search page, enter the address of the location you would like to view. Figure 6-9 shows a TerraServer image of Fenway Park.

You can also search by using Latitude/Longitude or Place. The Place search will attempt to find matches for what you enter, including Sites of Interest. For example, entering `fenway` will return both Fenway, Massachusetts, and Fenway Park.

The Results Page should contain a set of links to various images of the location. You will most likely want to download either the Urban Areas or Aerial Photo image type.

Download a Series of Images

Once you have made your selection, you will be presented with the satellite image of your location. You should quickly notice that when you click on the image, you are able to zoom in closer to point where you click. You can use this feature to continue zooming in until you have exhausted all of the

Figure 6-9. TerraServer image of Fenway Park

available images. You can also use the Navigate pane on the left side of the page (shown in Figure 6-10) to zoom in on the location by changing the resolution. You might increase the size of the current image by selecting from the Map Size pop-up button.

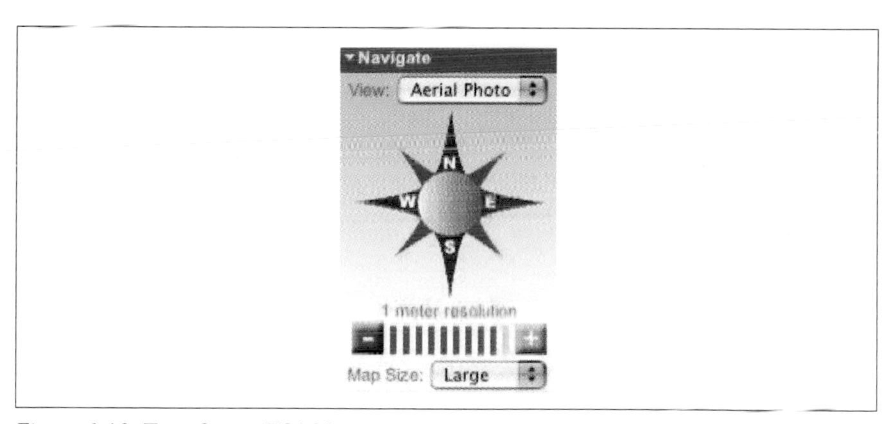

Figure 6-10. TerraServer USA Navigate pane

You will want to download the largest possible image available. Using the Navigate pane, you can select Large from the pop-up menu. The ideal image will have a resolution of at least 720×480 (NTSC) or 720×576 (PAL).

By following this process, you will be able to collect a series of images, which you can use to simulate a satellite zooming in on your location:

1. Click the Download button.
2. Save the image to your computer.

3. Click the Back to TerraServer link.
4. Zoom Out one level.

After gathering all of the available satellite images, you will need to import the images into your editing system.

Create the Zoom Effect

While using your editing program, import the images, like so:

Avid
: File → Import

Final Cut
: File → Import → Files...

Movie Maker
: File → Import

Premiere
: File → Import → File

iMovie
: File → Import

Most editing systems also allow you to import files by dragging-and-dropping them to specific windows within the application.

To simulate zooming in on your location, you will want to start with the image that shows the largest land mass and is farthest from your location. To do this, place each image in succession on your timeline, as shown in Figure 6-11.

Hacking the Hack

How you proceed from here is your creative decision. Here are a few ideas to get you going:

- Make the zoom more pronounced by slowly increasing each image's size/scale.
- To simulate a lens clicking into place (think microscope), insert a three-frame black transition between each image and add a dissolve.
- Overlay a Fog effect over one image to simulate flying through clouds.

Also, just because you can download images of the exact location where your scene was shot does not mean you have to. If you want to establish your child's baseball game at Fenway Park, go right ahead!

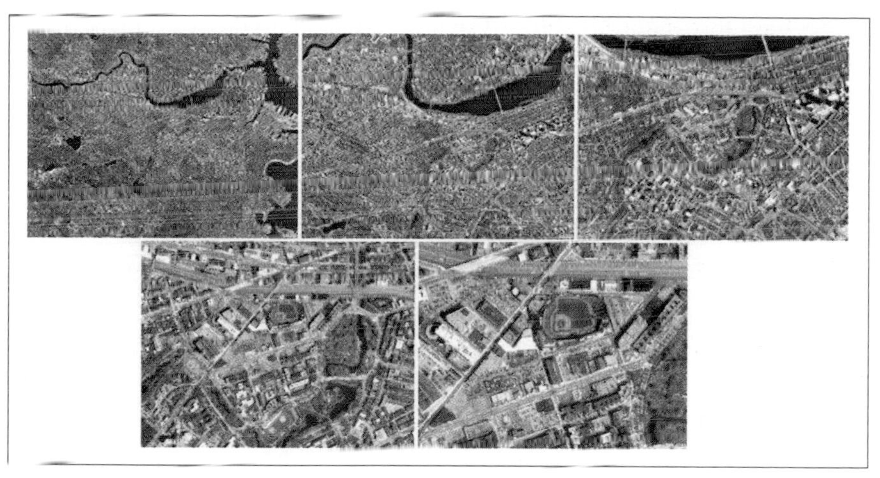

Figure 6-11. Five images, zooming in on Fenway Park

Remove an Unwanted Object

Sometimes objects, such as boom microphones, briefly appear in your raw footage. Here's how to get rid of them.

You can't always capture exactly what you want. Accidents happen and objects sometimes creep into your scene. Figure 6-12 shows an object intruding onto the video at the top edge. This is especially common when using a boom microphone **[Hack #53]** or when a glitch occurs in your video. With some layering in your timeline, and a few adjustments to the image, you can remove the offending object from your scene.

Figure 6-12. A boom microphone dipping into the picture

If you have the opportunity to cut to a different shot and avoid the object, I recommend at least trying the cut to see if it is acceptable to you. If it is acceptable, go about your business and concentrate on the storytelling process. Choosing to remove an object should be a last resort, prior to reshooting a scene.

Removing an object is completely variable from shot to shot, and is an artistic process. No single set of instructions can possibly cover every circumstance. Your creativity and patience will go a long way when attempting this hack.

Performing a Quick Fix

If your object is near an outer edge of your video, you can remove it easily by scaling your video slightly larger. Scale your image only for the amount of time that is necessary. Scaling will soften your picture, and the more you scale it the softer it will become. Therefore, you should scale your image only to a point that you can tolerate. Figure 6-13 shows a frame of video scaled to remove the boom.

Figure 6-13. Scaling to remove the offending object

If you would rather not scale your video, or if the image is too far into the frame to allow you to scale, you will need to alter the video and apply some effects in order to remove the object.

Determining a Replacement

You also need to determine how you are going to fill the area you are going to remove. This will require you to be familiar with your footage, because you need to determine if you can steal footage from another shot in your video or if you will have to create a fill. Keep in mind that you are attempting to remove an area of video and fill it with something that looks the same.

Using other takes. If you have shot your problematic scene using a tripod, or if you have captured more than one take, you can more than likely find a shot that has a *clean* section of video where you need to fill. In other words, if you have a shot in which a boom has crept in from the top right of the frame, and you have additional footage where the boom is not in the frame, you can use that additional footage for your fill. Figure 6-14 shows a problematic shot and the footage that will be used to fix it.

Figure 6-14. Finding a shot that has useable "fill" footage

Once you've found some footage you can use, you can begin the process of removing the object. Keep in mind that you can reposition your fill footage underneath your problematic footage. For example, if a boom mic has lowered into frame and there is a wall behind the boom, you might be able to shift the fill footage horizontally so that a clean section of wall shows through.

If after searching for good fill footage, you can't find a shot you can use, you'll have to get a little extreme and create your own fill footage.

Creating footage. Creating footage is a creative endeavor, and the approach will be different for every situation. In general, however, there are a couple of approaches you can take.

First, you can try to return to the location you shot your original scene and try to capture a *clean* shot. Keep in mind that the lighting conditions of the location will most likely have changed, so you footage will probably not match exactly.

You can also try using a photo-manipulation application, such as Adobe's Photoshop (*http://www.adobe.com*; $89.99 for Photoshop Elements, $649 for Photoshop CS), to create an image (or series of images) to use as fill. If you take this approach, it will help if you have at least a reference to your video. To create a reference, you should export a frame of your problematic video:

Avid
: Mark In and Out points for one frame → File → Export → Windows/Macintosh Image

Final Cut
: Mark In and Out points for one frame → File → Export → QuickTime Movie...

Premiere
: File → Export → Frame...

Then, import the frame into your photo application. From there, you can develop your fill footage, using the frame of video for reference.

Layering Your Timeline

To prepare for the object removal, you will need to at least two tracks of video in your timeline. If you have only one video track, you should add another:

Avid
: Clip → Add Video Track...

Final Cut
: Sequence → nsert Tracks...

Premiere
: Sequence → Add Tracks...

Once you have at least two video tracks in your timeline, place the scene on the topmost track. Figure 6-15 shows a timeline with two tracks and the topmost layer occupied.

The positioning of the object will determine how you remove it.

Removing the Object

Place your fill footage on the track just below your original footage. If your object is near an outer edge of your video, you will be able to remove it easily by cropping the appropriate edge until the image is no longer in view. If

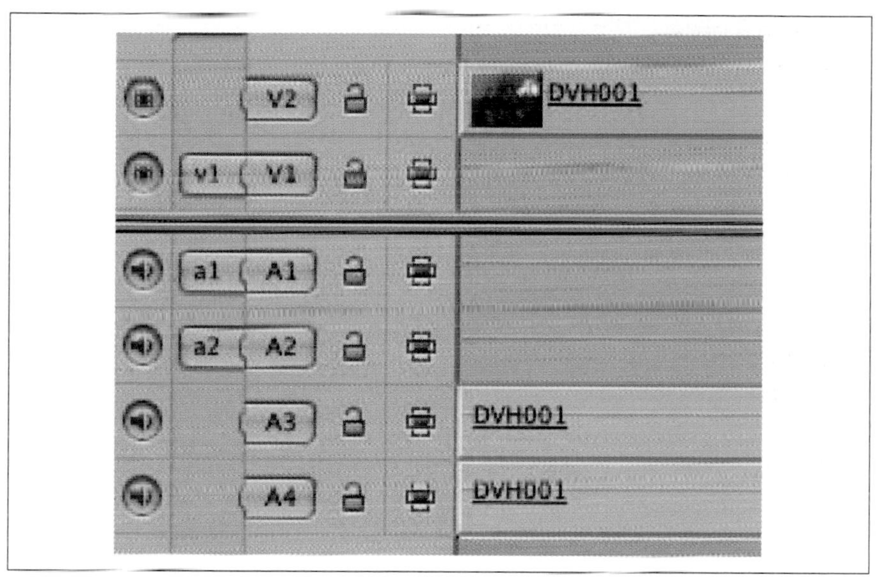

Figure 6-15. Placing your video on the topmost video track

the object is not near an edge, you will need to create a matte. Either way, you should first select your original clip and begin manipulating it.

Cropping your scene. The result of cropping your image is that you lose a portion of it. Don't fret when you first do this to your footage, because your fill footage is going to take care of the void. Here's how to crop your footage:

Avid
: Effects Palette → Image → Mask

Final Cut
: Motion tab → Crop

Premiere
: Effects → Transform → Crop

Once you have access to the crop settings, simply crop the edge in which your object is located. For example, if a boom mic has dropped into from the top, increase the crop setting for Top until the mic is no longer visible. If you discover while cropping that you are starting to remove important content from your scene, take a different approach and try using a matte. However, you should be aware that creating a matte can be a time consuming process.

Creating and using a matte. There are many different types of mattes, and your choice of matte will depend on your situation. For the most part, you should look to use a multipoint *garbage* matte, because it will provide precise control over the size, shape, and location of the matte. You will be able to find matte effects or their equivalent:

Avid
Effects Palette → Blend → Picture-in-Picture

Final Cut
Effects → Video Filters → Matte

Premiere
Effects → Video Effects → Keying

When you have located a matte you would like to use, you will need to shape it using the parameters supplied by the specific effect. Your goal is to remove the object (obviously) while keeping as much of the original image unaffected. Unless the object just momentarily pops into the scene, such as a video glitch, you will probably have to use keyframes to apply some movement to the matte.

Filling the Void

After successfully removing the object, you need to place your fill footage. If you have shot using a tripod, or from a static position, your fill footage should simply drop in and work. No fuss. No muss.

If you have not shot from a static position, you will need to finesse your fill footage until it fills the void created by the cropping. To accomplish this, you will probably have to reposition your footage:

Avid
Effects Palette → Image → Resize

Final Cut
Motion tab → Basic Motion → Center

Premiere
Effect Controls → Fixed Effects → Motion → Anchor

Once you have lined up your fill footage, your shot should be complete—minus the offending object. Figure 6-16 shows the original image, the cropped image, and the final image using the fill footage.

Given the fact that mistakes and technical glitches happen, at some point, an object will probably appear in your video that you'd rather not be there. Armed with a little knowledge, some creativity, and a lot of patience, you can remove the unwanted object from your scene.

Figure 6-16. Cropping and layering an image to remove unwanted objects

Create an Invert Effect Using Movie Maker

Windows Movie Maker includes an invert effect that isn't really available for use. Here's how you can get it out of the depths of Movie Maker and into your project.

In some instances, you might want to *invert* your video. Inversion causes your image to reverse its colors, kind of like a photo negative. Whether to create a unique effect, or as a means to accomplish a more complex effect, inversion can be quite useful. The Microsoft engineers realized this fact and created a Negative effect for use in Movie Maker to accomplish inversion. However, it is not available "out of the box."

I want to thank Rehan, of RehanFX (*http://www.rehanfx.com*), for clarifying the lack of an inversion effect within Movie Maker, and this particular implementation.

Creating the Effect

Movie Maker allows you to create your own effects. Using a text editor, such as WordPad, you can create effects using XML. To create the invert effect, open a new document and type the following:

```
<TransitionsAndEffects Version="1.0">
<Effects>
<EffectDLL guid="{B4DC8DD9-2CC1-4081-9B2B-20D7030234EF}">
<Effect nameID="62864" iconid="18" comment="Negative">
<Param name="InternalName" value="Standard" />
<Param name="Invert" value="-1.0" />
</Effect>
</EffectDLL>
</Effects>
</TransitionsAndEffects>
```

When typing XML, or any computer code, you should be very careful and exact. Most code will not work if something is out of place, such as a forward slash (/) or even a quote ("). If you discover that your effect isn't loading correctly, go back and check that you have typed everything *exactly* as shown.

Save the file as *invert.xml* (as a Text Document file) to *C:\Program Files\Movie Maker\Shared\AddOnTFX*, as shown in Figure 6-17. If the *AddOnTFX* directory doesn't exist, create it. When Movie Maker starts, it looks inside this directory for additional effects and transitions to use.

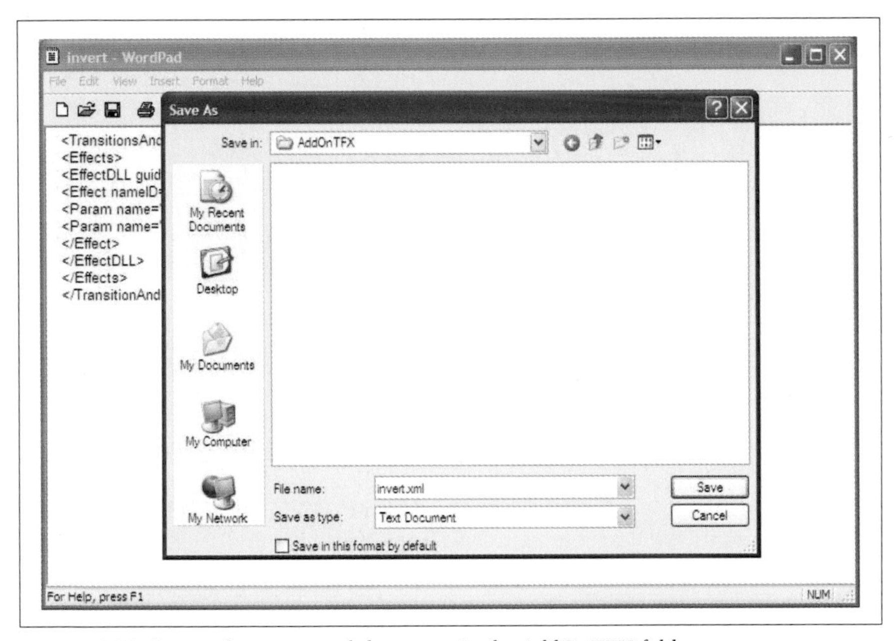

Figure 6-17. Saving the invert.xml document in the AddOnTFX folder

Loading and Using the Effect

After saving the file, launch Movie Maker. If it is already running, you will need to quit and restart it (the application, not the computer). Once Movie Maker is running, choose Tools → Video Effects, or click on the Video Effects item, to make sure the Negative effect is available. Figure 6-18 shows the Negative effect as it should appear in the Video Effects pane of Movie Maker.

If the effect is available, you will be able to select it and apply it to your footage. To do so, simply drag-and-drop it onto the clip to which you would like to apply the effect. Yup, it's just like any other effect in your palette. Figure 6-19 shows an image with the Negative effect applied.

Figure 6-18. The Negative effect, available to use

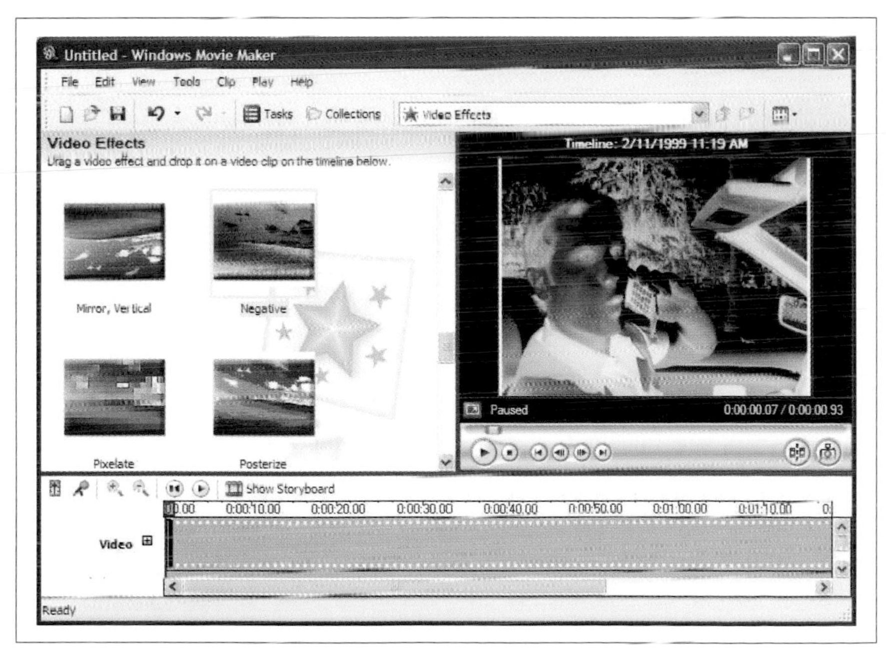

Figure 6-19. Making video look like a film negative

Hacking the Hack

If writing code is your thing, there is a world of creative effects and transitions waiting to be unleashed. Some companies exist solely based on creating effects and transitions for Movie Maker, and many people make a living writing them as well. If you would like to explore the wonderful world of creating your own effects, head over to the Microsoft Developers Network Library for Movie Maker Effects and Transitions at *http://msdn.microsoft.com/library/default.asp?url=/library/en-us/dnwmt/html/moviemakersfx.asp*.

HACK #66 Turn Video into Matrix-Style Symbols

The Matrix Trilogy was a pop-culture phenomenon. The streaming green text symbols that appeared in the movie slowly found their way into screensavers, web sites, and even on T-shirts. Here's a way you can watch almost any digital video in a similar way.

In the movie *The Matrix*, certain people had the ability to look at green characters streaming down a black computer screen and interpret what they meant.* It was as if they could see images in the characters. Well, you don't live in the Matrix (or, maybe you do...who knows?), but you can see images in streaming characters with a little help from your computer. Better yet, you can choose which images you'd like to see.

The conversion of video to a series of characters is often referred to as an ASCII (pronounced "as-kee") movie.

Downloading ASCIIMoviePlayer

If you are using Mac OS X, the QuickTime engine is buried deep in your operating system. To show this fact, some of Apple's engineers designed ASCIIMoviePlayer (*http://developer.apple.com/samplecode/ASCIIMoviePlayerSample/ASCIIMoviePlayerSample.html*) to play a QuickTime movie inside a Terminal window, which displays only text. The ASCIIMoviePlayer application takes the input of a video file and interprets its luminance as text-based symbols. It accomplishes this feat through the use of QuickTime.

After downloading ASCIIMoviePlayer you will have a folder containing four items: *ASCIIMoviePlayer*, *ASCIIMoviePlayer.pbproj*, *build* (a folder), and *qtplyr.c*. The one you are going to be most interested in is the file named *ASCIIMoviePlayer*, which is the actual application.

* For film geeks: the Wachowski Brothers did pay some homage to *Ghost in the Shell*.

The other files contain the pieces necessary for you to build the application from scratch. The actual source code for the application is in *qtplyr.c*, should you wish to see how it works or even enhance it. If you have installed the Developer's Tools, included with Mac OS X, you can double click *ASCIIMoviePlayer.pbproj* to work with the source code.

Running ASCIIMoviePlayer

To run ASCIIMoviePlayer, you need to run the Terminal application. If you are not familiar with Terminal, you can find it in the *Utilities* folder. The default color for Terminal is black text on a white background.

If you would like to change the color of the text and background in Terminal, choose Terminal → Window Settings → Color. There is a set of preselected color schemes, one of which is Green on Black. You can also customize the color combination, if you so desire.

To run a movie through ASCIIMoviePlayer, drag-and-drop the ASCIIMoviePlayer application to the Terminal window. Then, drag-and-drop a movie you would like to play (of course, you can type out the commands if you so desire).

When you drag-and-drop a file on the Terminal window, you will notice that the path of the file appears automatically. This is the expected behavior.

Before hitting the Return key, you should maximize the Terminal window by clicking on the Zoom button (the small green button in the upper-left corner of the window) Figure 6-20 shows a Terminal window about to run ASCIIMoviePlayer using a movie called *mymovie.mov*.

```
/Users/joshpaul/Applications/ASCIIMoviePlayer /Users/joshpaul/Movies/mymovie.mov
```

Figure 6-20. Terminal, ready to show a movie

Once you are prepared, press the Return key and watch the characters stream along your screen. You should be able to see your movie within the streaming characters. Figure 6-21 shows a frame from a movie playing in ASCIIMoviePlayer.

Figure 6-21. Did you take the Red Pill?

Alternate Approaches

There are other programs that are capable of converting movies to ASCII.

QuickASCII (*http://sourceforge.net/projects/quickascii/*; free, open source) is based on ASCIIMoviePlayer, but it has added optional color output and performance enhancements. QuickASCII runs on Linux, Mac OS X, and other Unix-like operating systems.

MPlayer (*http://www.mplayerhq.hu/homepage/*; free, open source) is a video player that is available for Windows, Mac OS X, Linux, and just about every other operating system on the planet. It is a capable media player in its own right, but if you add AA-lib (*http://aa-project.sourceforge.net/aalib/*; free, open source), you can get MPlayer to output ASCII movies by using the `-vo aa` argument. If you have trouble building AA-lib, there is an installer package at *http://sveinbjorn.vefsyn.is/aalib*. If you want to add color ASCII output, you can add the libcaca library (*http://sam.zoy.org/projects/libcaca/*).

Hacking the Hack

Sure, watching the video is fun for a while, but if you want to, you can integrate its output into your movie. For example, you can take an edited scene from your movie and dissolve to the continuation of the scene, only ASCII-fied. There are a couple ways to accomplish this effect:

- You can record your computer monitor with your digital video camera. If you use this approach, you will need to frame the image carefully, but you do that for all of your shots anyway, right? You might also have to change your camera's and computer monitor's settings in order to capture the image without flicker **[Hack #32]**.

- You can use a screen-recording program and capture the ASCII movie as a digital movie. I do not recommend this method unless you have very fast hardware and an abundance of hard drive space.

There is also a virtual community of people creating and working with ASCII art. There are ASCII webcams, ASCII pictures, and even ASCII cartoons. ASCII...who woulda thought?

Create a Surreal Scene

Using a layering technique, place people "inside" a surreal environment, such as a dream sequence.

There are limited genres and scenarios in which, as a movie producer, you can let your imagination run wild. One scenario in which this occurs is when you are attempting to convey a dream. Well, if you can alter reality in a dream, why not alter it in an unusual fashion? You can place your characters in a watercolor meadow, a crayon-colored office, or, if you're artistically challenged, a computer-rendered planet surface.

You can also follow NASA's lead and visit Mars! Images are available from the Spirit and Opportunity Rovers at *http://marsrovers.jpl.nasa.gov/gallery/all/*.

Creating Your Foreground and Background

You will be creating a complete environment where your characters will interact. In order to achieve this, you will need to create both a foreground and a background image to be layered in front of and behind your characters, respectively. Your foreground image should consist of props, such as desks, or cars. These props will be used to provide an illusion of depth.

When choosing a medium on which to create your foreground, choose a color you do not plan on using in your foreground artwork. This is because you will be transferring your artwork and adding it to your scene. In order to do so, you need to key out the color of the medium you have chosen. Otherwise, parts of your image will be made transparent along with your background. For example, if you have selected to create your foreground artwork on a sheet of green paperboard, you should not use any green in your actual artwork.

Some editing systems are capable of removing only blue or green backgrounds. Make sure you know the capabilities of your system prior to doing any production work.

Your background image should cover the entire surface of whatever medium you are using (or not, if that's your artistic vision). The background of your sequence should also provide some type of perspective, including a horizon line.

Using art supplies to make your images. If you are more artistic, paint, color, and draw your way to the environment of your dreams. Keep in mind that when you have finished, you will need to import your work into your computer. Therefore, you should create something that can be easily transferred to your computer.

If you have a scanner, scan your image at a high-resolution setting. Even though you will be using only a 720×480 image (720×576 for PAL), it is always better to have more detail available should you need to manipulate the image. If you do not have a scanner available, you can just as easily use your video camera to record your artwork. When using a video camera, make sure you both light the image appropriately and keep it framed. A good method is to capture the image using a tripod to stabilize your camera.

Using your computer to make your images. If you fall into the *artistically challenged* category, as I do, you can create your background using your computer. One program that is exceptionally adept at creating landscapes and environments is DAZ|Studio (*http://www.daz3d.com/studio/*; free). DAZ|Studio is an easy to use program that focuses on creating three-dimensional landscapes and comes with pre-built backgrounds and props. It is available for both Windows and Macintosh. Figure 6-22 shows a shot from the Fairy Forest setting within DAZ|Studio.

Figure 6-22. The Fairy Forest: a built-in scene

If you find you just can't create a background you are happy with, take a peek at the prebuilt offerings at DAZ. Many are reasonably priced and are quite impressive. I feel they offer a lot of production value, when you have the chance to use them.

Another program worth mentioning is Painter (*http://www.corel.com*; $399–$429), from Corel. Painter is a nice compromise between real-world painting/drawing and computer-aided creation.

Shooting Your Scene

In order to place your characters in the environment you have created, you will need to record them in front of a green screen **[Hack #70]**. Make sure your green screen and your characters are well lit. If you have the time while shooting your scene and have a semiportable editing system, capture some green screen footage and test how well it works with your background. By doing so, you will be able to correct any problems before they are recorded.

Compositing Your Scene

Once you have designed and imported your foreground and background images, and have digitized your green screen footage, you will have all of the elements necessary to composite your layers together and create your scene.

One the bottom layer of your timeline, place your background footage. On the next layer up, place the footage you shot in front of a green screen. Finally, place your foreground footage on the third layer.

The first step you will want to take is to key your foreground. This process is the same as keying out a green screen, with the difference being how you accomplish the key:

Avid
: Tools → Effects Palette → Key → RGB Key → (Select color to key out)

Final Cut
: Effects → Video Filters → Key → Color Key → (Select color to key out)

Premiere
: Effects → Video Effects → Keying → Chroma Key → (Select color to key out)

After keying your foreground, key your green screen footage **[Hack #70]** so that it, too, becomes a part of the scene. After accomplishing both the foreground and green screen keys, you will need to finesse your scene by moving the center of your layers around, so that your image composition is how you want it.

When you have successfully keyed the green screen and your foreground footage, and placed everything to your liking, the scene will look as if it was all recorded at the same time. You will probably have to make minor adjustments to the "look" of your scene before rendering your scene and considering it complete. Figure 6-23 shows a completed scene.

Figure 6-23. A composited scene consisting of three layers

When all of your effects have been placed and configured, and your scene is complete, render, sit back, and enjoy your dream.

Hacking the Hack

You can really be creative with this approach to creating a scene. For your background, create a long image horizontally. When you shoot your characters, have them walk from one side of the screen to the other. Then, when you composite the images together, add some movement to the background by changing its position relative to your characters.

Create multiple layers for use in your foreground. If you have more than one person in your scene, you can then place them at different layers within the scene along with foreground items between them. Add a shadow to the props and your characters **[Hack #71]** for more depth.

You can also apply different video effects to individual layers or your entire scene. Manipulating the saturation, brightness, and hue will help bring out the surreal nature of the scene you create. You can also resize your foreground props or even your characters!

Don't forget, it's your world; your viewers are just visiting.

Change a Scene from Day to Night

Using color-correction tools, you can make a scene that was shot during the day look like it was shot at night.

Not everyone has the experience or luxury to shoot a scene at night. If you find you need a nighttime scene but cannot shoot at night, there are a few color-correction tricks you can use to fake a nighttime look.

Although using color correction to change a scene from day to night is possible, you will get the best results if you can really shoot at night. The real thing is always better.

Using the Color Corrector

The primary difference between a daytime shot and a nighttime shot is the way colors are perceived. You will notice that at night everything takes on a blue-purple hue. This is the effect you are going to mimic.

Color correction deals with color by its very nature. If you plan on playing your movie on a television screen, you should make sure you color-correct using a television for reference **[Hack #44]**. You should especially be aware that NTSC is sometimes jokingly referred to as "Never The Same Color." This joke is in reference to the fact you can view an image on one television set and then view it on a second and see a completely different image, color-wise.

While using your editing application, locate the scene you would like to alter and select it. Then, open the color correction tool:

Avid
: Toolset → Color Correction

Final Cut
: Effects → Video Filters → Color Correction 3-way

Movie Maker
: Use Pixelan's SpiceFX Pack E1: Correction Effects (*http://www.pixelan.com*; $14)

Premiere
: Window → Workspace → Color Correction

iMovie
: Use GeeThree's Slick Effects: Volume 6 (*http://www.geethree.com*; $69.95)

There are a few key terms to be familiar with when color correcting:

Chrominance (a.k.a. chroma)
: The part of a video signal that includes the hue and saturation

Gamma
: The part of the video signal that includes the intensity and affects details

Luminance
: The part of a video signal that includes the brightness

In order to create the day-for-night effect, you will need to manipulate the Chroma, Gamma, and Luminance, in addition to altering the blacks, whites, and midtones. Depending on the features of your editing system or the plug-in you have chosen to use, your results will vary. Additionally, every shot is different, due to lighting, the camera lens used, and color composition of the scene, so you will need to adjust accordingly. Figure 6-24 shows the original image next to a day-for-night-effected image.

Figure 6-24. Mele Kalikimaka, during the day and at night

Reducing the Luminance and Chrominance

One of the first steps in creating the day-for-night effect is to reduce the luminance and chrominance of your scene. Obviously, by reducing the luminance, you will lose some of your scene's clarity. But, then again, when you are walking around at night, you can't see as clearly as during the day. Right?

You will also want to reduce the chrominance. This will cause some of the subjects of your scene to not stand out as much. As with all of the changes you are applying, you will need to experiment to see how much chrominance you need to remove. Figure 6-25 shows the image with the luminance and chrominance adjusted.

Figure 6-25. An image with altered luminance and chrominance

Reducing the Gamma

If you have the ability to reduce the gamma of your scene, you should do so. However, keep in mind that a little can go a long way when dealing with gamma settings. Figure 6-26 shows the image with a slightly adjusted gamma level, from 1.0 to 0.95.

Figure 6-26. Slightly adjusting the gamma level

Pushing the Blacks

In order to add a blue-purple hue to your scene, you can push your blacks more toward the blue range of the color spectrum. You do not want to go too far into blue, however, because your image will begin to take on an almost bizarre feel. Figure 6-27 shows the image with the blacks pushed toward blue.

Figure 6-27. Pushing the blacks toward blue, beginning to give a night look

Moving the Midtones

To push the image even more toward the blue-purple hue of night, push the midtones of the image toward blue as well. The initial effect might be a little startling, and your knee-jerk reaction might be to not push too far, so you may have to revisit your midtones a few times as you adjust your image to the desired look. Figure 6-28 shows the image with the midtones pushed toward blue.

Figure 6-28. A sense that it's getting later in the day . . .

Pushing the Whites

After you've put so much blue into your image, you'll probably be relieved to find out you get to play at the other side of the spectrum. When dealing with the whites, you will want to push them toward yellow. This seems a little strange, but it will bring out some of the whites within all of the blue. Figure 6-29 shows the image with the whites pushed toward the yellow side of the spectrum.

Figure 6-29. Pushing the whites toward the yellow side of the spectrum

Reducing the Saturation

Last, but definitely not least, you will want to reduce the saturation. By doing so, you will bring down the intensity of rest of the colors within the image. It will also bring your image back from the world of the overly colored. Figure 6-30 shows the image with the saturation reduced by over 50%.

Figure 6-30. Altering the color stauration

Color correction is a very powerful method of altering the look of your video. Whether you want to change a shot from day to night, or simply give your video a unique impression all your own, using color correction tools and techniques will give you the ability.

Create a Credit Flag

Most digital video software enables you to easily create text on screen. Some are even quite advanced and can create really cool titles. But sometimes, a more rudimentary approach can yield equally impressive results.

Most movies begin with an opening scene and a list of people's names and titles superimposed over it. This is known as the Title Sequence. Taking a simplistic method to creating credits (I call the method *credit flags*) will cause people to wonder, "How'd they do that?"

Preparing Your Flags

First, you will need to gather a few supplies: a large swatch of lightweight white fabric, a black marker, a wooden or metal pole, and a fan. You will also want a pair of scissors and way to attach the fabric to the pole, such as a stapler, push pins, or even tape. Figure 6-31 shows a sample set of the required materials.

Figure 6-31. Tools to create your credits

After you obtain the necessary materials, cut the cloth into squares large enough to write on using the marker. For example, when creating your personal credit flag, you will need a square of cloth large enough to write "Written and Directed by: Me!" in big, black letters. How you write and layout the letters is a matter of personal taste. You should create a flag for each person, or group of people, you would like to provide an on-screen credit.

Recording Your Flags

After you have created your flags, you are ready to assemble a credit flag and record your video. To create the flag, attach the square of cloth to your pole. Then stand the pole upright, with the fan aimed at the side of the flag.

You will need to set and focus your camera on the flag. Make sure that the flag fills the screen, even when being blown about by the fan. If you notice there are dark shadows on the flag while it is blowing around, you should provide additional light on it in order to reduce their appearance.

To provide a little more room for error, you might want to record your footage against a white background. If you don't have a fan, you can manually wave the flag, by holding the pole horizontally and rolling it back and forth.

After you have set your camera, begin recording. Capturing around 30 seconds per flag should provide more than enough footage. Repeat the process until all of the people important to your movie have had their flags recorded. Your final video will then need to be imported into your editing system. Figure 6-32 shows a credit flag, as imported into Adobe Premiere Pro.

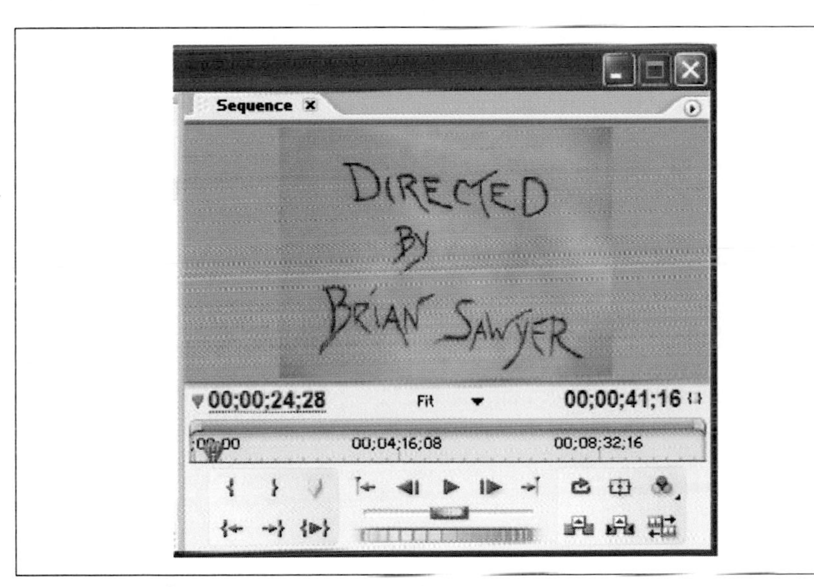

Figure 6-32. A flag, soon to become a credit

Matting Your Flags

After you have digitized all of your footage, open the footage in your Editor and apply an Inverse effect, making the letters white and the cloth black, as shown in Figure 6-33.

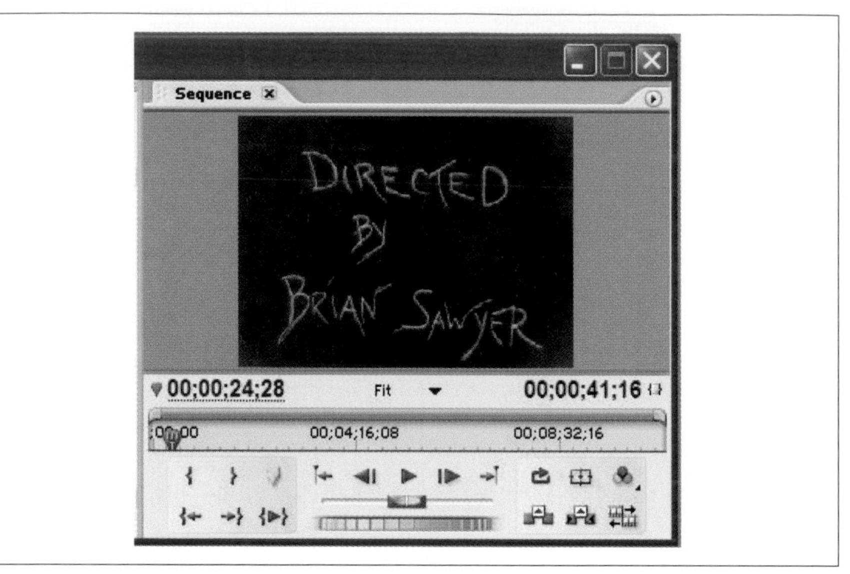

Figure 6-33. Not a chalkboard, just an inverted flag

Here's how to digititize the footage in various editing systems:

Avid
: Tools → Effects Palette → Image → Color Effect (Chroma Adjust → Invert)

Final Cut
: Effects → Video Filters → Channel → Invert

Movie Maker
: Add your own Invert Effect **[Hack #65]**

Premiere
: Window → Effects → Video Effects → Channel → Invert

iMovie
: Use the Negate B/W plug-in from cf/x (*http://www.imovieplugins.com/*; free)

If you find that your credit flag doesn't look "right," check your lighting set-up.

After applying the inverse effect to your footage, place the footage over your opening scene and apply a Matte or Key effect to it:

Avid
: Tools → Effects Palette → Key → Luma Key

Final Cut
: Effects → Video Filters → Matte → Image Mask

Premiere
: Effects → Keying → Luma Key (adjust Threshold and Cutoff settings accordingly)

iMovie
: Use the Chroma Key plug-in from cf/x (*http://www.imovieplugins.com/*; $3.50)

The Matte effect will remove the (now) black portion of the credit flag, allowing for the underlying footage to show through while the (now) white words remain visible, as shown in Figure 6-34.

Figure 6-34. The Directed By opening credit, as created

Once you have placed your credits in your timeline, render it and play. Your credits should now have an airy "feel" to them.

Hacking the Hack

For an additional "wow," apply some movement to the credits by making them move around the screen. Apply movement to have them bounce, fly on and off screen, and so forth. By matching the movement of your credits to the beat of some music, you will create a very organic feeling title sequence.

A variety of options can be applied to this method. Experiment, discover and, most of all, have fun.

HACK #70 Composite a Green Screen Image

Green-screen compositing allows you to create Hollywood-style effects.

Have you ever wanted to fly a spaceship? Or wanted to travel to Russia? Maybe you've just wanted to be a professional weatherman **[Hack #62]**.

Whatever your dream, you can probably make it happen by using a green screen **[Hack #14]**. The movie *Forrest Gump* used green-screen techniques successfully to create the illusion of Tom Hanks' character (Forrest) meeting historical figures such as John F. Kennedy, Richard Nixon, and Elvis Presley. By recording your scene in front of a green screen **[Hack #22]** you can create almost any illusion you can dream up **[Hack #67]**.

In addition to green-screen footage, you should also acquire footage you would like to composite it with. For example, in the case of meeting a historical person, you would want footage of that person. Before shooting your green screen, make sure you know what footage will be in the background of your shot. Figure 6-35 shows a frame from a green-screen shoot and a background image for compositing.

Figure 6-35. Two images that will soon become one

Starting the Composite

Compositing your footage requires you to use a *key* effect, which allows you to remove certain colors or portions of the video signal, such as luminance, from your final image. When using a key and multiple layers of video, the layer below the one that is keyed will show through. Figure 6-36 shows a green-screen image with and without a key applied.

Here's how to apply a green-screen key:

Avid
: Tools → Effects Palette → Key → Chroma Key

Figure 6-36. A green-screen image with and without a key applied

Final Cut
Effects → Video Filters → Key → Blue and Green Screen

Premiere
Effects → Video Effects → Keying → Green Screen Key

If you find that the specific key effect you have used doesn't work as well as you expected, you can try other key effects, such as like Color Key, Chroma Key, or RGB Key. Figure 6-37 shows how an initial composite might look.

Figure 6-37. Merging the two images in an initial composite

Unless you're really lucky (or really good), you will have to change the position of your actors within the frame.

Shifting the Shot

The next step in compositing your shot involves moving it to the location you desire. By moving the center point of the footage, you can change your actors' spatial relationship to the rest of the scene. This effectively enables you to place your actors in a more believable position.

Avid
: Tools → Effects Palette → Image → Resize (Position)

Final Cut
: Viewer window → Motion tab → Basic Motion → Center

Premiere
: Effect Controls → Motion → Position

Changing the position of your video is really a matter of trial and error. It might take you a short time to get the video to where you want it. Once you've placed your video, either play it (if your system is capable of real-time playback) or spot-check it by clicking in various places on your timeline. If there is a lot of movement in your green-screen image, you might need to reposition your video until it looks right. Figure 6-38 shows the composite after repositioning the green-screen image.

Figure 6-38. Moving the center of the image

Again, unless you're really lucky, you will need to make a few more adjustments in order to fool your audience into believing their eyes.

Adding the Finishing Touches

Depending on how you have lit your green screen, the colors and shadowing of your background, and the overall match of your footage, you might need to apply additional effects to achieve any level of realism. Some effects you might have to work with include saturation/desaturation, brightness and contrast, and general color levels.

If you are comfortable using color correction, that will be your most powerful tool. Additionally, if there are shadows in your background, you might want to add shadows **[Hack #71]** to the green-screen image to match. Figure 6-39 shows the image after slightly desaturating it and adjusting and the brightness and contrast.

Figure 6-39. Altering the saturation, brightness, and contrast

Getting a green screen to achieve a level of believability can take a long time. When working with the footage, you should take a break and walk away at least every hour. Otherwise, you might become too engrossed in the effect to think about it clearly.

HACK #71 Create a Shadow for a Green Screen Image

Compositing someone or something into a scene is an accomplishment. But to make it believable, you might need to add a shadow.

A green screen image is one where an image is captured with a background consisting of only one color, green. You can also acquire a blue screen image, where the background is blue, or just about any other color you can imagine. The essential ingredient is that your image is in front of a single, solid-colored background, and that your subject not have the same color as your background on it. It also helps if you are able to light and set the scene with compositing in mind **[Hack #22]**.

When you composite a scene using a *green screen*, you might discover there is no shadow being cast from your subject. If you discover the scene you are attempting to composite needs a shadow, because it just doesn't look *right*, there is an easy solution to your problem. Even better than that, the solution can be found within the problem itself.

This process works well for a shot of a single moving subject, such as a woman walking across a beach. It can also work just as well for more than one subject; however, the more movement there is between the subjects, such as dancing in a circle, the less successful this process will be, because your shadow won't cast *from* one subject *onto* the other.

Creating a Copy of Your Image

The first step in creating your shadow effect is to duplicate your video image. To do so, simply create another video layer on your timeline and drop a copy of your green screen video onto it. You will want to apply the same exact effect you used to initially composite your scene **[Hack #70]**. When prepared correctly, you should have two layers of video, with the same effects applied.

The best approach to duplicating a green-screened image is to copy-and-paste it into the timeline. After applying the necessary effects to key out the green screen on the original image, copy the entire clip, including the effects, and paste it to another (lower) layer on your timeline. This might require *shuffling* your footage onto newly created layers in your timeline.

Layering Video to Prepare for Shadow Creation

In order for the effect to appear correctly, you will want your *original* image on a layer above the shadow image's layer. If you do not do stack the layers correctly, the shadow will appear in front of your subject. Figure 6-40 shows the difference between having the shadow layer above the original layer and below the original layer.

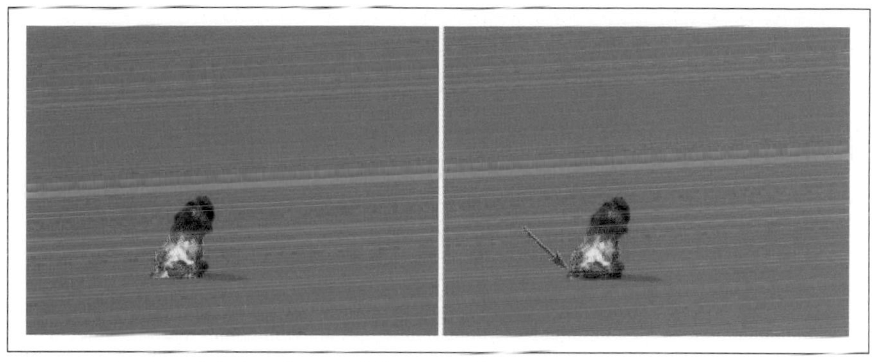

Figure 6-40. The subtle but distinct difference in layering

Create the Shell of the Shadow

Now that the image layers have been prepared, you need to remove the color from the image. By removing the color, you should end up with a black representation of your original image. To accomplish this:

Avid

Open the Effect Editor for the clip and choose Show Alpha.

With Avid, by changing the key to a moving matte, via the Show Alpha option, any additional effects you apply are independent of any other layers. After creating the moving matte, apply a paint effect to the clip, to *nest* the key effect:

Tools → Effects Palette → Image → Paint Effect

Then, drag a matte key on top of the paint effect, to provide the shadow:

Tools → Effects Palette → Key → Matte Key

Final Cut

Effects → Video Filters → Image Control → Brightness and Contrast

Premiere

Effects → Video Effects → Adjust → Brightness and Contrast

Shear the Shadow Off of the Subject

Once you have created the basic shadow element, you need to shear it away from your subject. If you don't shear the shadow image, your audience will never see it because it will always be behind your original image. To shear your image, you need to distort it:

Avid

Tools → Effects Palette → Blend → 3D Warp

Final Cut
: Effects → Video Filters → Perspective → Basic 3D (or use the Distort Tool)

Premiere
: Effects → Video Effects → Distort → Corner Pin

Getting your shadow to *fit* the scene takes time and patience.

There are no specific rules I can provide to shear your shadow realistically because every shot will require different settings. If there are other shadows in your background, use them as a reference point to help evaluate how well your settings match the scene.

Add Subtlety to the Shadow

Depending on your personal preference, as well as any preexisting shadows in your scene, you will possibly want to blend your new shadow. Blending the shadow is easy to do: just change its opacity. In order to add a little more realism, you can add a Blur effect to your shadow in order to soften it.

If you want to adjust any of the parameters on your effect, simply edit the effects you've applied until you obtain acceptable results. Again, if there are any preexisting shadows in your scene, you should attempt to match your self-created shadow as much as possible. Figure 6-41 shows a series of images from a completed scene with and without a shadow applied.

Figure 6-41. A completed scene with (top) and without (bottom) shadow applied

Hacking the Hack

There will usually be final, small details that you need to pay attention to. You will more than likely have to perform some minor adjustments to your subject image, including changing brightness and contrast to match your background. You might also have to do some color correction in order to have your original image *blend in* correctly.

When you've finished compositing your shadow, try toggling your shadow layer on and off, merely to remind yourself that you've created a shadow that follows your subjects... just like the real thing.

Alternate Endings Based on the Time of Day

An alternate ending is merely a different ending to a movie, as opposed to the one shown in the official release. Using some of the hidden features in QuickTime, available through Totally Hip's LiveStage Pro, you can provide an alternate ending to your movie based on the time the viewer is watching it.

You might remember the *Choose Your Own Adventure* books where, at the end of a chapter, you were able to choose one of a handful of options. Based on your choice, you would then jump to a specific page in the book, where the story would resume. The story would unfold based on the choices you made and you could read the book many times over, each time with a different outcome.

Reading is an easily adaptable medium for interaction, but video is not. People watch video passively; otherwise, they play video games, which are interactive. So, how can we adapt the passive medium of video to something that is interactive? One way is through time; depending on the time the video is watched, certain scenes will present themselves.

Using Totally Hip's LiveStage Pro (*http://www.totallyhip.com*; $449.95) you'll be able to create movies that can alternate scenes based on the time of day.

Planning for Passive Interaction

The first step in creating a passive-interactive video is to plan ahead. Just like the *Choose Your Own Adventure* books, you will need to plan for the occurrences you want to present and shoot the appropriate scenes. For simplicity, you should start with only one alternate ending, which means you need to shoot a total of two endings. As you get used to this type of video, you can create as complex a movie as you like.

You will not want to add the interactivity until after you have completed editing your movie. Upon completion, you should output three digital movies:

- The entire movie, minus the ending
- The original ending
- The alternate ending

You will compile these movies into one, final movie for distribution.

Creating Your Project

Start a new project in LiveStage Pro and choose Edit → Document Settings... to change some of the properties of the movie. Under the Playback tab, change the Transport → Controller: pop-up to Standard Movie. This will place the familiar time indicator at the bottom of the movie, along with a Play/Pause button and volume controls.

Under the Settings tab, change the width and height of the movie to 720 and to 480, respectively (720×576 for PAL).

If you would like to annotate your movie—with your name, some comments, or and disclaimers, for example—you can do so under the Annotations tab.

Importing Your Movies

Locate your complete movie on your hard drive and drag it onto your LiveStage Pro timeline. The video and audio tracks will be separated out and labeled Video 1 and Sound 1, respectively. You can change their labels by double-clicking on the label on the left side of the window. This will open a window, allowing you to change the name of the track, the height and width, and a whole bunch of other details (which do not pertain to this hack).

Adding the endings. In order to make the alternate ending work, you obviously need the two endings to become a part of final movie. Locate the original ending and drag it to the timeline. Then, change the name of the video track to `OriginalV` and the audio track to `OriginalA`. Repeat the process with the second alternate ending, but change the tracks to `AlternateV` and `AlternateA`.

The naming of the tracks is important, because that is how your movie is going to identify which tracks to play.

You will most likely notice that your movies are all *stacked* on top of each other. If you were to play your movie right now, all of the audio tracks would mix together and only the top-layer video track would be visible. That's not what you want, so you need to move the two endings to the end of your movie.

Moving the endings. To move the endings to the end of the movie, you need to enable the Track Manipulation Tool by clicking on the bottom icon on the left side of the timeline. Once you've enabled the tool, your tracks will turn grey-silver in color. You can then click on the video and audio tracks and drag them individually to the end of the movie. If you move slowly near the end of the movie, the tracks should snap into place. Figure 6-42 shows the Track Manipulation Tool as enabled and the alternate endings lined up.

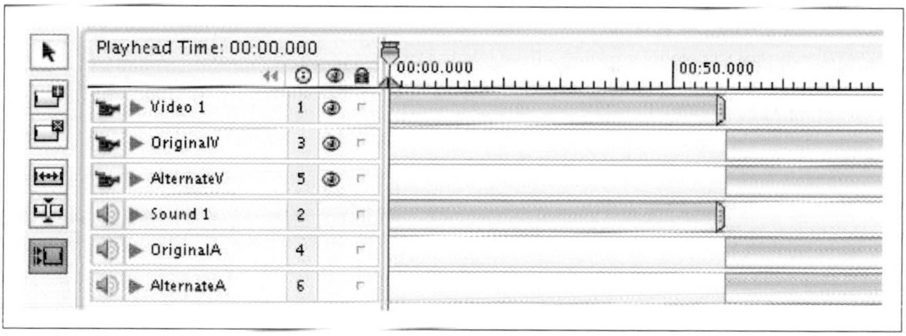

Figure 6-42. Moving the tracks with the Track Manipulation Tool

If you want to make sure your movies have lined up correctly, you can zoom into the timeline (just like most video editing programs) using the Zoom Slider at the bottom left of the window or by using the keyboard shortcuts to zoom in ([) or zoom out (]). After placing your endings, click on the Selection Tool, which is the arrow button at the top left of the timeline.

Enabling passive interaction. Okay, you've got everything lined up. So, how do you make the movie "know" what time it is and which track to play? You can either use a Sprite Track or a Text Track. Both track types will accomplish the same goal, but I like the Text Track for this task.

To add a Text Track, select Movie → Create Track → Text. You will then have a new track at the top of your timeline. If you cannot see the green-colored track, you need to move to the start of your timeline. You should double-click the green track, which will open the track editor.

Under the Track Events, on the bottom left of the window, you will see a list of events what can occur. Click on the Frame Loaded event and type the following script:

```
if (LocalHours < 17)
    TrackNamed("OriginalA").SetEnabled(True)
    TrackNamed("OriginalV").SetEnabled(True)
    TrackNamed("AlternateA").SetEnabled(False)
    TrackNamed("AlternateV").SetEnabled(False)
else
    TrackNamed("OriginalA").SetEnabled(False)
    TrackNamed("OriginalV").SetEnabled(False)
    TrackNamed("AlternateA").SetEnabled(True)
    TrackNamed("AlternateV").SetEnabled(True)
Endif
```

The first line checks to see if the local time of the computer is earlier than 5:00 p.m., or 17 military time (I know, I know, 1700—go with the flow). If the time is earlier than 5:00 p.m., then the original ending is shown. Otherwise, the second alternate ending is shown.

If you don't want the big, white box placed by the text track to show, simply click on the track name (most likely "Text1") and drag it down, below your video tracks. You can also select the Properties tab and enable Keyed Text. If, on the other hand, you want to get creative and add some text, go right ahead. Just make sure you change the Duration of the text if you do add some.

Compiling Your Movie

Now that you've assembled all of the pieces together, you can compile everything together and have your complete movie. To export your movie, choose File → Export → QuickTime Movie... and type the name of your movie. The process might take a while, depending on the length of your movie. That's it. You're done.

Hacking the Hack

LiveStage Pro is a powerful program and there is a lot you can do with it. Even within the limitations of enabling alternate endings, you can change the way your movie chooses the ending. For example, instead of using the hour, you can use the minute with the `LocalMinutes` (or the second using `LocalSeconds`) instruction instead of the `LocalHours`. You can also determine whether the viewer is on a Macintosh or Windows machine and even whether they are connected to the Internet.

If you want to *really* dive in, you can determine the viewer's IP address and, from that information determine their global location using something like GeoIP (*http://sourceforge.net/projects/geoip/*; free, open source). From there, you could figure out the current weather in their area using the Weather Underground (*http://wunderground.com*; $5 per year) and then choose an ending based on that information. If you want to go down that path, look at "What's Your Visitor's Weather Like?" (*http://hacks.oreilly.com/pub/h/999*) from *Spidering Hacks* by Kevin Hemenway and Tara Calishain (O'Reilly). The possibilities are nearly endless.

The really cool part is that this whole approach to alternate endings is merely scratching the surface of what you can accomplish using QuickTime and LiveStage Pro.

HACK #73 Alter a Video's Look After Editing

Even if you've completed editing your video, you can still alter its appearance.

If you've installed QuickTime Pro (*http://www.apple.com/quicktime/buy/*; $29.99), you have the ability to add a filter to a video without using any other application. In fact, QuickTime Pro adds 14 standard filters; plus, you can add more from third-party companies, such as Buena Software (*http://www.buena.com/*; free–$30). You also have the ability to cut, copy, and paste video together, as well as create skins **[Hack #76]**, create chapters **[Hack #49]**, and even convert a movie from one format to another **[Hack #29]**. But, I digress.

Getting to the Filters

If you have a video file that QuickTime can read, you can open the file and then select File → Export to begin the process.

QuickTime-supported file formats include: 3DMF (Mac OS 9 and Windows), 3GPP, 3GPP2, AIFF, AMC, AMR, Animated GIF, AU, Audio CD Data (Mac OS 9), AVI, BMP, Cubic VR, DLS, DV, FlashPix, FLC, GIF, GSM, JPEG 2000 (Mac OS X), JPEG/JFIF, Karaoke, MacPaint, Macromedia Flash, MIDI, MPEG-1, MP3 (MPEG-1, Layer 3), M3U (MP3 Playlist files), MPEG-2 (through a separate Playback Component), MPEG-4, M4A, M4B, M4P (iTunes audio), PDF (Mac OS X), Photoshop, PICS, PICT, PLS, PNG, QCP (Mac OS 9 & Windows), QuickTime Image File, QuickTime Movie, SD2 (Mac OS 9 & Windows), SDP, SDV, SF2 (SoundFont 2), SGI, SMIL, System 7 Sound (Mac OS 9), Targa, Text, TIFF, TIFF Fax, VDU (Sony Video Disk Unit), Virtual Reality (VR), and WAV.

From the Export window, choose Movie to QuickTime Movie from the Export pop-up menu. Then, click the Options... button, which brings up the Movie Settings window. Figure 6-43 shows the Export window.

Figure 6-43. Exporting a movie to a QuickTime Movie

In Movie Settings window (Figure 6-44), you have the ability to change the video's codec using the Settings button, change the video's look using the Filter button, and change the video's size or rotation using the Size button.

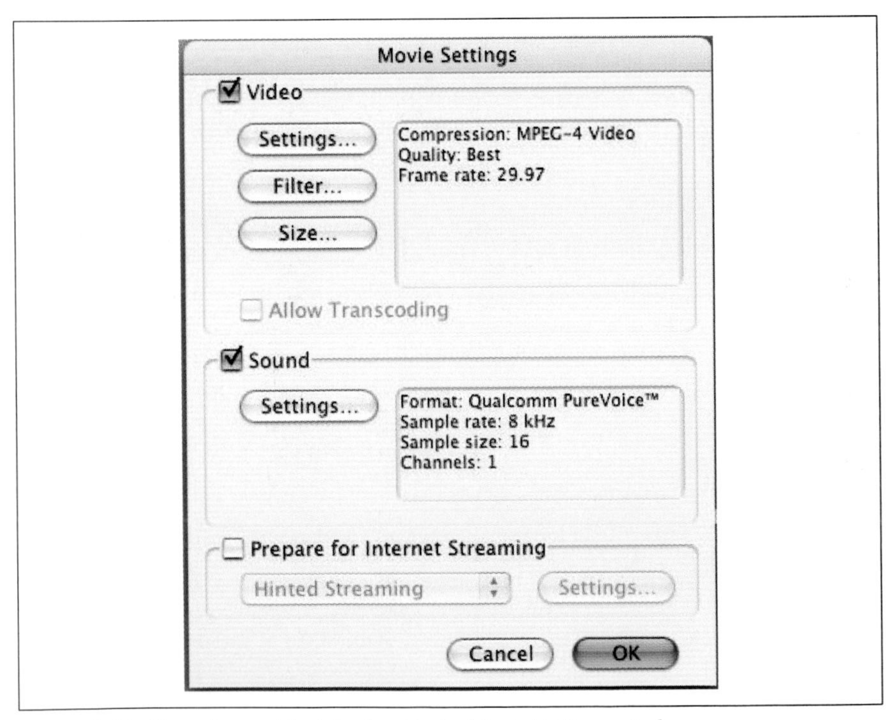

Figure 6-44. Changing a video's look with the Movie Settings window

You're only concerned about altering the video's look, so you should click the Filter... button.

Selecting a Filter

Clicking the Filter... button brings up a new window with the available filters, a preview of your footage (which reflects any changes you apply), and a set of controls. The availability and type of controls will depend on the filter you choose:

Adjustments

Alpha Gain
Alters the alpha channel in your image (requires an alpha channel)

Brightness and Contrast
Allows you to increase or decrease the brightness and contrast, separately

ColorSync
Alters the spectrum of color for your image to match a given ColorSync profile

HSL Balance
Enables you to change the Hue, Saturation, and Lightness

RGB Balance
Provides control over the Red, Green, and Blue channels

Blur

Blur
Lets you blur your image

Filters

Edge Detection
Locates and draws edges within your image

Emboss
Creates two offset greyish images, providing the illusion of an embossed image

General Convolution
Takes a 3×3 grid of pixels, and replaces the center pixel with a weighted average of the surrounding pixels

Sharpen

Sharpen
Brings out fine details within the image, especially around defined edges

Special Effects

Color Style
: Enables you to both solarize and posterize the image

Color Tint
: Provides quick options for creating a Sepia or X-Ray looking movie, as well as control over the light and dark colors for a unique tint

Film Noise
: Adds hairs, scratches, dust, and fading to provide a film-like appearance

Lens Flare
: Creates a lens flare at a point you specify

I personally like the Edge Detection filter, when set to the highest setting ("7-thickest"). I think it adds a surreal, almost cartoon-like feel to video. Figure 6-45 shows the Edge Detection filter set to "7-thickest" and to "Colorize result."

Figure 6-45. Edge Detection using the "Colorize result" setting

Another cool filter is the Film Noise effect. Not only does it allow you to add the appearance of hairs, scratches, and dust, but it also offers a type of film fading. I like the 1930s color film choice, because it *really* saturates the image. Figure 6-46 shows the Film Noise filter being applied to a movie.

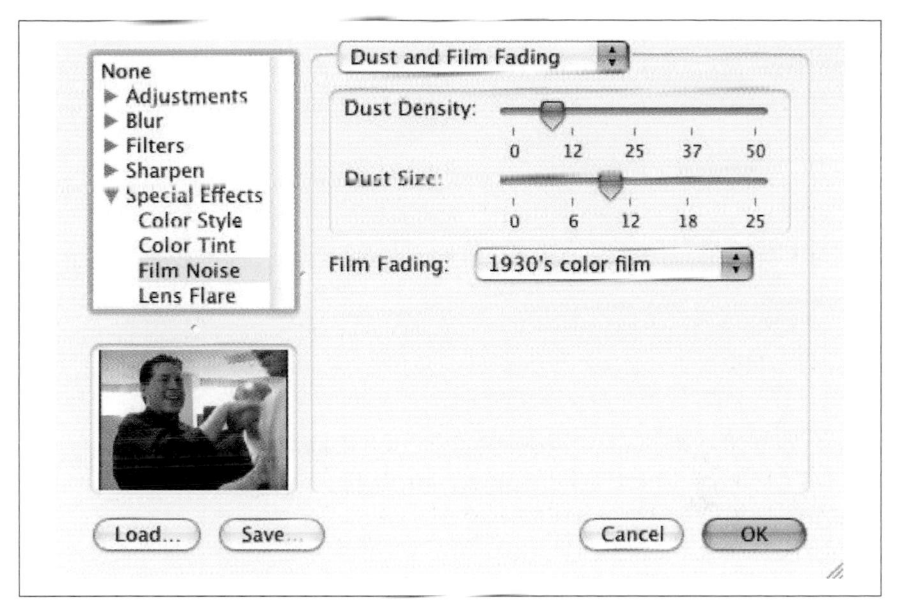

Figure 6-46. Film Noise, making the video look like a 1930s color film

When you have chosen the filter you would like to apply, and configured it to your liking, click OK and then Save. The export process can take a long time. Depending on the speed of your computer and the length of your movie, you might want to go grab a coffee.

Saving Movies for Reference

Because the export creates a new video file, and does not alter the original, you can select a new filter and repeat the process. The best way to become familiar with the filters and their effects is to apply each one individually to a short video. Then, save the video to your hard drive and keep the filtered videos for future reference.

Expose Only One Color

The movies "Pleasantville" and "Shindler's List" have scenes that appear in black and white while exposing one color, such as red. You don't need a big budget to use this effect in your movies.

It is rare you will ever want to remove color from a shot, or an entire movie. Yet, in some instances, it makes sense within the story you are telling. Using the following technique, you can create a black and white scene, while allowing one color to "pass through," known as a *color pass* effect. Figure 6-47 shows an original and completed image using the color pass effect.

Figure 6-47. Using yellow to help alert people to avoid physical harm

When planning on using this technique, it will help if you can limit the color you would like to expose to the object where you would like to draw attention. For example, if you would like to focus on a girl in a red dress, you should try to avoid having other red-colored objects in the frame.

Premiere includes a Color Pass effect, which makes creating this look ultra-easy. You just need to drag-and-drop the Color Pass effect onto your footage and select the color you would like to expose. Two steps and you're done.

Layering Your Video

In order to create the color pass effect, you will need to create two layers of video. One will contain the image as a black-and-white composition, while the other will contain just the image/color you would like visible. When combined, they will create single image with only one color visible.

The method you use to create two layers will depend on your editing system. For the majority of systems, you should be able to drag-and-drop your footage onto your timeline. You can then repeat the process; however, you should drag your footage either above or below your first layer.

Creating a Black and White Image

The first step in creating the color pass effect is to completely desaturate your lower-level track. This will create the base image for your composition as black and white and allow you to work more efficiently. To desaturate your image:

Avid
: Tools → Effects Palette → Image → Color Effect (Chroma → Saturation)

Final Cut
: Effects → Video Filters → Image Control → Destaturate

Even though you've now removed all of the color from your footage, you have affected only the lower track. Therefore, your image will still appear in your record monitor (a.k.a. Canvas or Program View) in full color. Don't worry; that'll change soon enough.

Choosing a Color

The next step is where you choose what color you would like to keep. You can do this in a number of ways, but using either a chroma key or a color key is the easiest. You should apply this effect to the top layer of your composition:

Avid
: Tools → Effects Palette → Key → Chroma Key

Final Cut
: Effects → Video Filters → Key → Color Key

Some systems enable you to simply click on a point of your image to select the color you would like to use. If not, you will need to use the controls provided to locate the color you want to expose. As always, don't worry about being exact because you'll be able to change any selections you make later in the process.

Exposing the Color

After choosing your color, you might notice that you accomplished the exact opposite of what you intended. Instead of exposing the color, you removed it. Fortunately, you can invert the key, thereby exposing the color and removing everything else. You should select the clip and effect/filter you've applied, and click on the invert checkbox.

After inverting your key, you may notice a few spots in your image where the color exists, but you don't want it to. For these spots, you can simply crop or mask them out of the image.

Hacking the Hack

Another approach is to maintain color throughout your image, but to emphasize a specific color. To accomplish this effect, instead of converting your image to black-and-white, you can reduce the saturation on the bottom layer and increase the saturation on the top layer. This will cause your chosen color to subtlety stand out from the rest of your image.

CHAPTER SEVEN

Distribute

Hacks 75–89

Distributing your final project is the most gratifying, and yet most nerve-racking, facet of digital video production. When you distribute your video, it means you've completed your project. That, in and of itself, is a huge accomplishment.

When you send off your video for someone to see, your head will swirl with questions. How will they react? Will they like it? Will they hate it? Should I have had Greedo shoot first?

Make a Window Burn Copy

If you want to make notes on your edit, it is best to make them while you're away from your editing system. Creating a window burn allows you to do so.

When you are editing with a computer, it's easy to make changes to scenes. It's so easy that you can overedit a scene, or an entire project, by continually making changes. A friend told me a story that describes the problem perfectly:

> Two kindergarten teachers were sitting in the Teachers' Lounge, talking over coffee. One mentioned how much better the other's students' finger paintings always looked than her own students'. She asked what she was doing wrong. The answer was unexpected: "I know when to take the paintings away."

Movie making is part art and part science. I don't mean to knock George Lucas, but how many different versions of *Star Wars: Episode IV* do we need? He is a good example of an artist who can't let his project go.

So, to overcome the temptation to make changes while you are viewing your project, you can make a copy of your project and view it away from your editing system. The copy should have a visual reference to your project's timecode so that you can make accurate notes. The visual reference is called a *window burn*, and the copy is referred to as a *window dub*. Figure 7-1 shows a frame of video with a timecode window.

Figure 7-1. A window showing the timecode of 01:30:04;11

The word *dub*, which is used often in the motion picture industry to mean *copy*, is short for the word *double*.

Creating a Dub with a Window Burn

The least time-consuming way to create a VHS dub with a window burn is to output your project to digital videotape and simultaneously copy the image to VHS. This method takes a little more time to set up, but if you are working on a project that is longer than 30 minutes, it is probably worth it. To configure your system to do this requires you only to place your CPU, camera or deck, VHS machine, and finally your TV in order, as shown in Figure 7-2.

If you are willing and able to connect your camera or deck to your television, you do not have to dub to VHS. Simply output to digital videotape and play back the cut on a television.

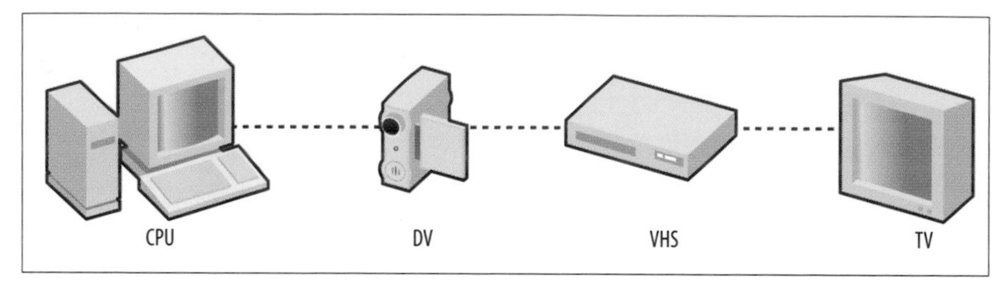

Figure 7-2. Setup to create a window burn copy from your computer

With this configuration, the video signal is going to flow from your computer to your digital video camera (or deck), then to your VHS, and finally to your television. You want the signal to end at your television because you want to make sure the signal has passed through every stage of your setup. If you do not see an image on your television, it means your video signal is not reaching a certain point in the chain and you need to check your connections.

Some digital video cameras will not pass a video signal coming in digitally and back out analog. If your camera or deck has this restriction, you will have to record to digital videotape and then to VHS.

You will want to turn on your camera's timecode overlay. When you have finished the dub, you will have a digital video dub and a window-burned VHS.

Some editing systems do not write the timecode of the timeline to videotape. Under such circumstances, you might want to output to a black and coded tape **[Hack #4]** instead. Doing so will make taking and addressing notes easier. For example, if your timeline starts at 01:00:00;00 but your videotape starts at 00:00:00;00, you will have to keep in mind that there is a one-hour offset to your notes.

Sitting Down, Relaxing, and Taking Some Notes

Once you have made your window dub, take a break from your project for at least a few hours, preferably until the following day. Your goal is to watch your movie as your audience might. You do not want to view it as the Writer, Producer, Director, or Editor.

When you finally sit down to watch your movie, make sure you have a pen and paper so you can take notes. While you are taking notes, you want to reference the timecode on screen. You do not need to be frame accurate; the reference is meant to help you quickly locate where you need to make changes when editing. For example, if you have a note of "00:35:47 - zoom in on candle," you will know to go to 00:35:47 in your project's timeline and add a zoom.

Knowing When to Take It Away

Following the window dub process in this hack will help you stay "at arm's length" from your project. However, the more often you repeat the process, the less you will benefit from it. Every time you watch your movie, you will find something you would like to change. In your opinion, your work will very, very rarely be perfect.

In order to succeed in creating a work of art, and not a swirling mess of paint, you need to limit yourself. You should go through the process only a certain number of times (I recommend two).

Hacking the Hack

If you prefer to watch your movie on your computer, you can create a digital movie with a window burn. Some editing systems provide a video effect that will create a timecode reference in a window. This effect, however, must be rendered on most systems.

To create a digital movie, apply the timecode effect to your timeline, render, and then export the movie to a digital file.

Skin Your Movie

Delivering a completed movie is an accomplishment. If you are distributing your movie online, you can get people to remember your movie from the others by wrapping it in a "skin"—a unique interface that will become a part of your video.

Most audio and video playback applications allow people to create *skins* that alter the appearance of the given application. However, QuickTime has a unique feature that enables producers to meld together a movie along with a unique interface. Using QuickTime Pro, a photo manipulation program, and a text editor, you can create a unique look for your movie and begin to build a memorable brand.

The skin of a QuickTime movie can consist of any image QuickTime can read. So, not only can you create an image in a program such as Photoshop, but you can also take a picture with a digital camera or even use a scanner to import one you've found. Whatever your choice, the process of creating the skin is the same. Skins can be of any shape, so don't be shy about using unusual shapes to help brand your movie.

Mark Your Image for Skinning

Once you have chosen the image you are going to use for your skin, it is helpful to edit the image and mark an area where your movie will display. Mark the area using either black or white. For example, if you are using an image of a television set, mark the screen with a black or white box. Figure 7-3 shows an image with a white box representing the display portion of the skin being edited in Stone Studio's Create.

Figure 7-3. Defining an area for the movie to display

Once you are satisfied with your image, save it in a QuickTime-compatible format, such as TIFF, JPEG, PICT, GIF, PNG, BMP, PDF, Photoshop, or even Targa.

Joining Your Movie with Your Image

After you've created your skin image and chosen a section in which to display your movie, you are ready to join your movie and your skin image together. While running QuickTime Player, choose File → Open Movie in New Player... and open the skin image. You need to copy the image as a movie, so choose Edit → Select All and then Edit → Copy.

Next, open your movie in QuickTime Player. To apply the skin, you need to paste it over your movie. However, you need to *scale* it to the entire length of your video. In order to do this, simply choose Edit → Add Scaled.

Getting Your Movie to Display

After adding the skin, you will probably notice that it has taken over your movie altogether. Fortunately, QuickTime allows you to alter video in a number of ways, including selecting a color to become transparent. To select a transparent color, open the movie Properties window by choosing Movie → Get Movie Properties.

In the first pop-up menu, select Video Track 2, and from the second pop-up menu, select Graphics Mode. Then, from the list of available modes, select Transparent. Finally, click the Color... button and choose the color you have used to mark the display within your skin (white or black, if you've followed the previous directions), or choose the magnifying glass tool and place it over the display section of your skin. Figure 7-4 shows a before-and-after image of the Transparent mode being applied.

Figure 7-4. Displaying your movie by using Transparent mode

Under most circumstances, you will have to resize and reposition either or both the movie and the skin.

Lining Up the Movie and Skin

From the Properties window, you can access the sizing controls by selecting the appropriate track (e.g., Video Track 1) from the first pop-up, choosing Size from the second pop-up, and then clicking on the Adjust button. While changing the size and position of an image, you will see a set of red marks related to the particular track.

By clicking and dragging the corner marks, you can resize the image, and by clicking and dragging on the image itself, you can reposition it. You can also sheer the image by clicking and dragging on the marks on the sides of the image. Figure 7-5 shows a movie being resized within a skin.

Figure 7-5. Sizing and positioning a movie image

After you are happy with the look of your movie/skin combination, choose File → Save As... and select the option to "Make movie self-contained."

Creating Masks for Your Skinned Movie

Even though your movie and skin are now merged, the QuickTime Player window is still visible. The next step in the process is an easy, yet vital part to display your movie properly. Within your graphic editing program, create two masks: one for the *window* area and one for the *drag* area.

The window mask will provide the outline for QuickTime to determine the form in which to present your movie. For example, if the skin is of a television, the window mask will be in the shape of a rectangle. The window mask should essentially be a completely black representation of your skin image.

The drag mask will provide QuickTime the information on where a viewer can click in order to drag the window around the screen. This mask should be a replica of your window mask, with the exception of providing white areas where your movie or its controls will show through. Figure 7-6 shows the window and drag masks for the television skin.

Figure 7-6. A window mask (left) and a drag mask (right)

After you have created your window and drag masks, save them as individual files. You should now have three files: your movie and skin as a self-contained movie, your window mask, and your drag mask. Place the three files inside the same directory on your computer.

Bringing It All Together

The final step in creating your skinned movie is to combine your three files into one standalone movie. To do so, type the following into a text editor:

```
<?xml version="1.0"?>
<?quicktime type="application/x-qtskin"?>
<skin>
    <movie src="myMovie.mov"/>
    <contentregion src="myWindow.tiff"/>
    <dragregion src="myMask.tiff"/>
</skin>
```

Your images can be in any QuickTime-compatible format. They do not need to be TIFFs, though that is the format used in this hack.

Replace `myMovie.mov` with the name of your movie file, `myWindow.tiff` with the name of your window mask file, and `myMask.tiff` with the name of your drag mask file. Then, save the skin text document to the same directory where your three files are located and name it using the *.mov* file extension—for example, *mySkin.mov*.

Open the skin text file using QuickTime Player. Your new movie should appear within the window you've designed. Figure 7-7 shows a completed movie using a QuickTime skin. Notice the absence of QuickTime controls.

Figure 7-7. A skinned movie being played

There are no longer any controls for the movie, so set the movie to Auto Play. Bring forward the movie Properties window by choosing Movie→Get Movie Properties and then select Auto Play from the second pop-up menu. Simply check the Auto Play Enabled checkbox.

Your movie is now compiled and configured. The last step is to save your movie, making sure to select "Make movie self-contained," and distribute it.

Hacking the Hack

How would you like to be able to present your video in the shape of an octagon? Or the shape of a star? Well, instead of surrounding your video with an image, simply create an outline of the shape with no considerable border. Figure 7-8 shows an odd-shaped movie being played.

You can also take a more traditional approach to skinning and include controls to handle the play, stop, fast-forward, and reverse functions. By using an application such as LiveStage Pro (*http://www.totallyhip.com*; $449.95), GoLive (*http://www.adobe.com/products/golive/*; $399), or Flash MX (*http://www.macromedia.com/software/flash/*; $499.95), you can create control buttons and add them to your movie, just like you add a skin. These programs unlock more potential than just adding control buttons, so you'll be able create truly interactive movies.

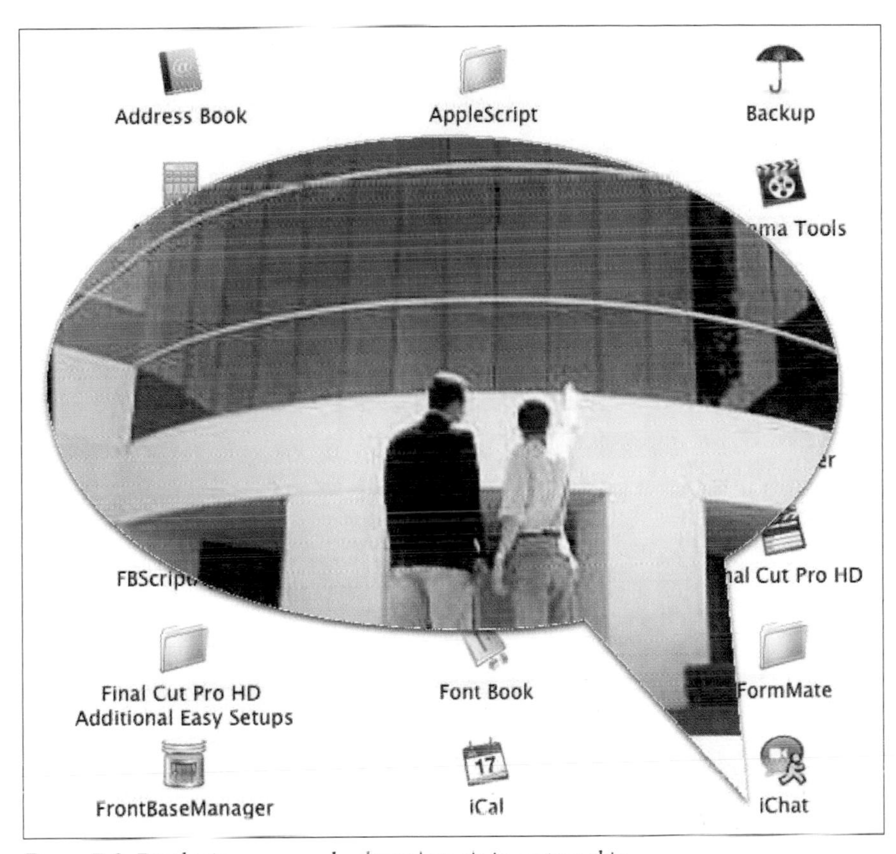

Figure 7-8. Displaying a uniquely shaped movie in custom skin

For example, with the television skin, you could create a toggle button that hides the video and mutes the audio, while still keeping the movie playing. When a user clicks the button, the television would seem to turn off, and when the user clicks the button again the television would appear to turn on, as the video would have progressed in linear time (because it continued to play).

HACK #77 Determine Which Codec to Use

Knowing how you plan to use or distribute your video will help determine which codec you should use.

When talking about digital video, you need to be clear whether you are talking about the DV format or digital video in general. Digital video covers a wide range of technologies, including DVDs, cell phones that can record and play video, streaming video on the Internet, and even television broadcasts via satellite and cable. Every form of digital video requires the use of a *codec*,

More often than not, when talking about editing digital video, the conversation is about DV. The DV format generally includes MiniDV, Panasonic's DVCPRO, and Sony's DVCAM. It does not include DVD.

Understanding Codecs

A codec is a *co*mpression/*dec*ompression algorithm that helps reduce the amount of storage, and computer horsepower, required to reasonably use video on a computer. Since uncompressed video takes up a lot of disk space and requires a lot of computer power, codecs have been created to help play video on a range of devices. Basically, a codec enables a device such as a computer or DVD player to play audio/video content. There are hundreds, if not thousands, of audio/video codecs available.

To overcome some of the frustration created by competing codecs, the number of codecs, and the fact that some hardware/software works with one codec and not another, various groups have attempted to create standards. Many of the standard codecs are also associated with video formats (sometimes called *containers*). For example, DV is both a codec and a format.

Some of the standards have been successful, primarily those set by the Moving Pictures Expert Group (MPEG). For example, the DVD movies you rent and purchase are encoded with the MPEG-2 standard, which is published by MPEG. The official MPEG website is located at *http://www.chiariglione.org/mpeg/*.

Can you imagine if each movie studio used a different codec for their DVDs? You might have five different DVD players in your house! Standards are a good thing for both producers and consumers.

Codecs are a vital part of delivering digital video. Every codec is targeted at a certain use, although some target multiple uses. For example, Avid editing systems have their own codec and are targeted to editing only on an Avid system.

Table 7-1 provides a compact list of the more common codecs and their typical uses, as well as their typically associated file format extensions.

Table 7-1. Common codecs, their recommended use, and file extensions

Codec	Used for	File format extension
MPEG-1	Video CD	*.mpg, .mpv, .m1v*
MPEG-2	HDV; DVD; professional acquisition	*.mpg, .mpv, .m2v*

Table 7-1. Common codecs, their recommended use, and file extensions (continued)

Codec	Used for	File format extension
MPEG-4	Internet streaming and download; consumer acquisition; computer playback	*.mpg, .mp4*
DV	Professional and consumer acquisition and editing; tape playback	*.dv*
3GPP/3GPP2	Cellular phone playback	*.3gp/.3g2*
H.264 (a.k.a. AVC, MPEG-4 Part 10)	Playback of Standard and High-Definition video via Internet; HD-DVD; satellite distribution	*.mp4, .264*
VC-1	Playback of Standard and High-Definition video via Internet; HD-DVD; satellite distribution	Undertermined at the time of this writing

Knowing where your video will be seen will help determine the codec you should use. According to Table 7-1, if you plan on delivering a video to a cell phone, you should use the 3GPP codec. While a 3GPP encoded movie might play on a computer, the quality of the image will be much lower than an MPEG-4 encoded video, which is targeted at such playback.Table 7-2 shows the resulting files sizes, and Figure 7-9 shows a variety of codecs set side by side and

Images in print versus those on a television or computer appear differently and *much* smaller. The images presented here are for reference only.

Table 7-2. Resulting file size for 20-second video clip

Codec	File size
DV	63.3 MB
MPEG-1	2.4 MB
MPEG-2	8.2 MB
MPEG-4	2.8 MB
3GPP	172 KB

Unfortunately, some questions start to arise when you plan on delivering to a computer. Whether via streaming video, Internet download, or CD-ROM, there are many applications available to play back video content. Many of them use proprietary codecs and most are not compatible each other. The major playback applications are QuickTime, Windows Media, and Real One.

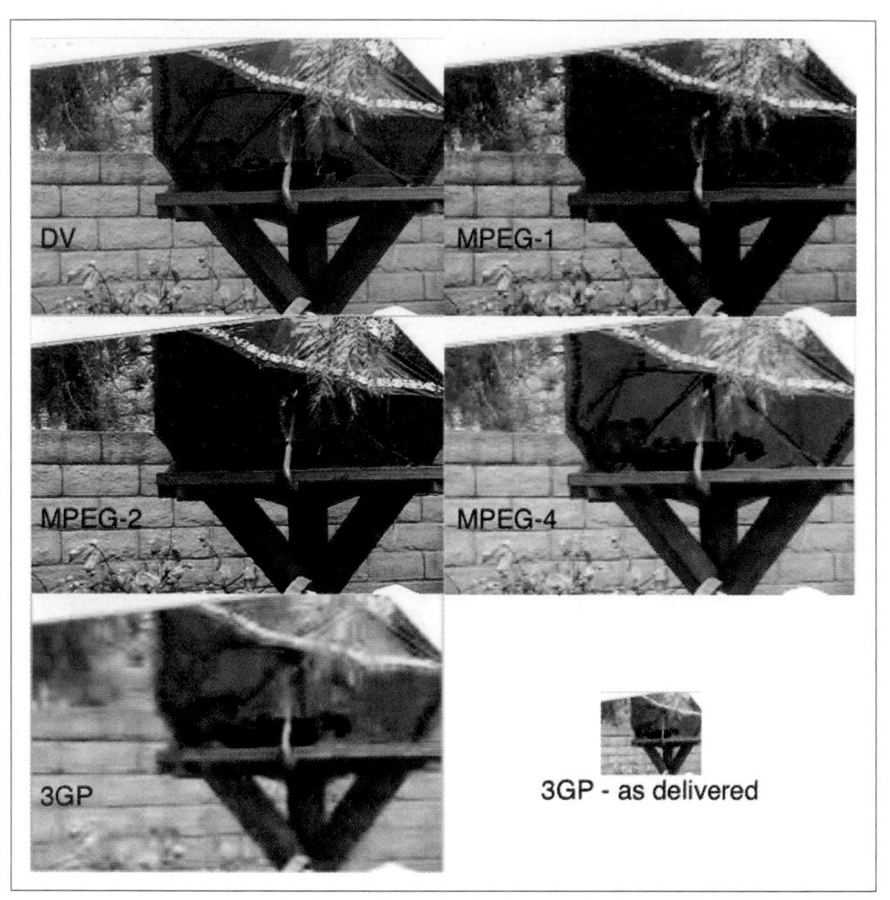

Figure 7-9. DV and MPEG codecs, side by side

Understanding the Major Video Players

QuickTime Player, Windows Media Player, and Real One Player are all applications that use codecs to play audio and video content and they are all available for both Macintosh and Windows computers. Each one has its benefits and drawbacks. Real is the only one that is available and supported on Linux.

The majority of the time, the videos that these applications are able play aren't compatible with each other (what else is new?). If one of these applications tries to open a video file that has been encoded for another player, the video file most likely won't open. This detail can be frustrating both as a consumer and as a producer. Table 7-3 shows the major media players, and the common file extensions.

Table 7-3. Major playback applications and their common file extensions

Playback application	Common file extension
QuickTime Player	*.mov*, *.qt*
Windows Media Player	*.wmv*, *.asf*
Real One Player	*.rm*, *.ram*

Much of this can become confusing, frustrating, and annoying, even if you've been immersed in digital video for years. To make it even more complicated, the landscape is continually changing. Fortunately, it is somewhat difficult to get yourself stuck, because you can usually convert one format and/or codec of video to another by transcoding **[Hack #29]**, should you need to do so.

QuickTime. If you plan on using the more robust capabilities of the QuickTime architecture, such as interactive sprites **[Hack #72]**, you will be required to deliver a QuickTime format file and your viewers will have to use the QuickTime player to view it. The QuickTime format offers a lot of different codecs out of the box, and many third-party vendors have added their own as well. You can download QuickTime at *http://www.apple.com/quicktime/download/*.

Some of the more prominent codecs include Sorenson (good for Internet video), Pixlet (good for high-quality video), and Photo-JPEG (good for low-resolution editing). If you choose a codec that is obtained from a third party, such as On2's VP3 (*http://www.vp3.com/vp3/quicktime/index.shtml*; free, open source), your viewers will be required to download and install it as well. For the most part, QuickTime will handle this detail for them.

QuickTime is installed along with the iTunes music application. If your viewers have installed iTunes, they have also installed QuickTime by default. There is also an undisputed rumor that iTunes uses QuickTime to create the interface to the iTunes Music Store (see comments at *http://weblogs.mozillazine.org/hyatt/archives/2004_06.html*).

QuickTime is an interesting beast, because it encompasses much more than audio/video creation and playback. Although the majority of people think of QuickTime as a media player only, it can be used it to create 3D scenes, truly interactive movies, and even utilities such as calculators. It is a robust technology with a vast and deep assortment of capabilities, which seem to be a hidden secret inside Apple.

Windows Media. Windows Media is the default media format for computers running Microsoft Windows. The format (now in Version 1.0) has improved exponentially since Version 7. It is also the basis for the VC-1 codec used in Internet, DVD, Standard and High-Definition, and satellite distribution of video. You can download Windows Media Player from Microsoft at *http://www.microsoft.com/windows/windowsmedia/default.aspx*.

The focus for Windows Media has been on audiovisual content and it is noticeable in the quality of the video it produces. Due to its focus, Windows Media does not offer much in the way of interactivity features, although it does offer chapter tracks **[Hack #49]** and the ability to open URLs **[Hack #84]**. If you plan on delivering to a majority of Windows viewers, Windows Media is a safe bet.

Real. Real was a golden child during the Internet boom of the late 1990s. Since then, it has begun to take a back seat to both QuickTime and Windows Media. Part of the slide might be attributable to the former cost of delivering Real-based content. However, if you plan on delivering to the widest possible audience, including Linux users, Real is the format of choice.

In mid 2002, Real announced it was going to give away the source code to its server, codec, and player products. The project is called Helix (*http://www.helixcommunity.org*) and, at the time of this writing, has almost 75,000 registered users. Helix is able to handle both Real video and QuickTime video.

You can download Real One Player from *http://www.real.com*.

Choosing a Codec

So, what in the world should you do? If you are planning to distribute your movie on a DVD **[Hack #79]**, you don't need to look much further than the MPEG-2 codec and a program to burn your movie to disk. But if you plan on delivering via the Internet, whether streaming or for download, your choice becomes more difficult.

As a ground rule, you should stick to standards when possible. For example, use MPEG-4 for videos you are going to distribute online. MPEG-4 videos *should* be playable on all of the players.

Here are a few questions to help narrow your selection:

- Do you have access to a streaming server?
- If so, which formats does it support?

- Do you have an interactive element within your movie?
- Are all of your viewers using a specific platform?

Answering these questions should yield some insight to your requirements. If your viewers are using Macintosh, choose QuickTime. If they are using Windows, choose Windows Media. If you don't know, try both and see which one you prefer. For example, look at which provides a better image. Or ask yourself, which is easier for you to produce and distribute?

Using an Alternative

There are a few alternatives to the mainstream players and codecs. Although the alternatives muddy the landscape even more, they also provide viable options. If your viewers are willing to download a new player and a new codec, these alternatives are good choices for online video distribution.

The players. The two video players gaining the most momentum are VLC (*http://www.videolan.org/vlc/*; free, open source) and MPlayer (*http://www.mplayerhq.hu/*; free, open source). Both VLC and MPlayer are the equivalent of a video Swiss Army knife, because they seem to open a majority of video files available on the Internet. They are also able to play DVDs and accept streaming video. VLC is available for Macintosh, Windows, and Linux. MPlayer is available for Macintosh and Linux.

The codecs. Three codecs gaining the most momentum are DivX (*http://www.divx.com*; free), XviD (*http://www.xvid.org*; free, open source), and 3ivx (*http://www.3ivx.com*; $6.95–$19.95).

All three codecs provide excellent quality video for the bandwidth they require, and all are variations of the MPEG-4 standard. They are all available on Macintosh, Windows, and Linux. The major difference between the three is the fact that DivX and 3ivx are much easier to approach from a user's standpoint, as their installation is uncomplicated.

In the end, your choice is just that: a choice. It is personal and your viewers will either find a way to play your content, especially if it is compelling and entertaining, or they won't. If you want to cater to the widest possible audience, keep it simple and use internationally agreed upon standards, such as those published by the Moving Pictures Expert Group (MPEG), the Society of Motion Picture and Television Engineers (SMPTE), and the International Organization for Standardization (ISO).

Play a Movie Off a CD in a DVD Player

DVD media is expensive. CD media is cheap. By compressing your movie using the MPEG-1 codec, you can burn your movie to a CD while still being able to play it in a DVD player.

It is somewhat gratifying to hand someone a DVD of your movie and have him play it on a DVD player. The simple fact is that the presentation of the disc implies technical capability and professionalism. But there are times when a DVD is excessive, such as when you are presenting a five-minute movie or handing out discs to a lot of people.

The Video CD (VCD) format was launched in 1993, but it never really caught on in the United States. A VCD can hold about 70 minutes of footage, so you shouldn't plan on putting an entire feature length movie onto one disc. The VCD format uses MPEG-1 compression, so the quality of a VCD is roughly equivalent to a VHS tape. The discs can be played on most DVD players, in addition to personal computers and VCD players.

Getting Your Movie onto a CD

The VCD format uses the MPEG-1 codec. You can easily convert your movie to another format **[Hack #29]**; however, simply converting to MPEG-1 and copying the resulting video file to a CD *will not* produce a VCD. A valid VCD needs to have specific files written in a particular way.

There are many applications that can create valid VCDs, and they are available for just about every operating system. These are the two most popular, for Windows and Macintosh:

Windows
: Nero (*http://www.nero.com*; $69.99 download, $99.99 boxed)

Macintosh
: Toast (*http://www.roxio.com*; $99.95)

Some DVD players require CD-RW discs, not CD-R, for VCD playback. So, if you discover the VCD you've created doesn't play, try using a CD-RW disc. In fact, you might want to use CD-RW discs in the first place because they don't become coasters if a burn goes bad.

Both Nero and Toast are capable of burning DVDs, SVCDs (super VCD), audio CDs, and more. A really nice feature of both applications is that they enable you to create menus on your VCD that act similar to those found on DVDs. As expected, each application has a somewhat different approach to creating a VCD.

Using Nero. Nero handles a lot more than just burning VCDs. The application suite includes a set of 20 applications that make achieving certain tasks, such as creating a music CD, very easy. The Nero web site (*http://www.nero.com/en/Tutorials.html*) has an extensive collection of tutorial files, including a 14-page tutorial on creating a VCD. Using the Nero Vision Express application, you can create a VCD like so:

1. Choose Make a CD → Video CD.
2. Click the Add Video Files... button.
3. Select your video files.
4. Click the Next button.
5. Set up your Menu.
6. Click the Next button.
7. Test the Menu you created, to make sure it works as you expect.
8. Click the Next button.
9. Click the Burn button.

Using Toast. Toast, like Nero, can handle much more than simply creating a VCD, such as creating password-protected CDs. Here's how to create a VCD using Toast:

1. Select the Video tab.
2. If the Disc Options drawer is not open, click the Disc Options button.
3. Select VCD.
4. Select NTSC or PAL, as appropriate.
5. From the Video Quality pop-up menu, choose the High quality setting.
6. Click the Add button.
7. Add the video files you want to use.
8. Click the Record button (the big, red one in the corner).

Playing a VCD

Playing a VCD in a DVD player will work most of the time. However, some older DVD players do not recognize the VCD format and therefore cannot play the discs. If this happens, you can still play the disc on a personal computer.

If you are using Windows, the Windows Media Player should play the VCD with no problems. However, if you do encounter problems, try using PowerDVD (*http://www.gocyberlink.com*; $49.95). Macintosh and Linux users should use the VLC application (*http://www.videolan.org*; free, open source) to play VCDs.

To play a VCD, insert the disc into your computer and locate the *MPEGAV* folder. Then, double-click the *.dat* file located inside. Windows users might have to instruct their computer to use Windows Media Player to open the file. Macintosh and Linux users might have to open the *.dat* file from within the VLC application.

If you've produced a video that is shorter than an hour, passing along your finished project on a VCD is a quick and easy alternative to creating a full DVD.

HACK #79 Distribute Your Movie on DVD

Burn your digital video to a DVD and play it in a standalone player.

Not even 10 years ago, the thought of creating and distributing your own DVD was a very expensive proposition. Today, it's an affordable, yet professional, way of distributing your video to clients and consumers. As long as you have the video content and a DVD burner, getting your video prepared and burned onto a DVD is easy.

> There is a technical difference between DVD-R and DVD+R discs. The DVD-R format is the older format and is compatible with most consumer players. If you use to DVD-R media, you will be more likely to have better playback results.

Preparing Your Video

How you prepare your video to be burned to a DVD will depend on the origin of your video and the application you will be using to burn the disc.

Exporting from your editing system. If you are exporting your video directly out of your editing system, and the DVD-burning application is supported by it, the process is going to be a lot easier:

Avid
: File → Export

Final Cut
: File → Export → QuickTime Movie

iMovie
: File → Share... → iDVD→Share button

Movie Maker
: Tasks → Finish Movie → Save to my computer

Premiere
: File → Export → Movie...

Export your video in the same format as your footage. This means that you should not change the resolution (width and height), the frame rate, or the codec from that of your footage. So, if you are working in NTSC DV (720× 480) at 29.97 frames per second, export your video using the same settings.

Converting from videotape. If your video is coming directly off of videotape, you have a couple of options. Your first option is to digitize the video using your editing system and then export it as mentioned previously. The other option is to convert it from analog to digital [Hack #37], while making sure your footage is converted using the highest resolution possible. Additionally, some DVD software applications include a feature to digitize your footage.

Burning Your Video

Once you have your video as a digital file, you need to convert the file for use on a DVD. This will require your video to be transcoded using the MPEG-2 codec. Some DVD software applications, such as iDVD on Macintosh and NeroVision Express on Windows, provide this feature. Others applications might expect you to do the conversion prior to creating your DVD.

Using iDVD on Mac OS X. Burning a DVD using iDVD (*http://www.apple.com/ilife/idvd/*; $79), takes little effort. First, launch the iDVD application. From the Start window, you can choose to Create a New Project, Open an Existing Project, or create a OneStep DVD, as shown in Figure 7-10.

Figure 7-10. Only three choices to make

Selecting OneStep DVD will import footage from your DV camera and automatically create a DVD from it.

To create a DVD from footage you already have on disk, as you should if you've exported it from your editing system, click the Create a New Project button. The Project window will then open and you will be able to select a theme from the Customize drawer. If the drawer is not visible, click on the Customize button located on the lower-left of the window. Figure 7-11 shows a list of Themes in the Customize drawer and the selected Sliding Panes theme.

Figure 7-11. Selecting a theme for the DVD

If your theme includes *motion menus*, as indicated by the icon of a person walking in the lower-right of the Theme's icon, then you can drag items from your Media list onto *Drop Zones*. When the DVD is finalized, the Drop Zones will become animated portions of your DVD's menu. They are not, however, a part of your movie. You can consider them more decorative than anything else. In Figure 7-12 a movie called *My Great Movie* has been dropped onto the first Drop Zone.

To add the video you would like to play on the DVD, simply drag it from your Media list onto the DVD window.

If you drop media onto a drop zone, it will not play as video on your DVD.

Figure 7-12. Adding video to a motion menu

You can change the name of buttons and menu items on your DVD by double-clicking on them and changing the text. Figure 7-13 shows the My Great Movie button highlighted and ready to be changed.

Figure 7-13. "My Great Movie" becoming an active button

If you want to check how your DVD will work after it is burned, click the Preview button. This will put iDVD into a simulation mode and provide a virtual remote control for you to interact with it. You will be able to click on the buttons you've created and experience what your DVD will be like. When you have completed previewing your DVD, just press the Preview button again. Figure 7-14 shows iDVD in Preview mode.

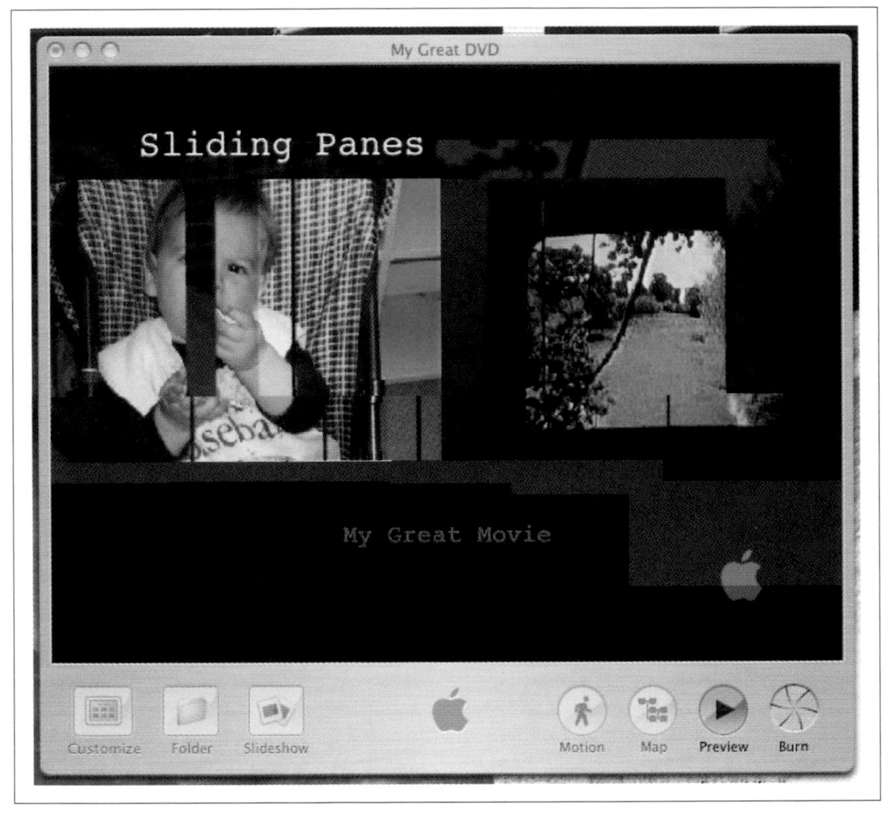

Figure 7-14. Previewing a DVD before burning it

If you'd like to make any changes before burning your DVD, make them and preview your DVD again (to make sure you like the changes). When you have finished making changes, click the Burn button on the bottom-right of the screen (or choose File → Burn DVD...).

When starting the process of burning a disc, iDVD will ask you to insert a blank DVD. Upon doing so, the application will handle all of the details of transcoding your footage and burning it to disc, as shown in Figure 7-15. Depending on the length of your footage and the speed of your computer, the process might take a while.

Figure 7-15. iDVD burning a disc

When iDVD finishes burning your disc, it will eject the finished DVD and then allow you to burn another disc or complete your session.

Using Nero on Windows. Nero Ultra Edition (*http://www.nero.com*; $99) includes a suite of applications to handle audio, video, photos, and general data. For creating DVDs, NeroVision Express makes the process of adding video, encoding it to MPEG-2, and burning the actual disc an easy, step-by-step process. Additionally, NeroVision Express can capture video directly from an attached camera.

After launching NeroVision Express, start the process of creating a DVD by choosing Make DVD → DVD Video from the menu on the right side of the window shown in Figure 7-16. If you would like to be able to edit your DVD at a later date, select the DVD Video (VR) option. However, choosing this option requires you to use a rewritable DVD and is not guaranteed to play in all DVD players.

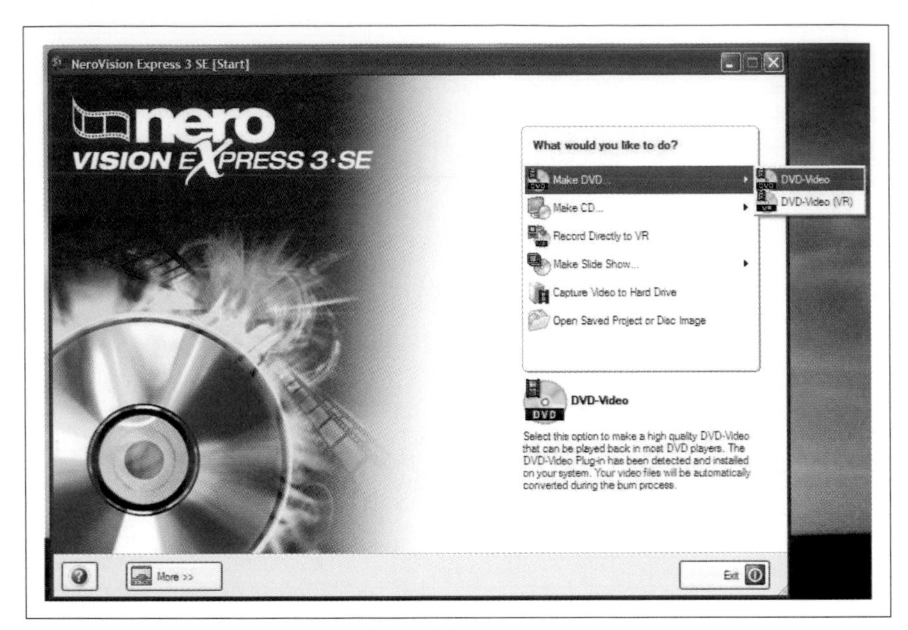

Figure 7-16. Starting to create a DVD with Nero

Next, choose Add Video Files..., as shown in Figure 7-17, to add the video files you would like to include on the disc. NeroVision handles the conversion to MPEG-2, so you can select most video file formats, including Advanced Systems Format (*.asf*), Audio-Video Interchange (*.avi*), MPEG-4 (*.mp4*), and Widows Media Video (*.wmv*), and many others.

Once you've added your video files to your project, you can edit the DVD menu that your viewers will see upon playing the DVD. When you select a template from the menu on the right, NeroVision will configure your menu and change the background, as necessary. You can download additional templates from Nero at *http://www.nero.com/us/NeroVision_Express_3_Template_Package.html* and *http://www.nero.com/us/NeroVision_Express_3_Template_Package_2.html*. Figure 7-18 shows the main template in its initial configuration.

Click the Edit Menu... button to change specific portions of the Menu, including the background, the font, and the buttons. You can even configure short portions of your video to play inside your buttons by selecting the Automatization menu item. Figure 7-19 shows a button being configured.

When you have finished configuring your menu, click Next to go back to the Menu window, where you should click Next again. NeroVision provides you the opportunity to test your menu's interaction by using a virtual remote control, as shown in Figure 7-20. Feel free to click around and press the remote's buttons to see how your DVD will work. If you discover you

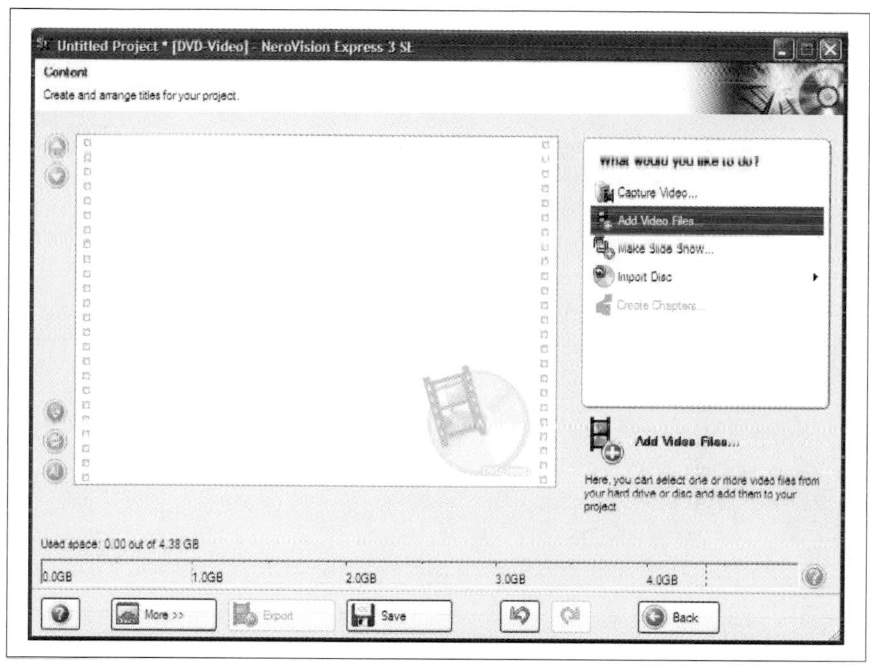

Figure 7-17. Adding one or more video files to the DVD

Figure 7-18. Selecting a template

Figure 7-19. Configuring a DVD Menu button

would like to change an item, simply click the Back button and make the necessary changes. Once you have the menu working as you like it, click the Next button.

With all of the personalization and configuration options out of the way, you can now proceed to burn your DVD. From the Burn to... item, select your DVD Writer from the options available. You might also find out the technical details of your video and how it will be burned to disc by clicking on the Details button on the left side of the window. If you would like to change any of the technical configuration, click on the More>> button on the bottom of the screen and then press the revealed Video Options button. Figure 7-21 shows the selection of a DVD Writer, along with the technical details of the video.

As soon as you are ready to burn your DVD, click the Burn button on the bottom right of the window. NeroVision will check your DVD writer for the availability of a disc. If one is not present, you will be presented a window asking you to insert a blank DVD. Once you've placed the disc, the application will begin to burn your DVD, as shown in Figure 7-22.

When NeroVision finishes burning your disc, it gives you the opportunity to burn more discs, save the current disc as a project, or start a new project.

Figure 7-20. A virtual remote control and a preview of a DVD

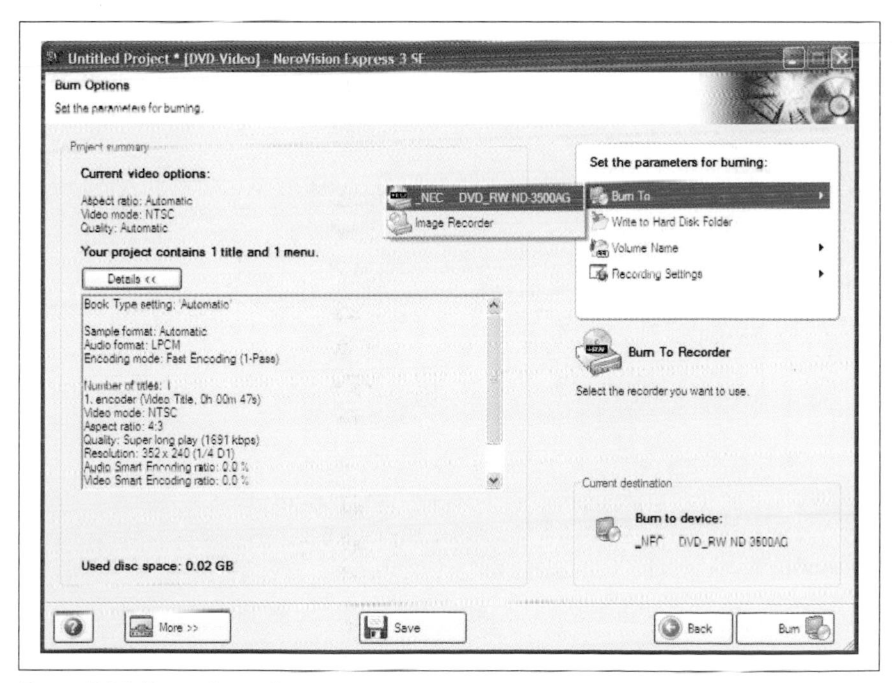

Figure 7-21. Preparing to burn a DVD

Figure 7-22. Burning a DVD

Using other applications. If the application you are using to create your DVD requires you to have your video in the MPEG-2 codec, you should transcode your video **[Hack #29]** appropriately. After doing so, follow the directions for your specific application.

Viewing Your Video

After you've burned your DVD, make sure that it plays in a DVD player. Although you can check it using the DVD player included with your computer, I recommend using a consumer DVD player (or multiple players, if you're so inclined). This extra step will guarantee that your DVD will work in a standalone player, which won't be as flexible as the player in your computer.

HACK #80

Stream a DVD

You can stream a DVD to one or more people, across the Internet.

Practically anyone who's worked with digital video is familiar with streaming video, even if that familiarity is just from watching a streaming video via the Internet. For the most part, streaming video involves either a live feed

[Hack #87] or a preprocessed streaming video file [Hack #83]. Using Video Lan Client (*http://www.videolan.org*; free, open source), also known as VLC, you can stream something somewhat unexpected: a DVD.

To stream a DVD, you will need the VLC application and a network connection. To view the stream, you will need a video player, such as QuickTime Player (*http://www.apple.com/quicktime/*; free), Windows Media Player (*http://www.microsoft.com/windows/windowsmedia/mp10/default.aspx*; free), or VLC. Of the three players mentioned, VLC will be able to view the widest range of streaming video.

Setting Up to Stream

Before launching VLC, place the DVD you would like to stream in your computer. Then, launch VLC and choose File › Open Disc... from the application's menus. VLC should automatically locate the DVD for you. If not, you will need to select it from the menu provided. Figure 7-23 shows VLC on Windows XP ready to open a DVD.

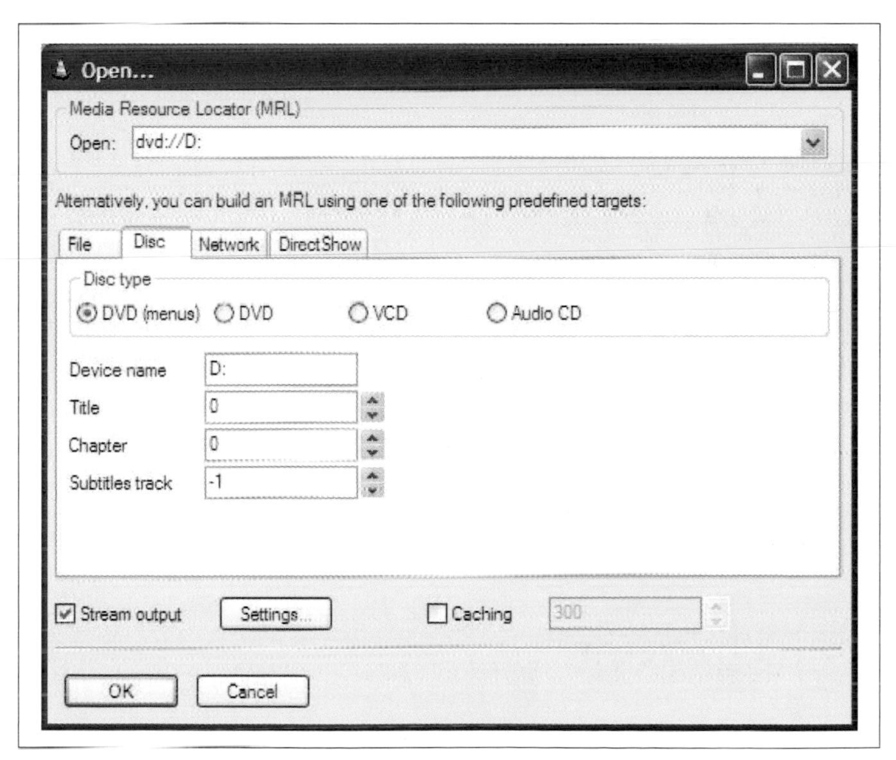

Figure 7-23. Opening a DVD using VLC on Windows XP

Some DVDs won't stream correctly unless you enable the DVD menus option. To do this on a Macintosh, check the "Use DVD menus" checkbox. On Windows, press the DVD (menus) radio button.

Just opening a DVD won't stream it, however. In order to stream the DVD, you need to enable the Advanced output option by clicking the checkbox. Doing so enables the Settings... button, which, when clicked, enables you to configure the stream.

You will probably notice that VLC can stream from Video TS (a file prepared for a DVD, but on a hard disk), DVD, VCD, and Audio CD.

Configuring the Streaming Options

In the Settings window, you'll have a wide variety of choices, many which can be confusing. Figure 7-24 shows the various advanced output settings.

If you want to watch the DVD on the computer that is going to stream it, select the "Play locally" checkbox. This will play the DVD on the computer as it is being streamed. Therefore, if you want to make sure the stream is occurring in the manner you expect, you should enable the option.

If you wanted to send the DVD's audio and video to a file on your computer, you would select the File radio button and then click the Browse... button to select where you wanted the file to be placed.

For the type of stream, choose the protocol that best fits your needs. Your choice of Type will also affect the options you have for your Encapsulation Method:

HTTP
: Streams the audio and video over the Hypertext Transfer Protocol, which many people are used to seeing in web addresses (i.e., *http://*). However, to view the stream, viewers will need a player that can translate it correctly. If you choose this option, you will have the widest range of Encapsulation Methods available: MPEG PS, MPEG TS, MPEG 1, ASF, and Raw.

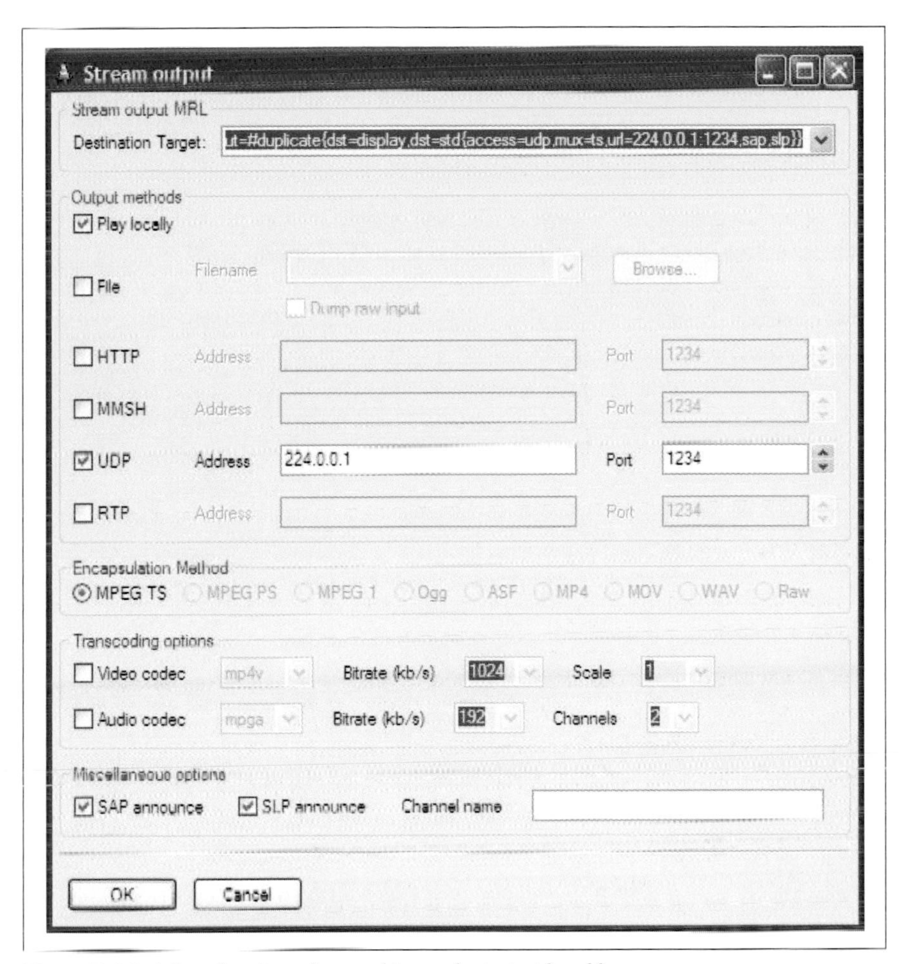

Figure 7-24. A lot of options, but nothing to be intimidated by

MMSH
: Enables viewers using Windows Media Player to view the stream. This method of streaming is not reliable, especially when not streaming from Windows. If you use this option, you must use the ASF Encapsulation Method.

UDP
: The User Datagram Protocol is lightweight and works well for real-time data, like audio and video streams. When using UDP you will have the option to stream to only one computer (Unicast) or to many computers (Multicast). When using UDP, you can use only the MPEG TS Encapsulation Method.

RTP

Using the Real-Time Transport Protocol is one of the better available options, because it is a standard that most media players can read. RTP was designed with a focus on audio and video communications. The RTP type will require you to use the Raw Encapsulation Method.

After you have selected a Type and an appropriate Encapsulation Method, enter an IP address. For most situations, simply entering your computer's IP address will suffice. If you are going to use UDP, you have two options. First, you can Unicast, where the stream is sent to only one specific viewer. In this case, you will need to know the viewer's IP address. Your other option is to Multicast, in which case you need to make sure you select and enter an appropriate IP address, between 224.0.0.0 and 239.255.255.255.

If you plan to stream across a router, choose an IP address above the 224.0.2.0 block. Also, change the Time To Live setting to something higher than 1. For an in-depth explanation of multicast addressing, point your web browser at *http://www.tldp.org/HOWTO/Multicast-HOWTO-2.html#ss2.1*.

Here's how to find out a computer's IP address:

Macintosh

System Preferences → Network → Select the active network connection

Windows

Start → Settings → Control Panel → Network → Connections → select the active network connection

When entering the IP address, take note of both the Address and the Port number, because you will need the information to view the stream.

You can use the Transcoding and Miscellaneous options for transcoding your video and Stream Announcing for more sophisticated situations. Transcoding allows you to convert one video codec and file format to another **[Hack #29]**. Stream Announcing allows your viewers to automatically open your stream in a media player, and should be enabled if you have selected to use UDP.

As soon as you've completed setting your streaming options, click the OK button. Then, click the OK button on the Open... (Windows) or Open Source (Macintosh) window. VLC will then start streaming your DVD. If you have selected to "Play locally," then the video will open in its own window.

Viewing the Stream

How your viewers configure their media players to connect to the stream will depend on the selection you made in your Stream Settings, as well as the specifics of the media player they are using:

QuickTime Player
: File → Open URL in New Player...

VLC
: File → Open Network...

Windows Media Player
: File → Open URL...

In the URL or Network section, a viewer should enter:

HTTP
: `http://`*`ip_address:port_number`*

MMSH
: `mms://`*`ip_address:port_number`*

UDP

Unicast
: `udp://localhost:`*`port_number`*

Multicast
: *`udp://ip_address`*

RTP
: `rtp://`*`ip_address:port_number`*

After entering the appropriate information, the DVD should begin playing. Figure 7-25 shows a Mac OS X client viewing a UDP streamed DVD from Windows XP.

VLC offers a lot more than just streaming DVDs. As mentioned previously, it can be used to both view and distribute streaming videos, to save streaming videos **[Hack #90]**, and it can open and play just about any type of digital video (including some proprietary ones) you throw at it. If the Swiss Army made a media player, it would be VLC.

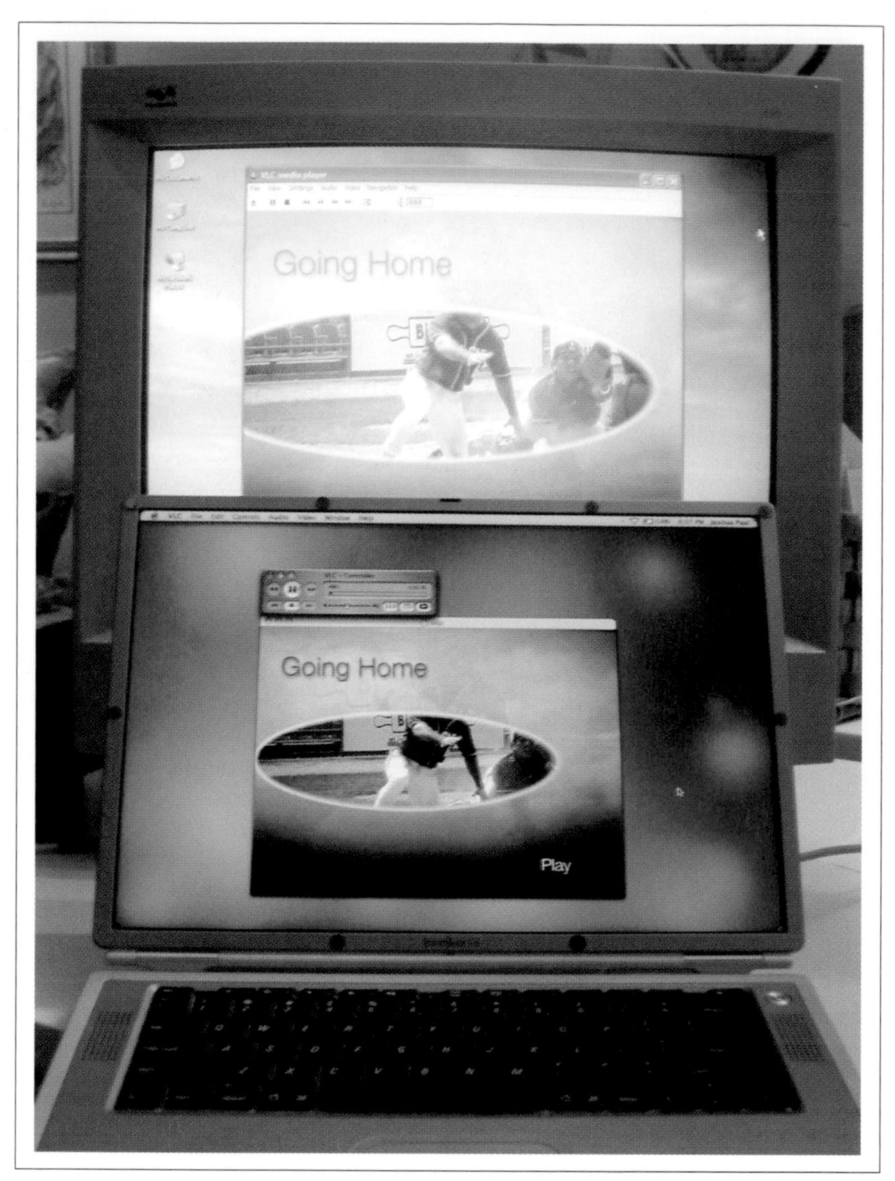

Figure 7-25. Streaming a DVD from Windows XP to Mac OS X

Vlog Your Life

Web logs, also known as blogs, are text-oriented web sites people that update on a daily or semiregular basis. Vlogs are the same concept, only they are video-oriented.

Blogging is a Web phenomenon. From teenagers to Presidential candidates, people from all walks of life have been given the opportunity to easily express and discuss their views, stories, and daily lives with the world in writing. With the availability of broadband Internet connections, digital video cameras, editing systems, video phones, and free blog sites, video logging (a.k.a. *vlogging*) is another outlet for your video projects.

Peter Jackson ran a vlog during the production *King Kong*. It can be found at *http://www.kongisking.net/kong2005/proddiary/*.

Starting a Vlog

A vlog is technically no different than a blog, so if you're already blogging, you're ready to vlog. If you don't have a blog, you can sign up for an account at a variety of sites. Here are some of the more popular blog sites:

Blogger
: Free: *http://www.blogger.com*

TypePad
: $4.95–$14.95 per month: *http://www.typepad.com*

Radio Userland:
: $39.95 per year: *http://radio.userland.com*

There are also *many* other blogging sites on the Web, which you can find by doing a search using the term `"blog host"`. Each blog site runs slightly differently. Some sites allow you to post from a web site only, while others offer desktop applications. Your choice is really a matter of what appeals to you.

If you have your own server, you can host your own vlog. There are many blogging software applications on the market, and a search for `"blog software"` will yield plenty of them.

Once you've signed up for a blog, you should become familiar with the way it works. At the very least, you should post one message, just to make sure you know how.

Posting a Vlog

To post a vlog, you need some video. Any video will do. Some people are posting unedited video, in small 10-second clips . . . off their cell phones! Others are posting edited pieces of an artistic, political, or simply experimental nature. Feel free to express yourself and be creative.

You can find more information on video blogging at *http://www.videoblogging.info/*. That site offers a few tutorials, a list of resources, links to recently posted videos, as well as a thriving and passionate community.

Once you have the video you would like to share, you need to compress your video for the Web **[Hack #82]**. If your blog service provides enough drive space for you to store video, use the provided space. Otherwise, if have a server where you can upload your video, use that server to host your video.

If you don't have somewhere to host your video online, you can use the Creative Commons archive at the Internet Archive (*http://www.archive.org*) or at Our Media (*http://www.ourmedia.org*). The process of placing video in the Internet Archive is made easy by the free ccPublisher application (*http://sourceforge.net/projects/cctools*; free, open source). You need an account with the Internet Archive, but the application steps you through the entire process, from selecting the video on your hard drive, to signing up for an account, to selecting a copyright license, to actually uploading the video. Figure 7-26 shows ccPublisher helping to select a video for uploading.

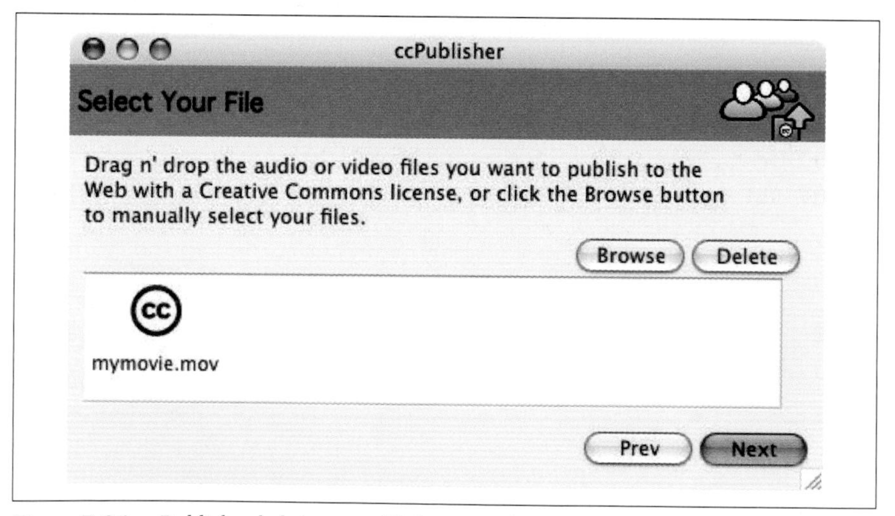

Figure 7-26. ccPublisher helping to publish your video

After you've uploaded your video, take note of its new URL. To post your vlog, use the tool provided by your blog service (i.e., web site or desktop application) and type a brief description or synopsis of your video where you would normally type your blog text. If you would rather link to your video, instead of embed it, add the following HTML to your post:

```
<a href="http://your.domain.tld/path/to/your/video.mpg">Click to watch</a>
```

Replace *your.domain.tld/path/to/your/video.mpg* with the actual URL of your video.

Viewing a Vlog

Viewing a vlog requires only a web browser and the appropriate media player. It's just like viewing any other video you might find on the Web. The only thing that is different is the way it is distributed.

However, applications are also available that can subscribe to people's vlogs and download the video automatically. If you are using Mac OS X (a Windows version is being developed), you can use ANT (*http://www.antnottv.org*; free, open source), a desktop application that can subscribe to vlog feeds. ANT then automatically downloads any new video that is posted overnight. You can also comment on the videos that you view, which essentially provides feedback to the producer of the video. Figure 7-27 shows ANT viewing a video by Eric Rice.

If you would like to allow people to subscribe to your feed using ANT, and your blog doesn't provide an RSS 2.0 feed, you can use FeedBurner (*http://www.feedburner.com*, free to monthly fees) to create the proper feed for you automatically.

On the Web, there is MeFeedia (*http://www.mefeedia.com*), which collects information on vlog *feeds*, which are basically broadcasts, and links to them. When you sign up for an account, you can subscribe to a lineup of vlogs you would like to track and view them at your leisure.

With vlogging, you have the ability to distribute your videos to anyone with an Internet connection and a media player, and you can view daily videos from people around the world. Have fun!

Figure 7-27. ANT playing a video called 300mph

HACK #82 Host Your Video on a Web Site

Putting your video onto a web site is a fun and easy way to distribute it.

With the growth of broadband Internet connections, most people now have access to fast download speeds. Given this fact, it is reasonable to expect your viewers watch your movie in a web browser, or download it to their hard drive to watch it using a media player. However, the size of an uncompressed digital video is still too large to be reasonably distributed online. Therefore, you need to compress your video for online distribution.

This hack covers the creation of video suitable for *downloading* from a web site. If you are planning on placing a video that is longer than 15 minutes, you should evaluate using streaming video **[Hack #83]**.

Exporting Your Video

To compress your video for use on the Web, you can export it from your editing system:

Avid
: File → Export → Options → QuickTime Movie

Final Cut
: File → Export → Using QuickTime Conversion...

Movie Maker
: Tasks → Finish Movie → Send to the Web

Premiere
: File → Export → Movie → Settings button

iMovie
: File → Share → QuickTime → Compress movie for: → Web

Some editing applications have presets, so all you have to do is select the preset and let the system do the rest. These applications do allow you access to more *expert* settings, if you want more control over the compression, but you might need to look for them. Other editing systems do not offer any presets and require you to configure the compression manually.

Configuring the Compression

If you need, or want, to configure the settings for compression, you will want to be familiar with the available codecs **[Hack #77]** and be able to select a file format. You should be aware that there are a lot of codecs available. Sticking to the major standards, and those that are widely distributed, will bring you more success with your viewers.

How you manually configure the export settings will vary, depending on which editing application you are using. As a general rule, you should attempt to keep your resulting file size as small as possible. To do so, you can do the following:

- Export your video at half-resolution. For example, if you are editing NTSC DV footage (720 × 480), scale your video to 360 × 240.
- Reduce the frame rate to 12 or 15 frames per second.
- Use a codec targeted at online video distribution, such as MPEG-4.

Creating the Web Page

Enabling people to view your video on a web page requires the video to be embedded in the page. The file format you have chosen for your video will determine the HTML you need to type for your web page. To start your page, type the following:

```
<!DOCTYPE HTML PUBLIC "-//W3C//DTD HTML 4.01 Transitional//EN" "http://www.
w3.org/TR/html4/loose.dtd">
<html>
```

```
    <head>
      <title>My Movie Page</title>
    </head>
  <body>
```

Then, add your embedding information.

For QuickTime files:

```
    <object id="Player" classid="clsid:02BF25D5-8C17-4B23-BC80-D3488ABDDC6B"
width="360" height="240" codebase=http://www.apple.com/qtactivex/qtplugin.cab
type="video/quicktime">
    <param name="src" value="./myMovie.mov">
    <param name="controller" value="true">
    <embed src="./myMovie.mov" width="360" height="240" controller="1"
pluginspage="http://www.apple.com/quicktime/download/">
    </embed>
  </object>
```

For Windows Media files:

```
  <object id="Player" classid:"clsid:6BF52A52-394A-11d3-B153-00C04F79FAA6"
width="360" height="240" codebase="http://activex.microsoft.com/activex/
controls/mplayer/en/nsmp2inf.cab#Version=6,4,7,1112" type="application/x-
oleobject">
    <param name="url" value="./myMovie.avi">
    <param name="showcontrols" value="true">
    <embed type="application/x-mplayer2" src="./myMovie.avi" width="360"
height="240" controller="1" pluginspage="http://www.microsoft.com/Windows/
MediaPlayer/">
    </embed>
  </object>
```

For MPEG files:

```
<object data="./myMovie.mpg" width="360" height="240" type="video/mpeg">
</object>
```

Finally, end the web page with:

```
  </body>
</html>
```

After typing out your video page, save it to your hard drive and name it with the file extension of *.htm* or *.html*.

Replace any information you deem necessary, such as the `src`, `url`, or `data` attributes that refer to your video file. For example, if your video file is named *MyGreatMovie.mov*, replace both instances of *`myMovie.mov`* in the HTML with `MyGreatMovie.mov`. You can also change the video's width and height, as well as whether the viewer should be able to control the video.

The ./ in front of the video file's name instructs the viewer's web browser to locate the video file in the same location as the HTML file. You can replace this with the specific path to where your movie will be located on the web server where you will upload your files, or you can leave it alone and simply upload both your HTML file and your video file to the same location.

Placing Your Files on the Server

Your server administrator will be able to provide you instructions on how and where to upload your files. There are many different methods to upload files, and every administrator handles access to the server differently. That said, you will have only two files to upload to the server: your video file and your HTML file.

After uploading your files, open a web browser and point it at your new web page. Within a few minutes, you should be watching your video—and you and your viewers will be able to do so from anywhere in the world.

HACK #83 Encode a Video to Stream on the Internet

Video files tend to be quite large, even when compressed, and can take a long time to download. By streaming a video, your viewers can get near-instant satisfaction.

The Internet is a great way to distribute your video. Cable modems, satellite dishes, and digital subscriber lines (DSL) have all brought broadband Internet connections to the people who want them. But even with a broadband connection to the Internet, downloading large files can still be a time-consuming process. Video files of any decent length, even when compressed, are much larger than even graphics-intensive web pages.

To overcome the hurdle of time-consuming downloads for video, various companies and organizations created methods to send video files in small pieces. These small pieces can be displayed immediately. The process of sending the pieces is called *streaming video*.

In order to stream video, three pieces of technology need to be in place. First is the video client, which the viewer will use to watch the video. Second is the streaming server [Hack #85], where the video will be distributed to the viewer. Third is the video file itself, which must be formatted or *encoded* in a specific way.

Creating a Streaming Video

There are a number of ways to create a streaming video. The amount of time it takes to encode video is different for each process and might also depend on the speed of your computer. However, as a general rule, you can expect "live" encoding to be the fastest and lowest quality, exporting from an editing system being in the middle for both speed and quality, and encoding from an uncompressed video file to be the slowest to process but the best in quality.

Encoding a live video. I use the term *live* video to represent two forms of video. The first is an actual live event, such as a sporting event, where action is occurring while you are encoding. The second is merely playing video from a camera or deck. Both types can be encoded using the same solution. However, this hack does not cover how to broadcast a live event **[Hack #87]**.

Microsoft and Apple have both made encoding software available for free. Microsoft's application is called Media Encoder and can be downloaded from *http://www.microsoft.com/windows/windowsmedia/9series/encoder/default.aspx*. Apple's application is called QuickTime Broadcaster, which is available for download at *http://www.apple.com/quicktime/download/broadcaster/*. Both programs work quite well and offer decent results, considering they are encoding (and often compressing) video in real time. Figure 7-28 is an image of Microsoft's Media Encoder about to broadcast and archive a stream and Figure 7-29 shows QuickTime Broadcaster.

The quality of your video will depend on your streaming server's upload speed. Do not exceed your server's upload speed when configuring your video's data rate. I have personally had good results at 500Kbps for a data rate with both applications.

Encoding and saving to disk is especially useful when you need to place the video online as soon as possible, because the encoding occurs in real time. However, because the encoding is done is real time, you will not get the best results possible.

To send video to either application, connect your camera or deck to your computer using an IEEE-1394 (a.k.a. FireWire or i.LINK) connector. After connecting your video source, launch the encoding application and configure your settings appropriately. Make sure you have set the application to hint the resulting file for streaming, as well as to save your encoded file to disk. You will need the saved file in order to create a video on-demand later on in the process.

Figure 7-28. Media Encoder ready to encode a video at 548Kbps

There is a bug in QuickTime Broadcaster that occurs when recording for long periods of time to a disk larger than 2GB in size. If you plan on capturing a video that is longer than 15 minutes in length, you should either partition a disk to have a 2GB partition or create a 2GB disk image using Disk Utility. I have used the latter option successfully in a production environment.

Exporting from your editing system. Most editing systems have a method to export video to a file on your computer. Depending on the operating system, the editing software, and possibly the plug-ins you've installed, your options for exporting a streaming video will vary. Make sure that your system allows you to export a *streaming* video. It will also be helpful if you know which codec you plan to use **[Hack #77]**.

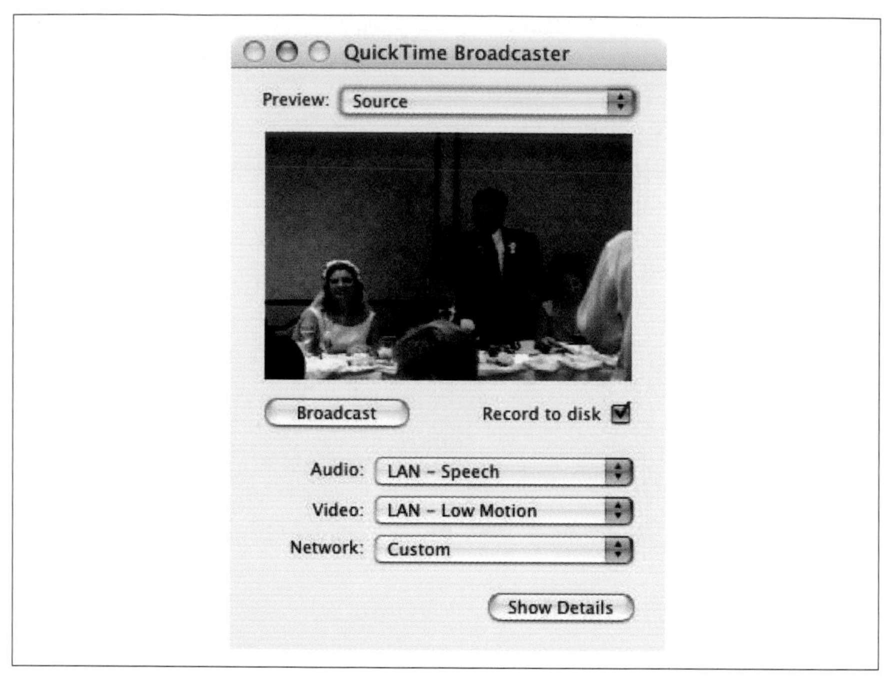

Figure 7-29. QuickTime Broadcaster set to Record to disk

Avid
: File → Export

Final Cut
: File → Export → Using Compressor...

Movie Maker
: Tasks → Finish Movie → Send to the Web

Premiere
: File → Export → Movie → Settings button

iMovie
: File → Share → QuickTime → Compress movie for: → Web Streaming

Some editing systems will export based on your timeline's current In and Out points as a default. If you are trying to export your entire movie, you need make sure your marks are set properly. You probably don't want to find out you've waited for your system to export your movie for the last three hours, only to discover you left off the first five minutes of your timeline.

Encoding using Discreet's cleaner. If you want the best quality streaming video you can achieve, and you have the time to wait for your system to carefully encode your video, use an application specifically designed for video encoding. Discreet's cleaner (*http://www.discreet.com*; $549) is such a program and provides great results. Additionally, cleaner offers a wide variety of options for encoding your video, including unlocking some of the more hidden features of digital video, such as URL references **[Hack #84]**.

To get the best quality video out of cleaner, capture your video at the highest resolution possible. This will most likely be uncompressed DV. Once you have captured your video, launch cleaner and drag-and-drop your video file onto the Batch window.

A really nice feature of cleaner is the ability to set an In point and an Out point. This allows you to change your settings, encode just the portion of video between your In and Out points, and then view the result. In practice, this enables you to tailor your settings to each individual movie, because you can make a change and see if your quality is better than before you made the change. Figure 7-30 shows the In and Out markers along the timeline as green and red marks, in addition to showing the current settings for the encoding process.

Following the In/Out approach, here's an example workflow:

1. Double-click on the movie for which you would like to change the settings.
2. Click on the Settings tab and make any desired changes, such as reducing the frame rate.
3. Locate a point in your video you would like to have as the start of your video tests.
4. Mark your In point (Edit → Set In Point).
5. Locate a point in your video you would like to have as the end of your video tests. Your Out point should not be more than 20 to 30 seconds from your In point.
6. Mark your Out point (Edit → Set Out Point).
7. Encode your test (Batch → Encode).
8. Locate the encoded movie on your computer and open it in a video player.
9. Repeat until happy or tired.

By following this workflow, you will consistently have a higher quality movie at the lowest possible data rate. If you are really patient, or possibly a masochist, you can change your In and Out points, encode, and view another section of your video. Depending on the outcome, you can continue the process or accept your results.

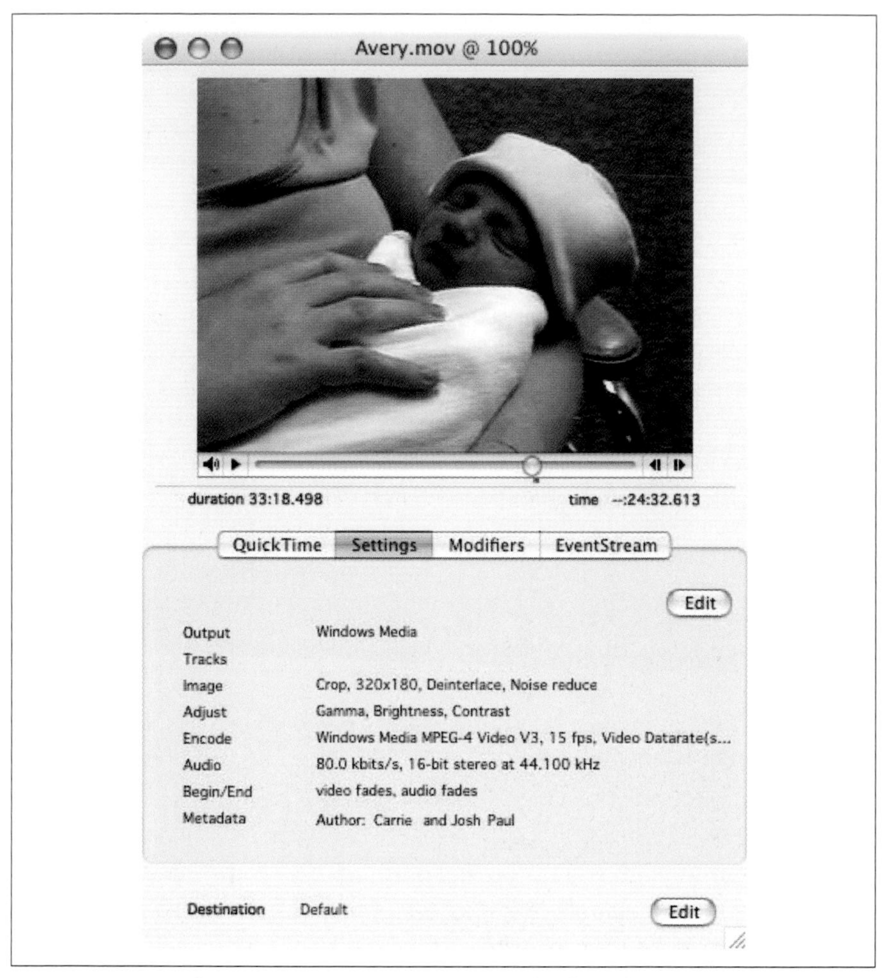

Figure 7-30. In and Out points appearing as green and red marks, respectively

When you are finally happy with your settings, make sure you set your In and Out points to the *actual* start and end times for your movie. Encode and enjoy.

HACK #84 Take Donations via PayPal

Make your video open a web page that accepts donations.

QuickTime, Real, and Windows Media can all forward viewers to a URL. By using Discreet's cleaner (*http://www.discreet.com/products/cleaner*) application, you can easily open a web browser from any of these players and send the viewer to a PayPal donation page. Whether people actually donate or not is a whole 'nother story.

The latest version of cleaner for Windows has dropped the EventStream feature. If you are using Windows, and would like access to EventStream, you should try to locate an older version of cleaner. You might also want to pressure Discreet to bring the feature back.

Importing Your Video

Once you've installed cleaner and launched it, you need to import your movie to the batch process. To accomplish this, you can either drag and drop your movie file to the Batch window, or select the menu item by choosing Batch → Add Files.... After you have added your movie, double-click on the icon that represents your movie or select the movie and then the menu item (Windows → Project).

In your Project window, jump to the end of your movie (or the point in your movie where you would like to seek a donation), as shown in Figure 7-31.

Figure 7-31. Jumping to the end of your movie

I usually *back off* the exact end of the movie by about one second, simply because I do not trust the video file to always act appropriately.

Once you have found the point where you would like to ask for the donation, select the Event Stream tab or select the menu item (Windows → Event Stream).

Creating the Event Stream

Once the Event Stream window is open, you need to create an Event by clicking the Add button. Then, select the Open URL item from the pop-up menu. You will then be able to fill in the URL for where you would like to send your viewers, as shown in Figure 7-32.

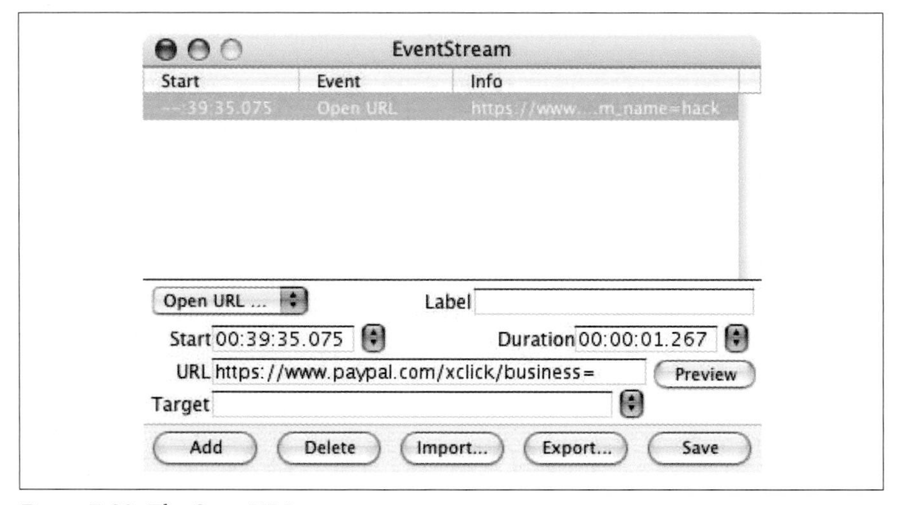

Figure 7-32. The Open URL event

Creating the URL

PayPal has a simple URL for donations:

```
https://www.paypal.com/xclick/business=<username>%40<domain>.<tld>&item_
name=<your_item>
```

The `%40` is the equivalent to an `@` symbol, so replace the *`<username>`* with the part of your email address that comes before the `@` symbol, the *`<domain>`* with the part of your email address that comes between the `@` symbol and the last dot (`.`), and the *`<tld>`* with the last section of your email address following the last dot. Lastly, change *`<your_item>`* to your description of the donation; something like `Donation` should be sufficient.

Wow! That is a little confusing, so here is an example using the email address of *video@domain.com*:

```
https://www.paypal.com/xclick/business=video%40domain.com&item_name=Donation
```

After entering the URL, check to make sure it works as you expect by clicking the Preview button. If you have typed the URL correctly, your web browser should open and take you to a PayPal page, allowing a viewer to fill in an amount she would like to donate. If you receive an error page, or if the page does not show the correct information (e.g., your email address is wrong), go back and make sure the URL you have entered is correct.

The interface cleaner offers to enter URLs can be frustrating to use, because the text field you type the URL into is too short. Basically, there is nowhere on screen you can see the *entire* URL you have entered.

To overcome this limitation, I recommend typing the URL in a simple text editor, such as WordPad or TextEdit, and then copying and pasting it into the Event Stream window.

When you have the Open URL event working as you like it, click the Save button.

Compressing and Distributing

Now that your Event Stream is in place, you need to select the settings for how you would like to compress your movie. There are almost infinite compression settings available, so for brevity, I will jump ahead, knowing you will select a setting appropriate to your distribution requirements **[Hack #77]**. However, make sure to use either the QuickTime, Windows Media, or Real file format. Once you've configured your settings, start the Encoding process and wait for it to finish.

Once the encoding process has completed, locate your movie file and double-click it. Your movie should automatically open in the associated movie player. After your movie opens, you should make sure the URL is opened at the correct time.

If you have a long movie, don't be shy about jumping ahead to test the outcome. Your viewers might just do the same at some point.

Happy! Happy! Joy! Joy! Get your movie out there **[Hack #86]** and hope people help support your digital video habit (and bandwidth costs).

Hacking the Hack

PayPal offers a wide variety of parameters for you to configure via a URL. One parameter allows you to specify how much money you are requesting. To specify this parameter, alter the URL with the *<amount>* variable:

```
https://www.paypal.com/xclick/business=<username>%40<domain>.<tld>&item_
name=<your_item>&amount=<amount>
```

In addition to specifying the amount of the donation, you can forward people who make a donation to your own "Thank You" web page, allow people to send you a personal note, and/or even request their address (maybe to send them a full-quality DVD?!).

If you plan on distributing Windows Media or QuickTime files only, you can seek donations using either Windows Media File Editor or QuickTime Pro, respectively.

Using Windows Media File Editor. Using Windows Media File Editor, which is a part of the Windows Media Encoder 9 Series (*http://www.microsoft.com/windows/windowsmedia/9series/encoder/default.aspx*), you can add *script commands* to a video. One such command will open a URL in the viewer's web browser.

After launching Windows Media File Editor, choose File→Open, and select the video that you would like to embed with your PayPal URL. After the video is open, move the time indicator to end of video and click the Script Commands... button. This will open the Script Commands window, where you should click the Add... button.

From the Type pop up, in the Script Command Properties window, choose the URL type. Then enter the PayPal URL in the Parameter text box. When you are finished, click the OK button. Then choose Save and Index... from the File menu.

Using QuickTime Pro. You can easily add your PayPal URL to the end of your QuickTime video by following the steps outlined in Apple's "HREF tracks" tutorial at *http://www.apple.com/quicktime/tools_tips/tutorials/hreftracks.html*.

Set Up an Internet Television Station

Using the free Darwin Streaming Server, create an Internet TV station, complete with a schedule.

Not many of us can afford our own television station. But if you are willing to broadcast your programming over the Internet, you can have your own station up and running in less than an hour. Oh, and it won't cost you anything. Once you have it installed and configured, you will be able to broadcast QuickTime, MPEG-4, and 3GPP movie files.

Downloading the Darwin Streaming Server

Broadcasting video from your computer requires streaming software. Apple has made its Darwin Streaming Server available for free. In fact, it's open source, which means if you know how to program a computer, you can make changes and enhancements to the way the streaming server works.

For people who don't know how to program a computer, or who would rather just install the software and get on with creating their Internet TV station, Apple provides prebuilt installation packages. Packages are available for Mac OS X, Windows 2000 and 2003 Server, Red Hat Linux 9, and Solaris 9. If you want access to the source code, you will need to register and agree to the Apple Public Source License (APSL).

You can download the software at *http://developer.apple.com/darwin/projects/streaming/*.

Installing the Server

If you do not have Perl installed on your computer, download and install it before installing the streaming server. Although the streaming server doesn't require Perl to run, the Administration and set-up tool does.

You can download Perl for Windows from ActiveState (*http://www.activestate.com*).

After downloading the software, or building it from the source code, you need to install it. The prebuilt packages will install everything necessary for you to be up and running quickly. If you are building from the source, follow the included directions.

Configuring the Server

After installing your server, you will need to configure it. You can do so by using a web browser and connecting to *http://hostname:1220*, where *hostname* is the name of your streaming server. The web page will then step you through the configuration process.

If you're administering from the server itself, you can connect to your installation at *http://127.0.0.1:1220*. This is quite useful if you've installed the streaming server on your laptop, for instance.

Setting your passwords. Initially, you will be asked for a username and password. The combination you use will be your Streaming Server Administrator's password and will be required for any changes you wish to make in the future. The username and password combination does not have to be the same as the username and password you are using on your computer. In fact, for security reasons, it should be different.

The next page presented will ask for another password. This password is the one you will use to send an MP3 audio stream to the server. The password should be different than the Administrator's password you entered on the previous page.

Performing the initial configuration. If your server has a valid secure sockets layer (SSL) certificate and you would like to securely administer the server, you can enable SSL by simply clicking the checkbox. Figure 7-33 shows the SSL configuration page.

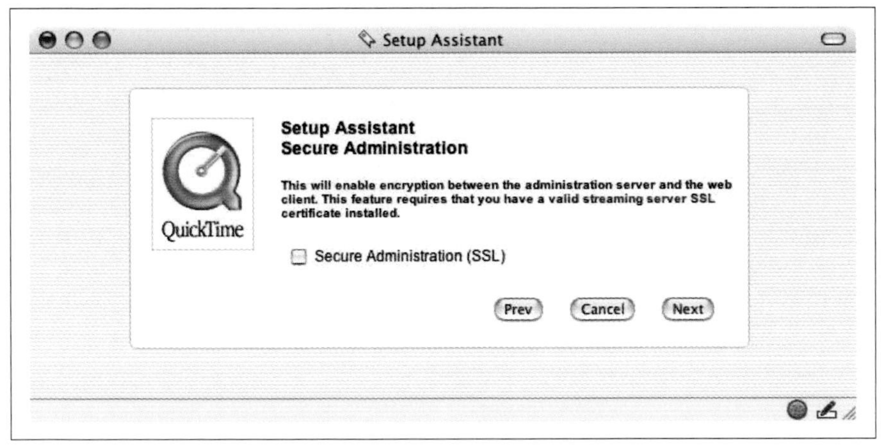

Figure 7-33. Enabling SSL (only if you have a certificate)

The Media Folder is where you will need to place your media files. I recommend using the default setting, although you can stream files from anywhere on your computer. Take note of where the server will expect your media files to exist. Figure 7-34 shows the default Media Folder configuration.

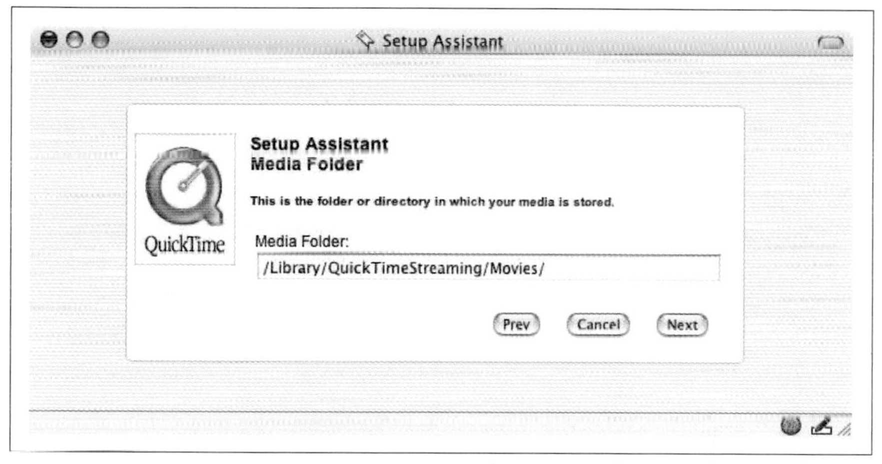

Figure 7-34. Streaming movies and music from the Media Folder

If you change the location, make sure the directory you choose allows the streaming server access to the files by assigning the appropriate permissions. The Streaming Server needs read/write access.

The streaming server can send data over port 80, which is the standard port for serving web pages. Enabling streaming media over port 80 will allow people who are behind restrictive firewalls to view your content (so long as they can view web pages). If the computer you have installed onto also serves web pages, then *do not* enable this feature. Figure 7-35 shows the configuration option to enable streaming on Port 80.

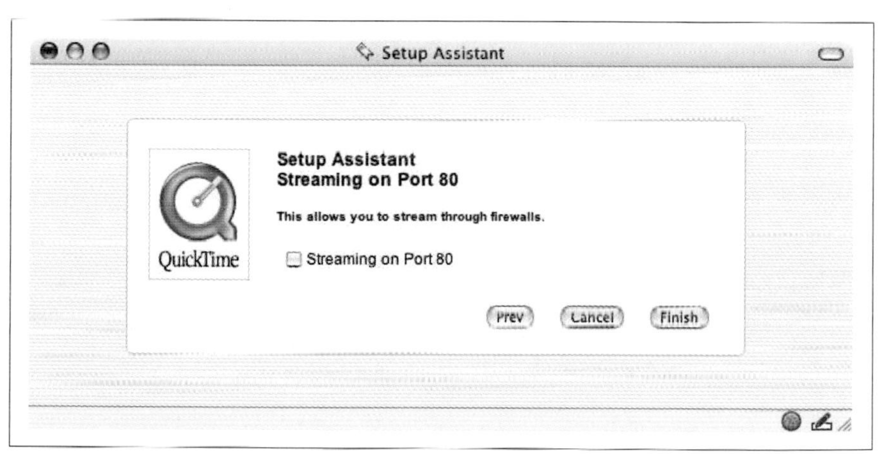

Figure 7-35. Enabling streaming on Port 80 (only when necessary)

When you have finished the initial configuration, you should see "Server is Running" at the top left of the web page. If you see "Server is Idle," you will need to manually start the server. Hopefully, this will not be the case. Figure 7-36 shows QuickTime Streaming Server running.

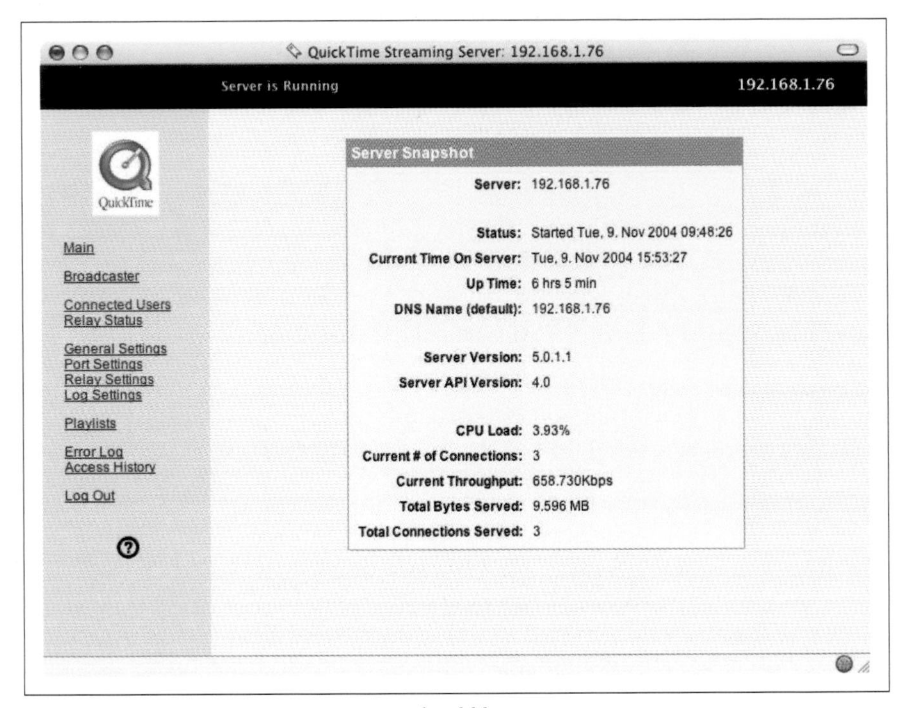

Figure 7-36. Server is Running . . . as it should be

At the time of this writing, a method to start/stop the server is not a part of the web-based Administration tool. If you need to start the server manually, you will need to run a Perl program:

- Mac OS X: */usr/sbin/stramingadminserver.pl*
- Windows: *\Program Files\Darwin Streaming Server\streamingadminserver.pl*

To run the program from the command line:

- Mac OS X: open the Terminal application, located in the Utilities folder.
- Windows: open the Run... item from the Start menu, and then type `command`.

Then, type:

- Mac OS X: `perl /usr/sbin/stramingadminserver.pl`
- Windows: `perl \Program Files\Darwin Streaming Server\streamingadminserver.pl`

To test your installation, launch your video player and open the URL *rtsp://hostname/sample_300kbit.mp4*. An animated QuickTime logo should play almost immediately. Figure 7-37 shows the 300kbit sample movie playing from a streaming server.

Figure 7-37. A streaming movie

You do not need QuickTime Player to view your broadcasts. There are sample movies in a QuickTime-specific file, but you will be able to stream MPEG-4 videos, in addition to QuickTime and 3GPP formats.

Creating Your Schedule

Now that the streaming server is up and running, you need to add some movies to broadcast. Locate the directory you selected for your Media Folder, as noted earlier. Then, find the movies you would like to broadcast and place them into the directory you've selected as your Media Folder.

Using the Administration tool, click on the Playlists link. You should notice there is already a Playlist called *sample*. Click on the New Movie Playlist link. You will then see a list of all the movies you have placed in your Media Folder. In order for your files to stream correctly, they need to be encoded for streaming **[Hack #83]**.

Configuring the playlist is rather easy. Your playlist's Name and Mount Point should be the same, except that the extension *.sdp* should be added to the name of the Mount Point. The Play mode offers three options: Sequential, Sequential Looped, and Weighted Random. In order to add movies to your Playlist, click and drag a movie from the Available Content list (list on the left) to the Playlist (list on the right). Figure 7-38 shows a playlist, named "monday," being created.

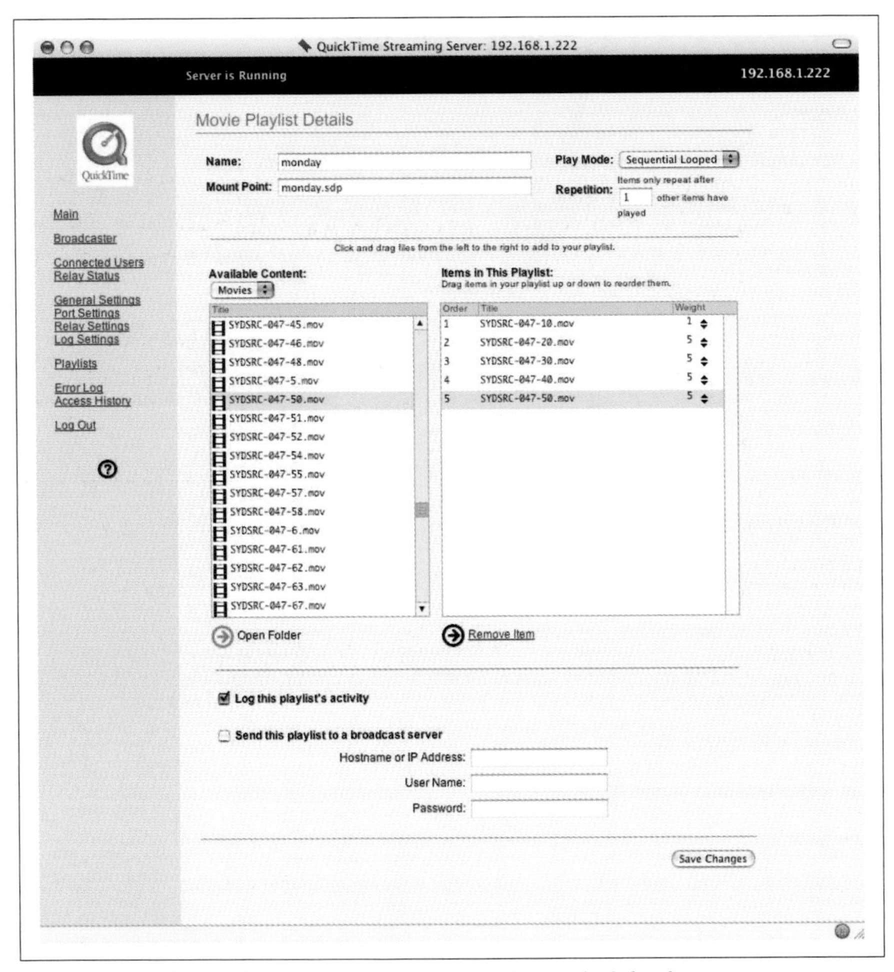

Figure 7-38. All available content displayed in the list on the left side

Selecting the Sequential mode will simply stream the movies in your playlist from the first movie in the list to the last. If you would like your playlist to continue broadcasting until you stop it, you have two options. The first option is to use the Sequential Looped mode, which will start at the first movie in your list and play in order to the last. It will then start at the top again and continue this process until you stop it.

The second option is to use the Weighted Random mode, which will broadcast a random selection from your playlist. There are two criteria that are examined when using this mode. The Repetition setting limits the selection to movies that have not been broadcast yet, until the number of movies broadcast exceeds the number of repetitions you've entered.

Make sure you don't set the Repetition number higher than the total number of movies in your Playlist.

The other variable you can set is the Weight. You accomplish this by clicking on the up and down arrows next to a movie's title in the playlist. Higher weighted movies will play more often than lower weighted movies. You can assign a weight between 1 and 10 to each movie.

Upon saving your changes, you will return to the Playlists page. Your newly created Playlist should be listed, however its Status will be Stopped. In order to start broadcasting, click the Play button. The status should then change to Playing. Figure 7-39 shows the "monday" playlist as available and Playing.

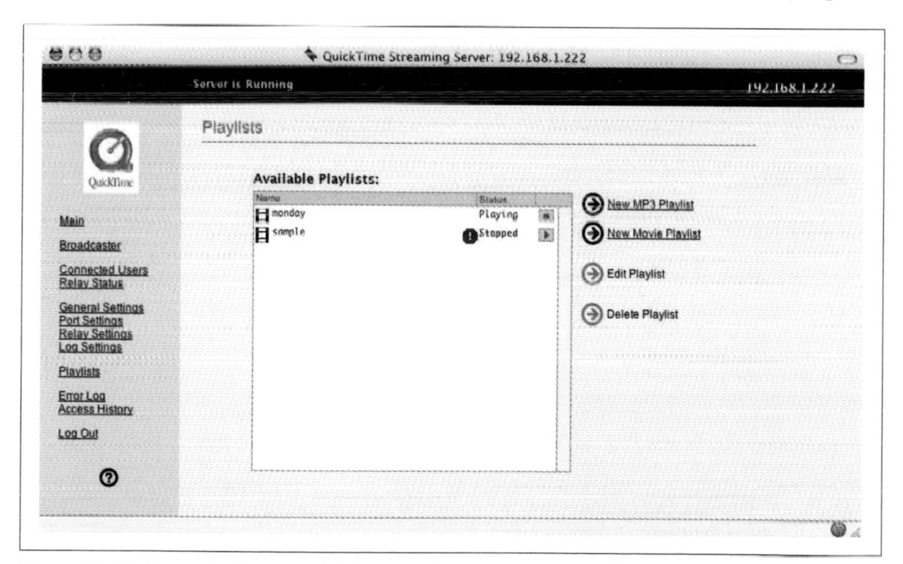

Figure 7-39. Two playlists available: one playing, one stopped

To access your broadcast, launch your player application and open the URL *rtsp://hostname/myPlaylist.sdp*, where *hostname* is the name of your streaming server and *myPlaylist* is the name you assigned to your playlist.

Allowing People to Tune In

Because your broadcast is streaming, people need to know where and when to *tune in*. You can distribute the URL to your broadcast via email or web page. If you plan on allowing people to view your content from a web page, you can either create a basic link or embed the broadcast into the page.

I feel that embedding your movie in a web page is too restrictive, because your viewers will not be able to resize the movie.

You can create a basic link to view the broadcast like this:

```
<a href="rtsp://hostname/myPlaylist.sdp">Click here to tune in</a>
```

When a visitor clicks the link, his preferred video player will open your broadcast. He can then sit back, relax, and watch your programs.

The final hurdle is to figure out what you are broadcasting and at what time. Your broadcast will begin when you click the play button on the Playlists page, so you need to plan accordingly. It will be important to know how long your movies are and in what order they are playing (unless, of course, you have selected Random Weighted mode).

For example, if you have five 30-minute movies playing, and you began broadcasting at 10:00 am PST, your schedule will look something like this:

Highway America - 10:00 a.m.
Gone Fishin' - 10:30 a.m.
Science Today - 11:00 a.m.
The Finer Things - 11:30 a.m.
Come 'n Get It! - 12:00 p.m.

If you have selected the Sequential Loop mode, *Highway America* will begin playing again at 12:30 p.m. (after *Come 'n Get It!* ends), *Gone Fishin'* will begin at 1:00 p.m., and so on. This is because your playlist is in a "loop" and will continue playing in order, from first to last.

HACK #86 Use BitTorrent to Distribute Your Video

Using BitTorrent, a peer-to-peer network application, you can have other people distribute your movie for you.

Digital video files can be quite large. If you plan on distributing your movie online, you might wind up paying more than you expect, or you might even have your distribution stopped due to exceeding your bandwidth allocation.

BitTorrent is a peer-to-peer (P2P) network application created by Bram Cohen. What sets BitTorrent apart from other P2P applications is the method it uses to distribute files. In most P2P applications, when you request to download a file, that file is obtained from only one machine. By downloading from one machine, you are limited to that machine's connection to the Internet, whether it is fast or slow, reliable or not.

BitTorrent, in contrast, attempts to download a file from many machines at the same time. This *torrent* method allows people to obtain files both faster and more reliably, because the requested file is delivered in pieces. Some machines will have faster connections than others, but combined, they can all deliver a specific file faster than a single machine by itself. Additionally, the more popular a file is, the more quickly it can be distributed.

BitTorrent uses *.torrent* files, which are usually small, to represent a complete file that exists in the network. Once someone has obtained a *.torrent* file, she can then download the complete file from the network. This is akin to browsing a movie rental store and taking the DVD cover to the front of the store so that the clerk can give you the actual DVD.

Given the design of BitTorrent, it is a terrific method to distribute video files. To quote the BitTorrent web page:

> The key to cheap file distribution is to tap the unutilized upload capacity of your customers. It's *free*. Their contribution grows at the same rate as their demand, creating *limitless scalability* for a fixed cost.

Downloading a BitTorrent Client

The official BitTorrent client is available for Windows, Mac OS X, and Linux and can be downloaded from the official BitTorrent web site at *http://bittorrent.com/download.html*, but there are also many other BitTorrent clients available on the Internet. One client of particular interest is Azureus (*http://azureus.sourceforge.net*; free), which includes a tracker (explained later) and is expandable through third-party plug-ins.

Azureus is available for Windows, Mac OS X, and Linux. Installation instructions for all three operating systems are available on the Azureus web site. You need Java installed on your computer; download it for free from Sun (*http://www.java.com/en/download/manual.jsp*).

After installing Azureus, you're ready to download your first torrent.

Downloading a .torrent File

There are essentially two steps to downloading a file using BitTorrent. The first is to obtain a *.torrent* file. The second is to open the *.torrent* file in your

BitTorrent client (in this case, Azureus), which begins downloading the specific file.

To find *.torrent* files, you can use a BitTorrent *tracker*. A tracker keeps a list of available torrents and distributes the *.torrent* files required to obtain the complete files they represent.

Two of the better-known trackers are the etree.org Community BitTorrent Tracker (*http://bt.etree.org*) and Legal Torrents (*http://www.legaltorrents.com*).

Once you have located a tracker to use, simply select a file you would like to download. The file you get from the tracker should be a *.torrent* file. Using Azureus, open the torrent file and your download will proceed. Figure 7-40 shows a download in progress of movie called *Panorama Ephera*.

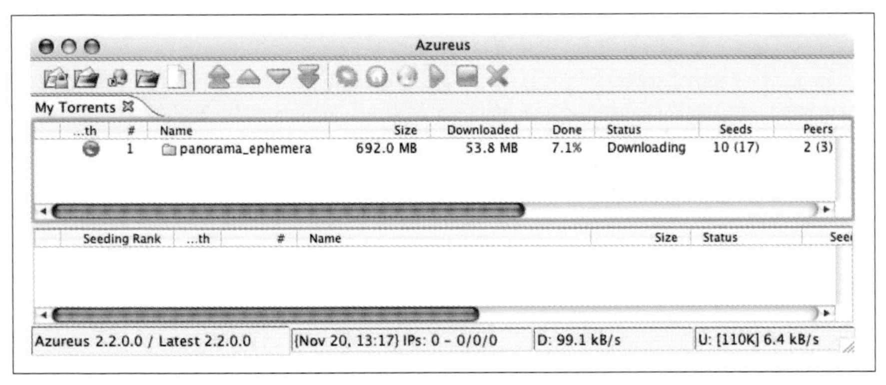

Figure 7-40. Downloading a movie in Azureus

When using BitTorrent, you are *cross-loading*: both uploading and downloading files. If you restrict your uploading, your download will be extremely slow. BitTorrent is a "tit for tat" system.

Depending on the size of the file you are downloading, your Internet speed, and the speed of the people who are sharing the file with you, your download might take some time to complete. You will more than likely want to work on something else while the download occurs—perhaps something like creating a torrent of your own to share with the world!

Sharing a File

If you want to make a file available in the BitTorrent network, you need to create a *.torrent* file. Unlike other P2P networks, which allow only one file, a *.torrent* file can be composed of many files. Most BitTorrent applications make this process painless, and Azureus is no exception.

The easiest way to create a torrent and share a file is simply to choose File→Share→File.... Depending on the size of your movie, the process might take a while. During the process, a window will display the current status. Upon completion, the *.torrent* file will be made publicly available through Azureus' tracker.

Once you have begun the process of sharing your file, you might notice that your movie's Status is listed as Queued in the My Torrents tab. If you find your movie is queued for a long period of time, right-click on it and select Force Start. This will force your movie to begin *seeding*, which is the process of injecting your movie file into the BitTorrent network.

If your computer is behind a firewall, you need to open ports 6969 and 6881–6889. Otherwise, people will not be able to contact Azureus to download the *.torrent* file or the actual movie. You will also want either a static IP address, or a dynamic DNS pointing to your computer. A dynamic DNS can be obtained from Dynamic Network Services (*http://www.dyndns.org*) or No IP (*http://www.no-ip.com*).

Sharing a file this way both creates the *.torrent* file needed for your viewers to obtain the movie and adds the *.torrent* file to the built-in tracker. Figure 7-41 shows the Azureus BitTorrent Client Tracker in action, as someone downloads a file.

Azureus : Java BitTorrent Client Tracker: 2.2.0.0/0.8.7

Azureus: BitTorrent Client Tracker (2.2.0.0/0.8.7)

Torrent	Status	Size	Seeds	Peers	Tot Up	Tot Down	Ave Up	Ave Down	Left	Comp	More
tall_guy_short_guy.mov	Running	573.1 MB	1	1	63.0 MB	61.5 MB	107.5 kB/s	105.1 kB/s	512.0 MB	0	...

Figure 7-41. Azureus enabling a viewer to download a movie

People can access the tracker at *http://your.host.name:6969*. They can then download the *.torrent* file and load it into their BitTorrent client in order to download the actual movie. If you choose to, you can also download the file for yourself and then email it out to your friends, family, and groupies.

Using a Public Tracker

If you would prefer to use a public tracker to announce your file, you need to create a *.torrent* file for the tracker to use. Azureus makes creating such *.torrent* files easy.

Using the creation wizard. While running Azureus, choose File→Create a Torrent. The creation wizard window will present a set of options for your torrent file. Figure 7-42 shows the first section of the creation wizard.

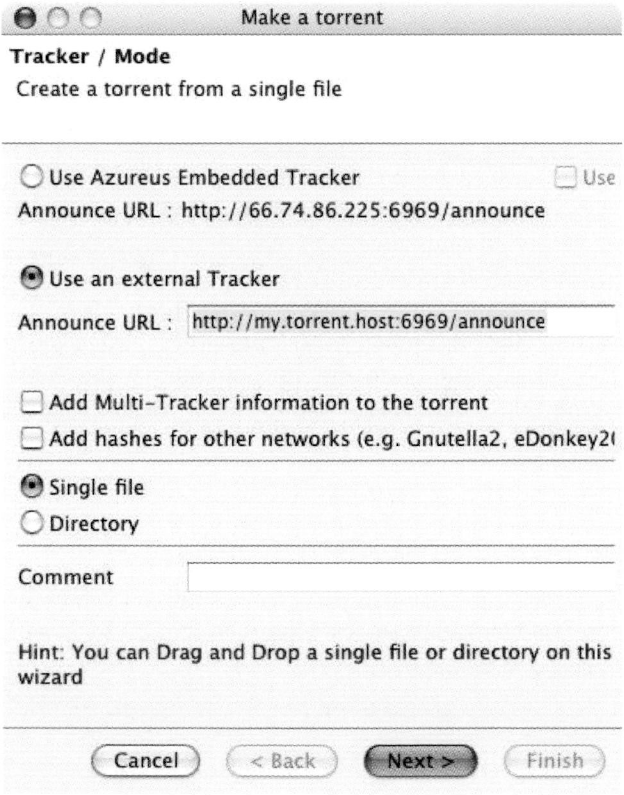

Figure 7-42. Creating a torrent file

If you would like to have multiple trackers handle your *.torrent* file, check the appropriate checkbox. If you select this option, you will be able to add additional trackers in the following window.

If you decide you would like to add more trackers *after* you've created your *.torrent* file, you can use Azureus's Add Tracker feature by right-clicking on your file and choosing Tracker → Add Tracker URL. There are many public trackers available, at no charge, to help announce your *.torrent* file to the community.

Unless you plan on using the other P2P networks, such as Gnutella2 or eDonkey2, you can safely deselect the "Add hashes for other networks" checkbox.

If you plan on creating a *.torrent* with multiple files, select the Directory radio button. Otherwise, keep the "Single file" radio button selected. Unless you plan on breaking your movie into multiple, smaller movies, you should keep the "Single file" button selected.

Finally, if you would like to add a comment to your *.torrent* file, you can enter it in the Comment text field.

When you have entered your information, click the Next button.

Selecting a file. Obviously, if you want to share a movie, you need to select the movie so that Azreus can create the torrent. Upon selecting your movie, click the Next button. You then have the option to configure the Piece Size for distribution. It is recommended that you allow Azureus to figure the size for you automatically. Figure 7-43 shows the File section and configuration window.

When you click the Finish button, Azureus begins the process of creating your *.torrent* file. This can take a while, depending on the size of your file and the speed of your computer. The final *.torrent* file will be substantially smaller than your original movie, which will help to distribute your movie quickly. Figure 7-44 shows a *.torrent* file with the file size of 24KB created from a 573.1MB original file.

Once you have created your torrent file, you should upload it to the tracker you entered in the first screen.

The small *.torrent* file is not your movie. It is simply a locator for your movie, so people can download it.

Figure 7-43. Finalizing the creation of a .torrent file

Figure 7-44. Displaying the small size of a torrent file

You still need to share the entirety of your movie by seeding it. Make sure Azureus remains running for a while after you've announced your torrent. This will allow Azureus to seed your movie to other people so they can continue to distribute it.

See Also

- BlogTorrent (*http://www.blogtorrent.com*; free, open source) is an easy-to-install tracker and client, distributed by Downhill Battle (*http://www.downhillbattle.org*).
- Prodigem (*http://www.prodigem.com*; free) is a tracker and hosting service for BitTorrent files.

Attend a Conference from Another Location

Sometimes you just can't make it to a meeting, a conference, or a lecture. Using Darwin Streaming Server and QuickTime Broadcaster, you can watch the action from a remote location.

Whether you are a student or a businessman, there are times when you just can't make it to a gathering. If you have a friend or colleague who can attend, you will be able to watch the action from a distant location. Furthermore, if your friend chooses to, there will be an archive that you or your peers can watch at anytime.

Configuring the Broadcast

In order to broadcast the conference, your colleague will need to download and install the QuickTime Broadcaster application (*http://www.apple.com/quicktime/download/broadcaster/*; free). At the time of this writing, QuickTime Broadcaster is a Mac OS X–only application. Once your colleague has downloaded and installed, she should launch Broadcaster as she would any other application on her computer.

After launching Broadcaster, she will see a Show Details button on the lower-right side of the Broadcaster window. Clicking on the Show Details button will reveal a panel that allows her to change the settings for the Audio, Video, and Network configuration. When the configuration details are visible, the Show Details button toggles to read Hide Details, which, when clicked, does so. Figure 7-45 shows the Broadcaster window with configuration details available.

If you want to obtain the best results, your colleague will need to know the speed of her connection to the Internet. A good way to determine the speed is to use a Speed Test, available through DSLReports (*http://www.dslreports.com/stest*). Figure 7-46 shows a speed test in progress and Figure 7-47 shows the results of the test.

The speed test will reveal the *upload* speed, which is how fast data will flow from her computer onto the Internet. In order to optimize the broadcast, the amount of data being sent (the video broadcast) will have to be less than the upload speed.

Using the presets. Under each tab there is a Preset pop-up menu. This menu provides a list of preset configurations for Modem, DSL/Cable, and LAN connections. The Audio presets provide for either Speech or Music and the Video presets provides six options for low-motion or high-motion video across modem, DSL/Cable, and LAN connections. Additionally, you can

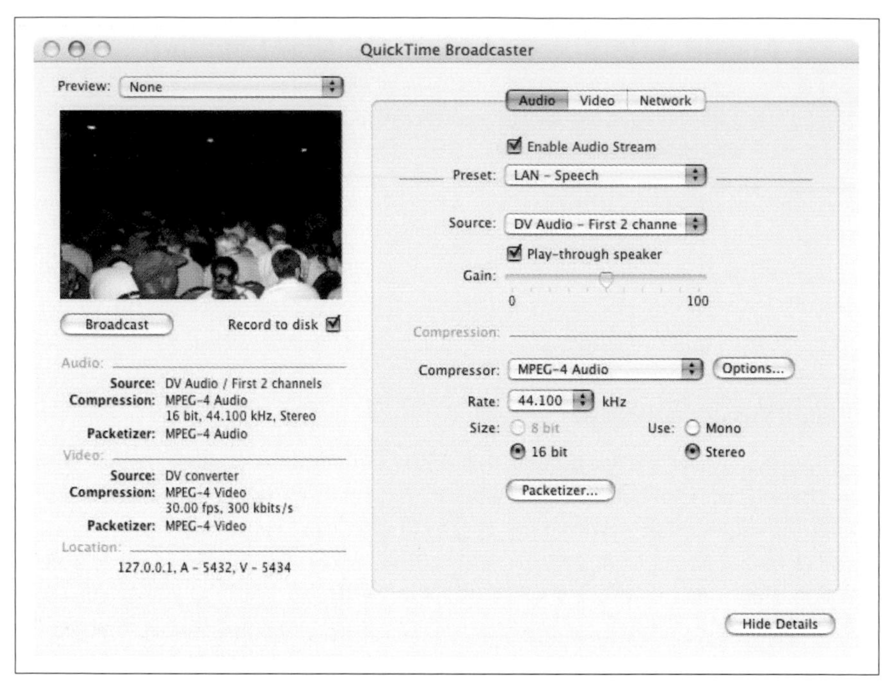

Figure 7-45. QuickTime Broadcaster configuration

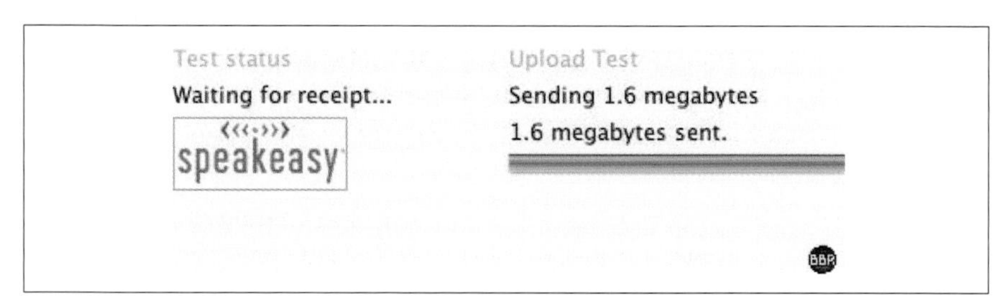

Figure 7-46. An upload test, in progress

create custom presets by altering the settings in a specific tab and choosing Preset: → Save Preset... in the pop-up menu.

Based on the upload speed of your colleague's Internet connection, use the following presets:

<56K
: Modem

>56K to 1MB
: Cable/DSL

>1MB
: LAN

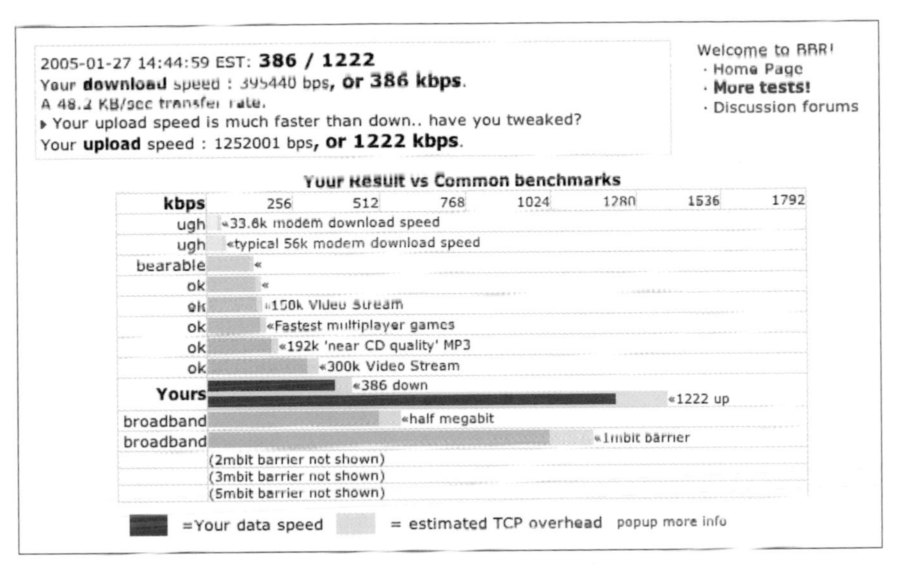

Figure 7-47. An upload speed of 1222 kbps, plenty for a video stream

For example, if the upload speed is 512K, your colleague should select the DSL/Cable preset.

Changing the audio settings. Clicking on the Audio tab in the Details panel reveals a variety of settings to optimize the audio portion of the broadcast. More often than not, the audio will be more important than the video, so these settings are very important. Your colleague should also make sure she is able to capture good audio where she plans to sit.

A good microphone, such as those available from Audio-Technica (*http://www.audio-technica.com/*) and Sennheiser (*http://www.sennheiserusa.com/*), will definitely help capture better quality audio and can be purchased from most quality distributors. In a lecture or conference setting (where a speaker is in front of an audience), you should use a shotgun microphone, while an omnidirectional microphone will help transmit better audio from a meeting. Figure 7-48 shows the Audio configuration tab in Broadcaster.

The most important settings in the Audio tab are the Compression settings. Within the Compressor pop-up menu, there are a wide variety of choices. If you are planning on viewing the conference with a program other than QuickTime, you should make sure the player is MPEG-4 compliant and that the Compressor is set to MPEG-4 Audio.

Selecting an audio compressor is a matter of personal choice. For most purposes, either Qualcomm PureVoice or MPEG-4 should provide good quality results. Depending on the selection, you can also modify the Rate, Size, and

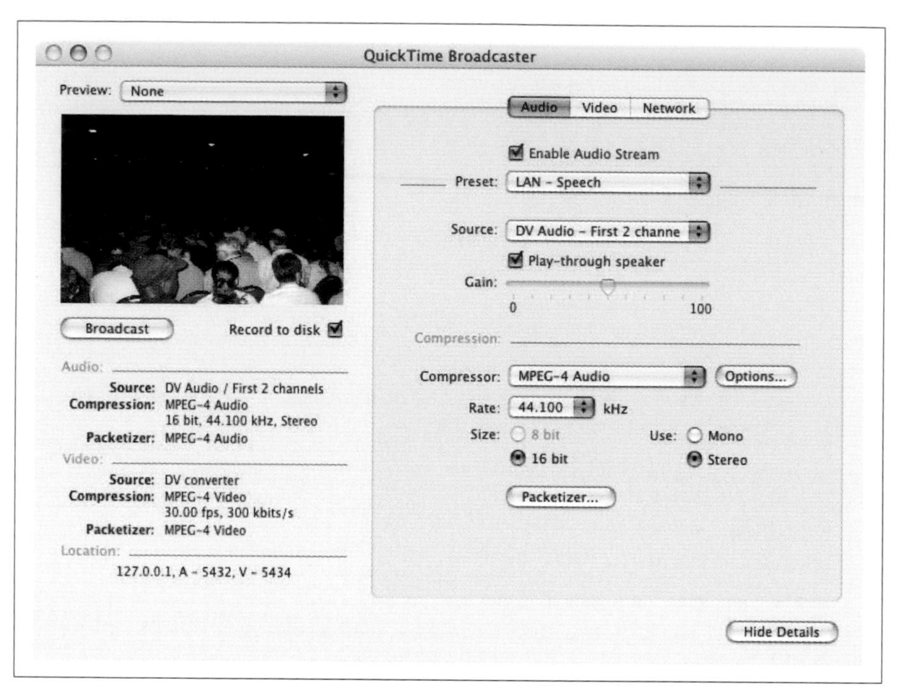

Figure 7-48. Using the LAN - Speech preset for audio.

Mono/Stereo settings. Setting these to 32kHz, 16 bit, and Mono, respectively, will result in respectable quality audio.

Changing the video settings. To change the video settings, select the Video tab. Your colleague can select any of the Presets, or customize her own setting. The Source pop-up menu will list the available video sources, in case there is more than one camera hooked up to the computer. Figure 7-49 shows the Video tab and its configuration options.

The Width and Height parameters provide control over the visual dimension of the video image. Reducing the size of the image by one-half to one-fourth will usually result in a visually acceptable image, while reducing the bandwidth required. For example, NTSC DV is 720×480, reducing the image in half would result in a 360×240 setting. For the best results, the settings should be evenly divisible by four.

Within the Compressor pop-up menu is a wide variety of *codecs* (*co*mpression/*dec*ompression algorithms). QuickTime has a standard set of codecs, but there are also third-party codecs available on the Internet. Of the standard set, Sorenson, MPEG-4, and H.263 will all provide an acceptable image. Some codecs allow for more control over their settings through the Options... button.

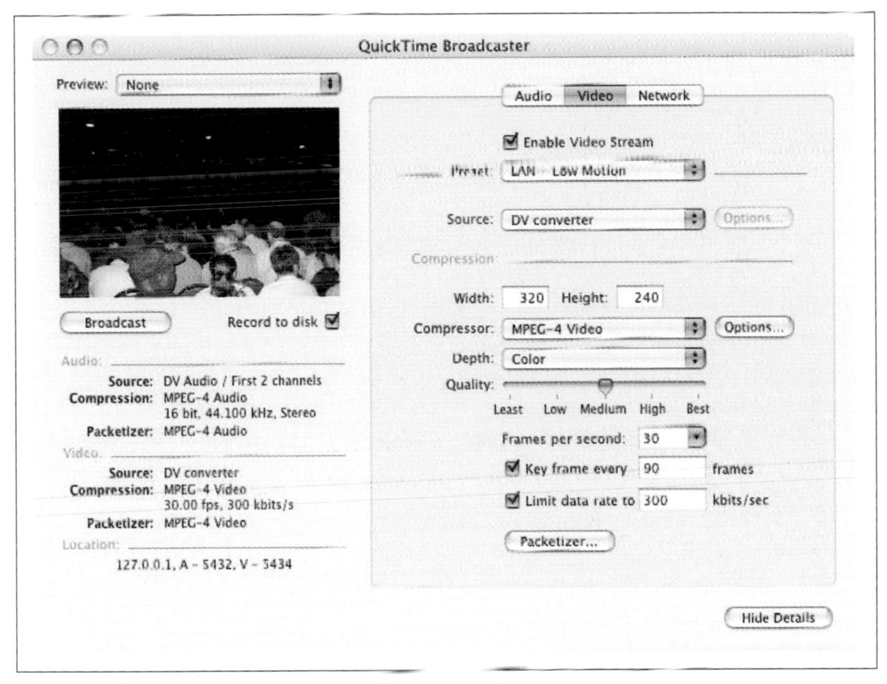

Figure 7-49. Broadcaster's Video configuration tab

The new MPEG-4 Part 10 codec, referred to as H.264 or AVC, promises a much better image at much lower data rates. If you have access to H.264, you should definitely give it a try.

The Quality setting ranges from Least to Best and will affect overall quality of the image, with the Least setting showing the most visible signs of compression. The higher quality settings will also result in a larger requirement for upload speed.

Changing the frames per second (fps) can help reduce the required upload speed. Most people can tolerate 7 to 8 frames per second, and 12 to 15 will provide the most tolerable video while keeping bandwidth requirements lower. If you have the upload speed to handle 30fps for NTSC or 25fps for PAL, then go for it!

Key frames are used by the codec to determine how often to sample a complete video frame's worth of information for use in subsequent calculations. The more often a key frame is obtained, the higher quality the video. However, the more key frames there are, the more upload speed will be required.

The key frame setting will depend on the frame rate of the broadcast, but a key frame every one-and-a-half to three seconds is a good average.

Finally, limiting the data rate provides the most control over the required upload speed. When entering a data rate, both the upload speed and the audio rate need to be taken into consideration (remember, the broadcast is both audio *and* video). I recommend being conservative when limiting the data rate, so if there is 1MB of upload speed, limit the data rate to 500kbps.

Changing the network settings. The Network tab is where the parameters of your streaming server are entered, along with *annotations*, such as copyright information. If you already have access to a streaming server, more than one person will be able to view the conference. If not, you will need to set up your own streaming server **[Hack #85]**, gain access to one, or allow only one person to view the gathering. Figure 7-50 shows the Network configuration tab.

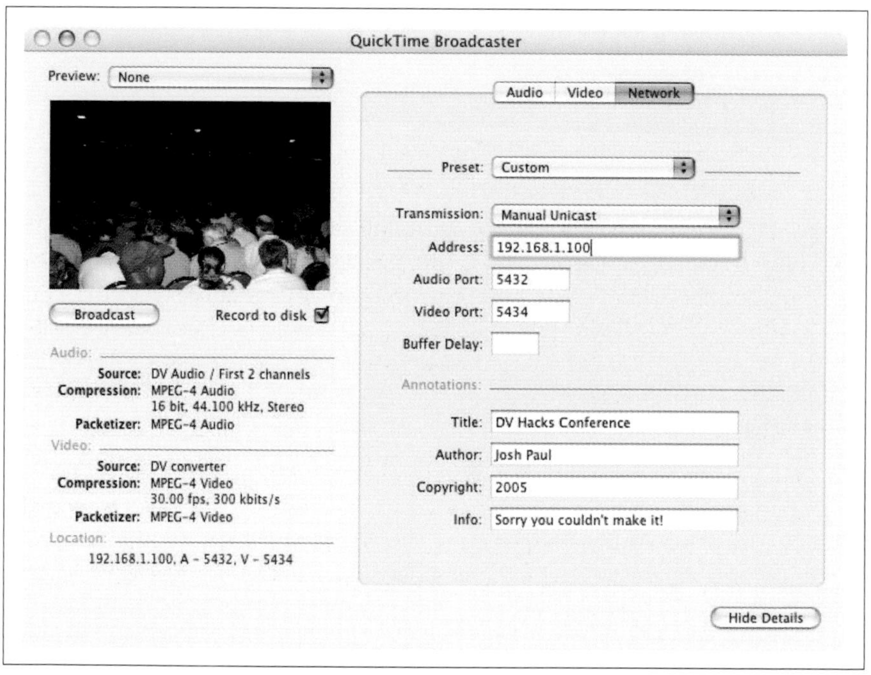

Figure 7-50. Network settings for Broadcaster

If your colleague is going to send the video stream to a single viewer (you), choosing Transmission: → Manual Unicast will enable you to access the video stream. In the Address field, enter the IP address of the *receiving* computer (yours), as shown in Figure 7-50. The Audio Port and Video Port settings can be changed, in case the receiving computer is behind a firewall, but the default values should work in most instances.

If your colleague is going to use a streaming server to distribute the video, she should choose Transmission: → Automatic Unicast (Automatic). For the Host Name setting, enter the domain name or IP address for the streaming server. The File parameter can be set to whatever you would like the broadcast to be named, for example *mystream*.

The Username and Password should be entered as you, or the streaming server's administrator, have set them up in the streaming server's configuration. Accounts can also be added to the streaming server, if individual accounts are required.

For both methods, the Buffer Delay setting provides a cache for the media, in seconds. For example, entering 7 in the Buffer Delay field would result in a seven-second delay between the time the audio and video are captured to the time they are broadcast. The delay allows for network slowdowns by making a store of media available.

Starting a Broadcast

After all of the settings have been configured, your colleague should click the Broadcast button to start broadcasting.

If you are a single viewer, you need to obtain an *.sdp* file from the broadcasting computer. A *.sdp* file can be created using QuickTime Broadcaster by choosing File → Export → SDP... and giving the resulting file a name, such as *mystream.sdp*. The resulting file can then be emailed or uploaded to a server for retrieval. After you have obtained the *.sdp* file, open it using QuickTime Player. The file contains the necessary information for QuickTime Player to accept the stream and begin playing it.

If you are using a streaming server, any clients who would like to view the broadcast can access it using the URL *rtsp://my.host.com/mystream.sdp*, where *my.host.com* is the name of the streaming server and *mystream.sdp* is the File name provided in the Network settings of QuickTime Broadcaster, followed by the *.sdp* extension.

Obviously, there needs to be communication between you and your colleague in order for the conference to be viewed successfully.

Archiving a Broadcast

If your colleague is using a digital video camera, the best way to archive the broadcast is to record the conference to tape while broadcasting. Then, after the conference has finished, digitize the recorded footage and compress it for the streaming server [Hack #83]. This approach will give the highest-quality results.

If your colleague is using a webcam, or if she forgot to bring a tape with her, she can select to Record to disk in QuickTime Broadcaster's preferences. By recording the broadcast to disk, the same video file that is sent to the streaming server will be preserved on her computer as well. She can then transfer the final video file to the streaming server.

A bug occurs in QuickTime Broadcaster when recording for long periods of time to a disk larger than 2GB in size. If you plan on capturing a video that is longer than 15 minutes in length, you should either partition a disk to have a 2GB partition or create a 2GB disk image using Disk Utility.

An archived broadcast will be available immediately upon placing the applicable file inside the streaming server's media directory.

Play Your Video on a Pocket PC

If you own a Pocket PC computer, whether it's a cell phone or an organizer, you can carry and play your videos anywhere.

The Pocket PC platform can be found in personal digital assistants (PDAs), cell phones, and palmtop computers. It's a versatile system that is capable of working with Microsoft Excel and Word documents, dealing with email, and browsing the Web. It can also play a movie, if it's encoded properly. The easiest way to encode for a Pocket PC is to use Windows Movie Maker.

Capturing Your Footage

There are two methods of creating a movie that can play on a Pocket PC, and which method you use will depend on what stage in the editing process you decide to create such a movie. If you are just starting your project and know that you will only be delivering your project to a Pocket PC device, you can select one of the Pocket PC profiles when you start capturing your footage. Figure 7-51 shows the Capture dialog box for Windows Movie Maker with the Pocket PC profile selected.

If you will be targeting other media, such as videotape, DVDs [Hack #79], or the Web [Hack #82], capture your footage at the highest resolution possible (for example, DV-AVI). By doing so, you will be able to *down-convert* your footage from its original form to a form of lesser quality. This is important,

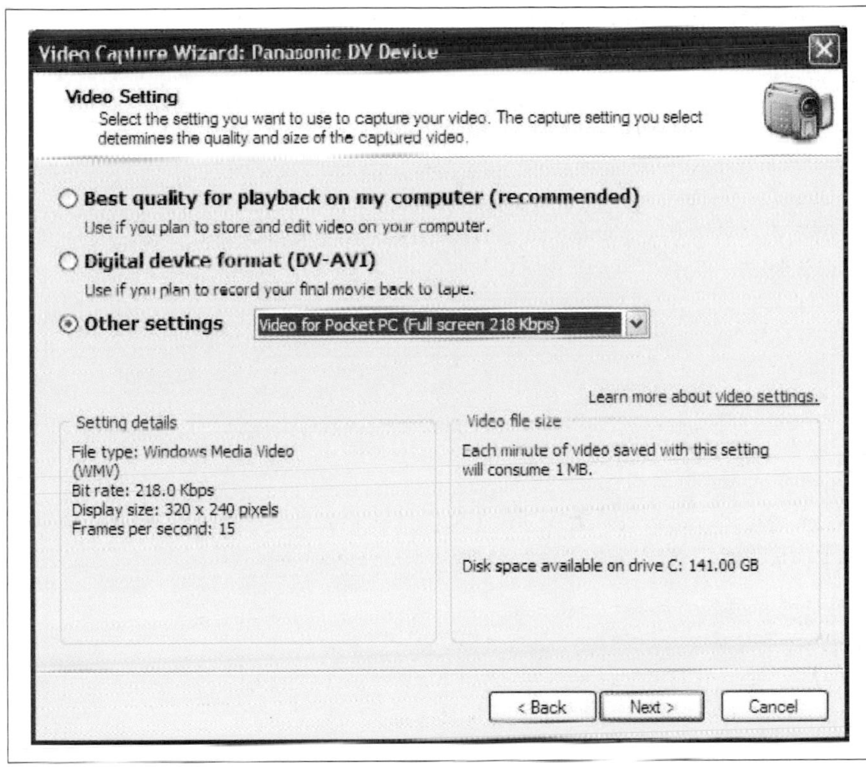

Figure 7-51. Pocket PC profile selected

because if you try to *up-convert* your footage, from a form of lesser quality to one of higher quality, you will get disappointing results.

Try to imagine a full glass of water representing your highest quality video. If you pour it into a smaller glass (down-convert), your new glass is still full of water, and if you were to hand it to someone, they wouldn't question why you gave it to them. On the other hand, if you have a small glass of water and pour it into a much larger glass (up-convert), and then hand the larger glass to someone, they might wonder why you've given them a partial glass of water. Did you drink it and want them to hold it? Should they go wash it? Ultimately, it is unsatisfying for them and doesn't fulfill your intention.

After you have captured and edited your footage, you need to export it.

Exporting for Pocket PC

From the main window of Movie Maker, choose Finish Movie → "Save to my computer." This will bring up a dialog box asking you to name your exported movie, as well as to select a location on your hard disk to save it. After providing the information, and clicking the Next button, you will have the opportunity to select the profile for saving your movie. If you initially captured your footage using the Pocket PC profile, you can keep the default setting and click the Next button.

One the other hand, if you've captured your footage at a higher quality, such as DV-AVI, you need to click the "Show more choices..." button and then select the "Other settings:" radio button. By doing so, you enable the drop-down list where you can select a profile. As a default, there are three Pocket PC options, so you should select the one that best fits your needs. Figure 7-52 shows the "Video for Pocket PC (Full screen 218Kbps)" profile selected.

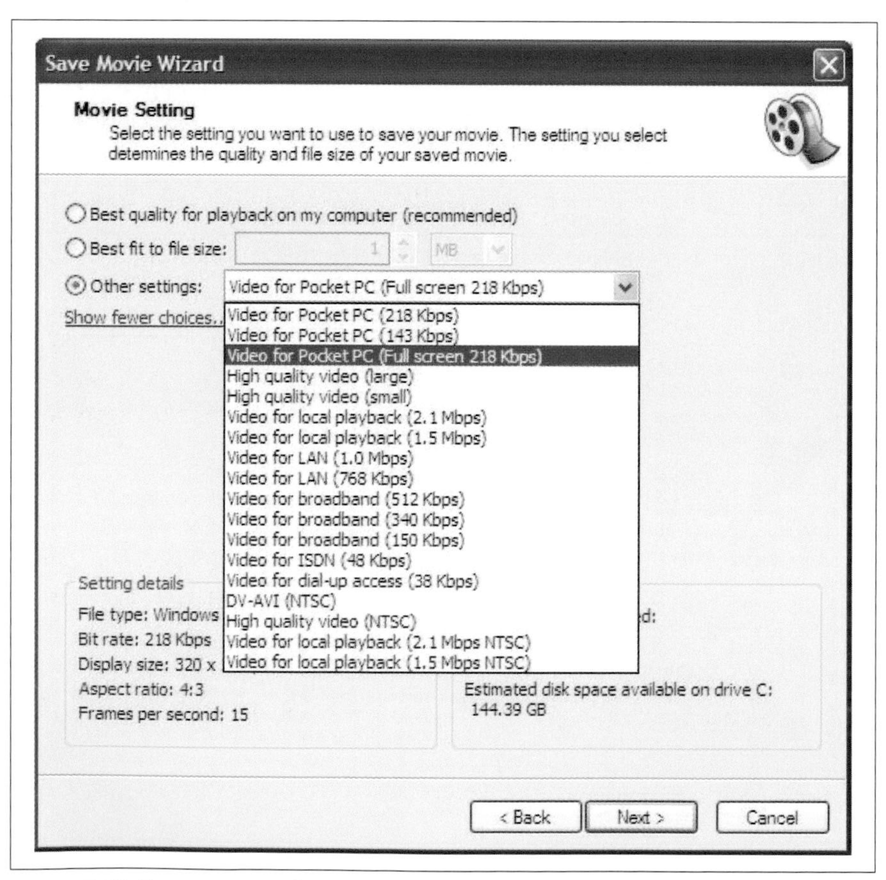

Figure 7-52. Exporting a Pocket PC video

After selecting your profile (if appropriate) and clicking the Next button, Movie Maker handles the details for you and begins processing your final movie. When it is complete, you have the option to immediately open the video in Windows Media Player, to ensure the quality is sufficient for you. If it isn't, you will need to go back and try one of the different Pocket PC profiles.

Selecting a Codec and File Format

If you don't have Windows Movie Maker, then you need to transcode your video **[Hack #29]** into a format that your Pocket PC can play. The standard Pocket PC distribution includes Windows Media Player, which can play practically any *.wmv* file.

There are also a lot of third-party media players that can open a wide variety of file formats and codecs, including MPEG-4 and QuickTime. Here are just a few:

- Beta Player (*http://betaplayer.corecodec.org*; free, open source)
- PictPocket Cinema (*http://www.digisoftdirect.com*; $39.95)
- PocketTV (*http://www.pockettv.com*; free–$49.95)

Transferring Your Video File

There are two primary ways to get a video file onto your device. The first is to use ActiveSync, which should have been included with your device. If it wasn't, then you can obtain it directly from Microsoft's Windows Mobile download site at *http://www.microsoft.com/windowsmobile/downloads/default.mspx*.

The other way is to transfer the video to your device is to use a memory card. By using a memory card reader, you will be able to copy the video directly to the card and then insert it into your device. The advantage to this approach is that it is much faster than using ActiveSync, especially for large files.

So, if you have that masterpiece you've always wanted to show Spielberg, you can throw it on your Pocket PC...and should you ever meet him, you'll be prepared.

Play Your Movie on a Cell Phone

Some cellular phones allow you to watch video on their tiny screens. Here's how you can convert videos from your computer and play them on your cell phone.

Digital video takes a lot of hard drive space. In its raw, uncompressed state, DV averages 14GB per hour of footage. In order to have video play on a cell phone, the digital video must be compressed into a smaller file.

Using QuickTime Pro

There are a lot of computer programs on the market that help you import and export digital video. One program that many people overlook is QuickTime Pro, from Apple Computer, which is reasonably priced. QuickTime runs natively on Windows and Macintosh and can run on Linux and Solaris when used with CrossOver (*http://www.codeweavers.com*; $39.95).

A nice feature of QuickTime Pro is that it can export movies that can play on cell phones. The format for these movies is either 3GPP or 3GPP2. Both formats are standards created by the 3rd Generation Partnership Project (3GPP; hence the format's name). Exporting to 3GPP was made available with the Version 6.3 update for QuickTime.

If your editing system is QuickTime capable, you can export your timeline to 3GPP using the QuickTime conversion engine.

Upgrading to QuickTime Pro. Even if you have installed QuickTime on your computer, you most likely do not have the Pro features enabled. To enable QuickTime Pro, you need to purchase a license key from Apple at *http://www.apple.com/quicktime/buy/*. After you've stepped through the payment process, you will receive your personal key to unlock the Pro features.

In addition to being able to create movies for cell phones, you'll also be able to do some minor editing [Hack #39], create skins [Hack #76], and play movies in full-screen mode. In order to enable the features, you will need to enter your personal key in the QuickTime Registration window.

Opening a movie. From within the QuickTime Player program, open a movie you would like to convert. For your initial tests, I recommend converting a short movie, because the process can take a while. Choose File → Open Movie in New Player... and then locate the movie from your computer.

Converting a movie. Once the movie is open, you want to export it. You can begin the process by choosing File → Export..., which brings up the dialog box shown in Figure 7-53. Name your movie whatever you want and then select Movie to 3G from the Export pop-up menu. Figure 7-53 shows the selected settings in the dialog box.

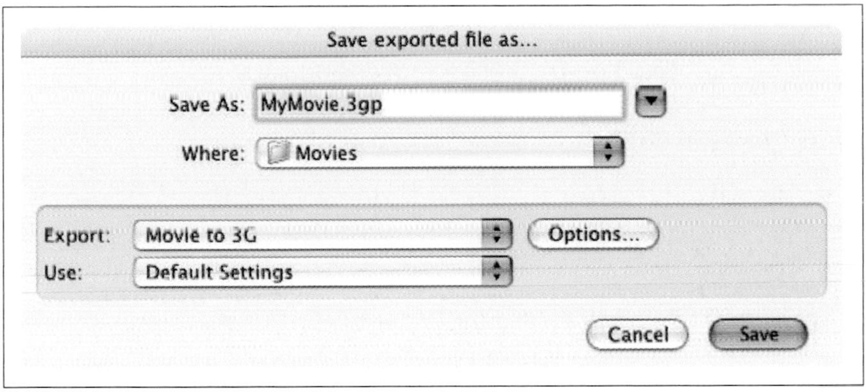

Figure 7-53. Exporting a movie file to a 3G-formatted movie file

Before clicking Save, you should click the Options... button, if only to see what's available to you.

Changing the default settings. The options for 3G export include settings for Video, Audio, Text, Streaming, and Advanced. Each of these will affect the final size of your movie file, as well as the file's conformance to the 3GPP standard.

If you change any settings and your file is no longer within the 3GPP specification, you will be able to tell by looking at the Conformance list item. Figure 7-54 shows the 3G Export Settings window.

Notice the File Format drop-down menu at the top of the 3G Export Settings window. This menu contains the following items: 3GPP, 3GPP2, 3GPP (Mobile MP4), 3GPP2 (EZmovie), and AMC (EZmovie). By choosing any of these menu items, you will enable or disable certain features. I recommend spending some time simply browsing around the tabs in this window as you change the File Format from one selection to another.

In particular, take notice of the Advanced tab, which allows you to place restrictions on the movie. For example, if you select one of the EZmovie formats, you can set the movie to play only on the handset where it is downloaded.

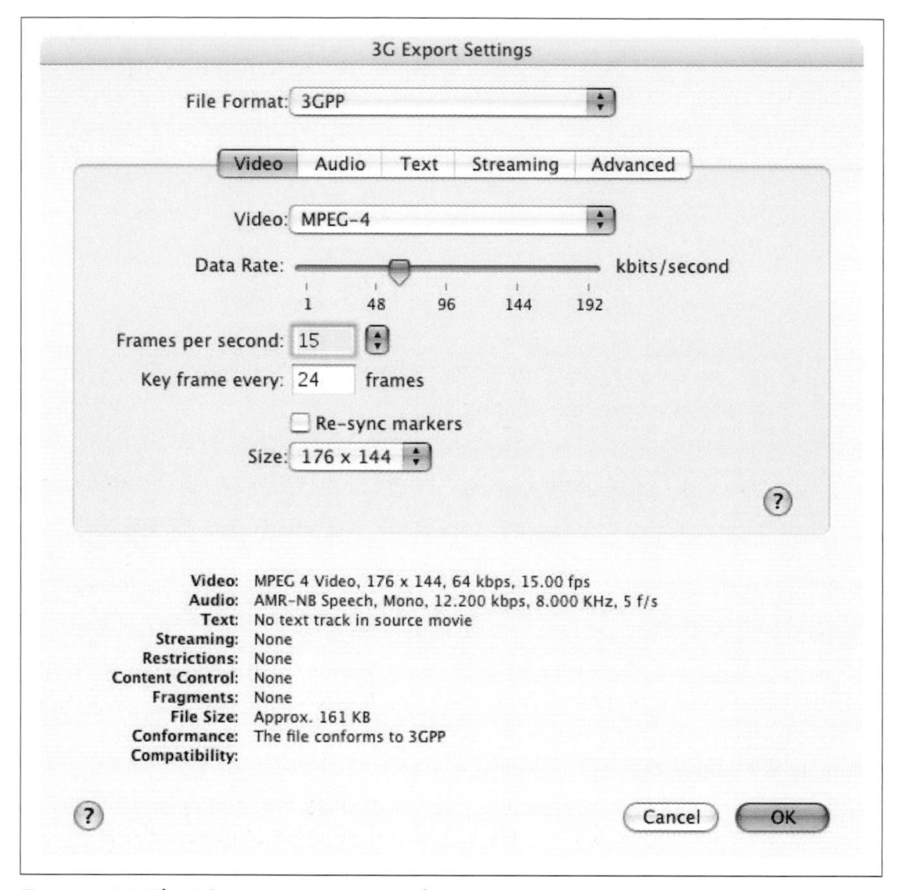

Figure 7-54. The 3G Export Settings window

If you want a higher-quality movie, the greatest impact will come from giving your file a higher data rate. The drawback of having a higher data rate, and therefore a higher quality movie, is the file size. Because you will be distributing this movie to a cell phone, you want to keep the file size small, so you will have to experiment with various data rates and determine what is an acceptable *quality-to-file size* ratio.

The new MPEG-4 Part 10 codec, referred to as H.264 or AVC, promises a much better image at much lower data rates, and is an international standard. Many cell phones can, or will soon, play H.264 video.

When you are finished, click OK. You will then be able to save your movie. You might want to experiment with the different settings to find out how they affect the file size, the quality of the image, and the quality of the audio.

Keep in mind that most cell phones do not have an abundance of storage. In my experience, people will tolerate a poor quality image if the audio is good, but not the other way around.

Your movies should be as small in size as possible, without making the movie unacceptable to watch.

Distributing Your Movie

So, how in the world do you get your movie onto a cell phone?

Email
: If your phone can receive email, you can send the movie to your phone using your desktop email client.

Disk
: If your phone has a memory disk, you can copy the movie to disk using the appropriate drive.

Bluetooth
: If your phone and computer both have Bluetooth, you can copy the movie wirelessly.

Web
: If your phone has web access, and you have a web site, you can make the movie available for download via your web site.

If you have a streaming server **[Hack #85]** that's capable of streaming 3GPP content, you can just place your 3GPP movie file on the server. You can then view your movie on-demand.

Playing Your Movie

After transferring your movie to your cell phone, launch your cell phone's movie player and play your new, very small (but viewable!) movie.

CHAPTER EIGHT

Random Fun

Hacks 90–100

Okay, so most of these chapters have been pretty serious and focused on creating, finishing, and distributing a video project. Although the hacks in this chapter can be used for real, functional, and even business purposes, they're primarily here to provide a little fun and inspiration. (Though that's not to say the other hacks in this book aren't fun and inspirational.) Enjoy!

Record a Streaming Video

VCRs and PVRs have enabled us to record live video content. Using an application called VLC, you can do the same with streaming video.

You can't always sit and watch a streaming video from beginning to end. When the stream is live, this becomes a problem, because you can't pause the stream and have it start where you left off. Using VLC (*http://www.videolan.org*; free, open source) you can save the stream and watch it at your convenience.

Finding a Stream

Every streaming video has an Internet address associated with it. In order for VLC to locate the streaming video, you need the entire address. This address often begins with either *rtp* or *rtsp* and is formatted much like a web URL.

The complete address will most likely include the type of movie you are downloading as a file extension. For example, an address of *rtsp://www.domain.com/some_movie.mp4* might be a complete address for a streaming video file named *some_movie.mp4*. Although not completely necessary, knowing the file extension will help you later in the process.

After you have the address for a streaming video, you'll need to configure VLC to open the stream and save it to your hard drive.

Opening a Stream

While running the VLC application, choose File→Open Network Stream... (Windows) or File → Open Network (Mac) to open a new window in which you can enter the necessary information for VLC to locate the streaming video. You will be opening a streaming video, so select the Network tab, if it isn't already selected.

First, select a protocol, depending on the address you are using. If the address starts with *rtsp://*, select the RTSP button. Then, enter the streaming video's address in the URL or Address text field. Although you might be able to enter the address without selecting a protocol, it might not always work. Figure 8-1 shows VLC being configured to open an RTSP stream from the address of *rtsp://192.168.1.102:554/sample_300kbit.mp4*.

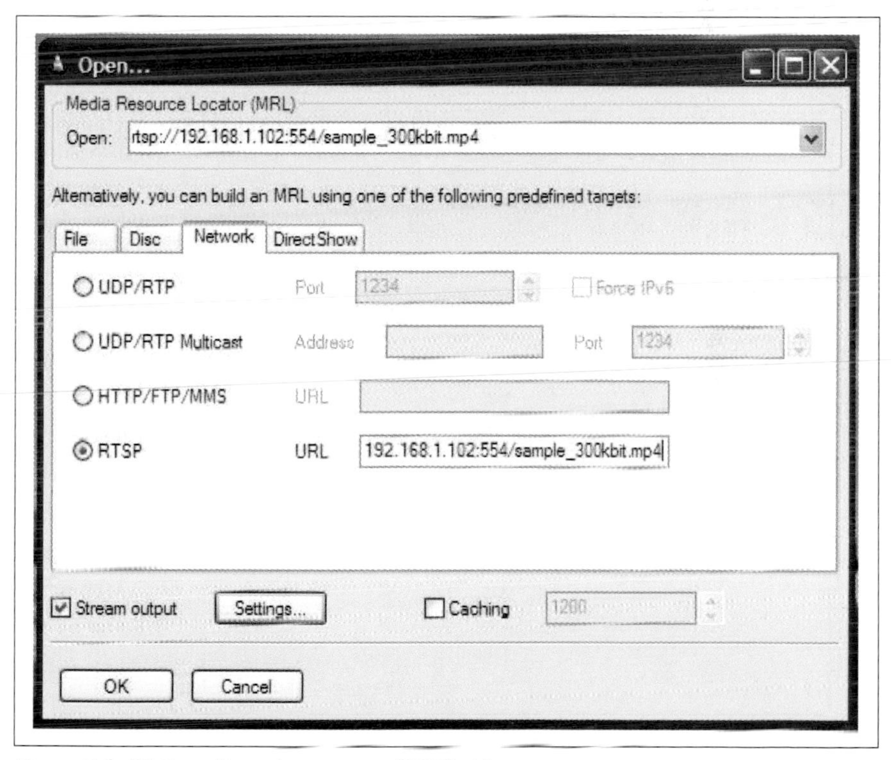

Figure 8-1. VLC configured to open an RTSP video

Once you've entered the address, you need to configure VLC to save the stream to a file. To do so, check "Stream output" (Windows) or "Advanced output" (Mac). This will enable the Settings... button.

Saving a Stream

When you click the Settings... button, a new window opens, allowing you to configure how VLC handles the incoming video. Because you want to save the video, as opposed to relaying it, you need to click the File checkbox/button. If you would like to view the video while it is being saved to disk (and I recommend doing so), also click the Play locally checkbox.

To save the incoming video, you need to set a filename. Using the Browse... button allows you to navigate your hard drive and locate where you would like the video file to be saved, as well as provide a name for the file. A good approach is to simply name the file the same as it is named in the address. For example, if the file is named *some_movie.mp4*, name your saved movie *some_movie.mp4*.

After setting the filename, select an encapsulation method that is compatible with the file format and codec of the streaming video. This can be a little tricky, so it will help if you are familiar with how to ascertain the file format from the file's extension [Hack #77]. Selecting the same type as the stream will ensure the best results. Figure 8-2 shows the available configuration options when using VLC.

When the configuration is complete, simply click OK and then click the OK button on the Open... window. VLC will proceed to process the stream and save it to your hard disk. If you have selected to "Play locally," the video will also play on screen.

Watching a Saved Stream

Watching a saved stream is exactly like watching any other video file. Depending on the configuration of your computer, VLC might not be the default application to open certain types of video files. If you find that a VLC saved file won't open in your preferred video player, try opening it using VLC. More often than not, it will open without any problems. Although other applications, such as Windows Media Player or Apple's QuickTime, can open some of the files you save, you will have the best results by opening VLC-created files using VLC.

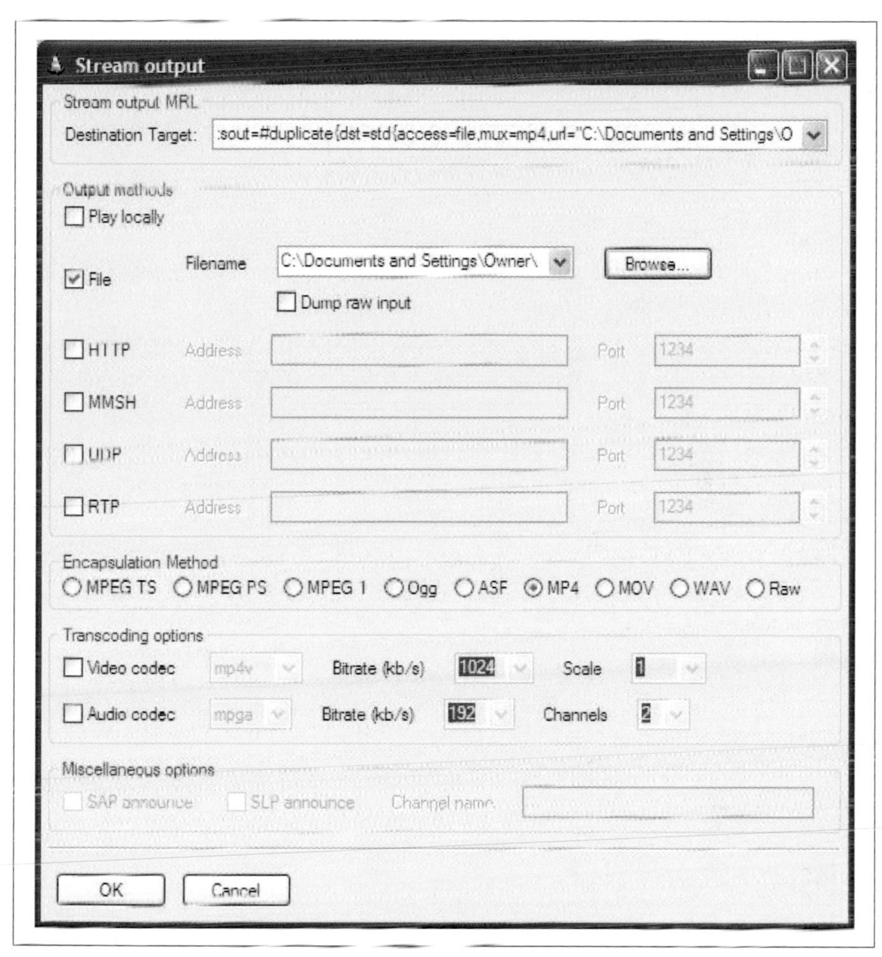

Figure 8-2. Configuring how VLC to save a movie

Create a Movie Using LEGOs

Using stop-motion software and LEGOs, you can create a movie on a very low budget.

Stop-motion movies involve taking inanimate objects, such as clay models, and patiently taking a picture, moving the model slightly, taking another picture, and so on until a complete movie has been produced. "Gumby and Pokey," "Wallace and Gromit," and "Celebrity Deathmatch" are examples of productions created using claymation.

Well, those of us who are artistically challenged refuse to be left out of the fun. We can make stop motion movies too, only we'll use plastic bricks instead of clay. Enter my favorite childhood toy...LEGOs.

Although I use LEGO bricks, you can substitute any other toy that consists of interlocking bricks to create a "brick film."

Gathering Your Tools

You should probably determine the genre of movie you would like to produce, since a cowboy will appear slightly out of place in a Science Fiction movie (unless of course, you've decided to make a SciFi Western). If you already have some LEGOs available, you might as well take them into consideration when deciding your genre.

A great side benefit of using LEGOs to create a movie is they are much easier to direct than human actors . . . and they don't eat as much.

Depending on your ambition and your budget, you should create a storyboard **[Hack #6]** to both clarify what you are going to be shooting and help you gather the necessary pieces for your production. You will also need a digital video camera, some stop-motion software, and a good location where you can produce your movie. Creating a stop-motion movie can be time-consuming, so finding a location that will not be physically disturbed is preferable.

Although stop-motion software is not essential, it is extremely helpful in producing this type of movie. Stop-motion software is commercially available for both Macintosh and Windows. For Mac OS X, Boinx software distributes iStopMotion. For Windows, Stop Motion Pro is available from, well, Stop Motion Pro.

Boinx
: Publishers of iStopMotion for Mac OS X (*http://www.istopmotion.com*; $39.95)

Stop Motion Pro
: Publishers of Stop Motion Pro for Windows 98 SE or ME, XP, and 2000 (*http://www.stopmotionpro.com*; $249)

Creating Your Environment

The location you choose to produce your movie can make the difference between success and failure. If your friends, family, and even pets are going to be wandering through while you are trying to shoot, your lighting will probably fluctuate and it is quite possible for your entire scene to be destroyed. Just imagine Fluffy, your beloved cat, executing a sneak attack on your lead actor and carrying it off to who-knows-where.

A background image can be a nice addition to your movie. In essence, you will be creating a diorama. If you are planning on shooting multiple scenes, your production value will go up dramatically if you change the background image when your scenes change.

Lights!

Just because you are shooting LEGOs, that doesn't mean you can avoid lighting your scene. You can use lamps from around your home, Christmas lights, nightlights, or any other light source you deem acceptable. Whatever your choice, once you set your lights for a scene, don't move them!

An exception to this rule is if you are trying to create a lighting effect, such as the sun setting.

Believe it or not, you can learn a lot from lighting LEGOs. Granted, your scenes are on a much smaller scale than working with people, but the principals of lighting still apply.

Camera!

After you have created your environment and collected your lighting, you should set up your camera. Whoa! Those little people sure do look big in the camera! You will most likely have to adjust your scene to take into account how your camera "sees" the environment.

You'll probably discover that what you *think* is going to appear on camera doesn't always translate from the real world to the on-camera world. You might have to position characters closer together, angle them slightly away from each other to suggest to the camera that they're facing each other, and even build things such as desks smaller than you initially anticipated. Oddly enough, all of these adjustments are similar to those you need make when shooting a movie.

Action!

Gather your patience, because making a stop-motion movie takes time. When making your movie, you are going to capture each frame...one...by... one, as shown in Figure 8-3. If you are going to play your movie on a television set, you will need to capture 30 frames for each second of action (25 if you will be recording for PAL). So, using basic math skills, you will discover that for a three-minute movie you will have to capture 5,400 frames of action. Think Zen.

Figure 8-3. Only three frames necessary to create this much action

Both software programs provide a feature called *onion skinning*. This feature allows you to see previous frames you have captured so that you can make sure that you do not capture the same frame more than once and you can keep your movements consistent. Onion skinning works by blending previous frames with the current frame, as shown in Figure 8-4.

Although it takes time to get used to viewing an onion-skinned image, the reference point it provides is well worth it.

Staying Inside the Lines

A nice feature offered by the software programs mentioned previously is that they offer an *overlay*. The overlay feature draws a semitransparent box over your image, roughly 15% from the outside edge of the frame. Stop Motion Pro refers to this overlay as TV Safe Zone and iStopMotion (correctly) calls it Action Safe.

The Action Safe area is where you are guaranteed your viewers will be able to see what is occurring on screen. If you plan on showing your movie on a television, you should take into account this area and not record any action that takes place outside its boundary. If you are planning on distributing your movie purely online, you won't have to concern yourself with this restriction. Figure 8-5 shows StopMotionPro with TV Safe Zone enabled.

Some television sets, primarily older ones, have a curve to the screen that causes images at the edge to be cut off and unable to be seen. Because each manufacturer designs their televisions differently, each television set will

Figure 8-4. Onion skinning to provide a reference to previous frames

Figure 8-5. TV Safe Zone, represented by the outside rectangle

have a different point where the image becomes cut off. Therefore, whether you're creating a stop-motion video or a feature length movie, a general rule is to stay inside of the TV Safe Zone area.

HACK #92 See Through Walls

A portable wireless video camera, combined with a portable TV, provides remote viewing for times when you really could use an extra set of eyes.

X10 Corporation's XCam2 (*http://www.x10.com/products/vk45a_how.htm*, $70) is a wireless color video camera that works by broadcasting its signal to a base station that's connected to a TV, VCR, or PC video digitizer. It's a handy device, but I've discovered a practical use that makes it almost indispensable for the home handyman.

I received an XCam2 battery pack for free (*http://www.x10.com/products/x10_zb10a.htm*; $20) when I bought the camera, but I hadn't put it to much use. The battery pack connects to the camera and turns it into a portable broadcaster because you don't have to find a power outlet for the transformer the camera normally requires.

> The XCam2's broadcast range is about 100 feet, but the frequency it uses (2.4GHz) is the same as microwave ovens, cordless phones, and WiFi equipment, and interference can reduce its range. Be prepared to do some channel-juggling (the XCam2 provides two channels to choose from) until you find the best selection for cohabitation with the rest of your gadgets.

I use the battery-powered XCam2 when I'm working alone and running wires for automation, such as a sensor wire for an alarm system, and I can't see what's happening on the other side of the wall or I don't know if my fish wire is going where I want it to. Sometimes, I mount the camera on a small tripod for easier positioning (*http://www.x10.com/products/x10_zt10a.htm*; $17).

At the receiving end, the XCam2 receiver is connected to a DC-powered, five-inch color LCD TV that I purchased online for about $130. The receiver and TV are in a large plastic project box in which I've drilled a hole to allow the television's antenna to protrude so that I can adjust it easily for best reception. A metal hanger on the box lets me hang it from a rafter for hands-free viewing.

Now, I have an extra pair of eyes that I use to view anything in the house, from anywhere I want. It's the greatest tool I've ever had. When you need to adjust the dampers from your HVAC system, tape some strips of tissue paper over the air vents in your house. To adjust a particular room, place the camera so that it can see the vents, and then watch the tissue to judge the amount of air that's coming through, all without having to leave the furnace area to check your work.

The same technique also is useful to identify circuit breakers and outlets. Put a lamp on the circuit you want to turn off, and then place the camera so that it's pointing at the lamp. Take the monitor with you to the breaker box and you will know when you have turned off the correct breaker.

In addition, I use the XCam2 when I need advice about how to best tackle a project or solve a problem. Instead of describing the situation, I just take my camera and transmit the images to my father, via the Internet, using X10's free XRay Vision software (*http://www.x10.com/products/xrayvision_software5.htm*) and a video-to-USB adapter (*http://www.x10.com/products/x10_va11a.htm*; $70). Now, I have both an expert available and a remote set of eyes!

—*Arthur J. Dustman IV*

Rental-Car Tips and Other Auto Hacks

A video camera can help you prove that you didn't put those dents in a rental car. And if you do get in an accident, it can document what really happened.

As you've probably figured out, many video camera hacks are also appropriate for camera phones. What makes your camera phone unique is that you probably have it with you in some odd situations, such as when you're renting a car or find yourself involved in a fender bender.

Rental-Car Checkout

At most rental-car shops, an agency rep walks around the automobile with you before sending you off into the world. This is a great time to pull out your camera phone and photograph or video any scratches or dents he points out. Be sure to get his hand in the picture to make an identifying element that is unique to that car.

Then, take a few more shots: one from the front, looking down one side of the car, and another from the back, looking up the other side. Be sure to get the license plate in at least one of these pictures. Also, get one full shot of the agency rep standing next to the automobile. Work quickly, so that you don't interfere with the regular flow of business.

When you return the car to the lot, take your basic shots again to prove that you returned it in good shape. Be sure to get a least one shot that shows where you are, and if there's an agent accepting the return, get him too. This documents that you returned the car in good shape.

Fender-Bender Documentation

Nobody likes being involved in a fender bender, but they happen all the time. After you've interacted with the other parties and have exchanged insurance information, pull out your camera phone and snap a few shots like the one in Figure 8-6.

Figure 8-6. Quick documentation of a fender bender

First, record the overall scene, including the placement of vehicles, damage incurred, debris in the roadway, skid marks, and anything else that looks important. Also try to shoot the point of impact, especially if one of the cars crossed a centerline or barrier.

You can even be a good citizen and take some pictures as a witness to an accident you were not involved in. Then, contact the police department and let them know what you have.

Timestamp

In any of these scenarios, send at least one shot via email to yourself to create a reliable timestamp of the event. With rental cars, you can prove that on this particular day this car was in good shape. As for the fender bender, by sending an email minutes after the event, you can help establish when the accident actually happened.

Nobody wants to have rental-car hassles or incur damage to their personal auto, but if trouble arises, your camera phone can help establish the facts and help resolve the situations quickly and accurately.

—*Derrick Story*

Save Your Presentations to DVD

With a little forethought, you can save your business presentations to DVD, as well as offer them online.

You've just made the greatest presentation of your career. The members of the board are applauding, and the Chairman is patting you on the back and offering you the hand of his eldest child in marriage. But there's a little nagging voice in your ear: you didn't videotape the presentation, so there's no record of your triumph.

Fortunately, it doesn't have to be that way. With a little planning, you can make a permanent record of your career triumph that you can send to friends, family, and people who couldn't make it. And you can also offer the presentation online so that your colleagues can relive the moment when you crushed that know-it-all from accounting with your devastating analysis of those third-quarter sales figures.

Solving Life's Little Problems

You're probably thinking that all you need to do is to stick a camcorder in the back of the room and hit Record. You could do that; the only problem is that the resulting video will look sub-par. Think about it: you're trying to record a screen in a darkened room filled with people, so it's not surprising that the results generally turn out to be disappointing.

If the PC that you are presenting from has a Video Out port, you could connect the computer to the camcorder, but again, the results are often disappointing; the quality of the presentation often suffers after it has been compressed and recompressed several times. If you use PowerPoint to make your presentations, Microsoft offers a free application, called Microsoft Producer (*http://www.microsoft.com/office/powerpoint/producer*; free), that can export them to a web page, but that application can't save a presentation to DVD.

So, in this hack, we'll make a high-quality video of the PowerPoint presentation on the PC and then combine this with an audio recording of the presenter. That way, the viewer gets to see the slides and hear the presenter talk.

Dealing with Audio and Video

First, let's handle the audio portion. If the presentation is to a live audience, connect the line output of the PA system to the audio input of your camcorder and record it. If this isn't possible, stick your camcorder somewhere close to the presenter, so the microphone picks up her voice. If you aren't presenting to an audience, you can record the audio using a separate microphone (or the one built into your camcorder) in a quiet room while you step through the presentation.

If you want to capture the presenter in video as well, set the camcorder to take a tight shot of her making the presentation, but don't worry about capturing the presentation itself. We are going to capture and add that in a bit.

Next, capture this video onto the PC and save the audio as a separate file, if the video-editing application you use can do this. If it can't, use an application such as VirtualDub (*http://ww.virtualdub.org*; free, open source) to save the audio as a WAV file. In VirtualDub, load the video and choose File → Save WAV. This strips out the audio from the video and saves it as a separate file.

Capturing PowerPoint to Video

To capture the presentation itself to video, we are going to use a screen-capture application; I use SnagIt from TechSmith (*http://www.techsmith.com*; $39). SnagIt is an application that can grab pretty much anything that appears on the computer screen. There's a fully functional 30-day demo available from the TechSmith web site, so you can try it out before spending any money.

Instead of using SnagIt, Mac OS X users can use Snapz Pro X from Ambrosia Software (*http://www.ambrosiasw.com/utilities/snapzprox/*; $69) to accomplish a screen capture.

Using SnagIt. The SnagIt application can capture a section of the screen and record it to a video file. Open your presentation application and set it to show the presentation in a window. From within PowerPoint, choose

Slideshow → Set Up Show and set the show type to Browsed by an individual (Window). This will stop the presentation from taking up the whole screen; we don't need it to do that. Start the presentation and leave the presentation application running in the background while you open SnagIt.

Once SnagIt is running, as shown in Figure 8-7, select to record a video of the screen-capture profile. Then, select Fixed Region from the input drop-down box in the Capture Settings section. This sets SnagIt to grab a certain section of the screen, but we need to tell it how much to grab. Click the input properties button (the three dots to the right of the Input menu) and enter the values of 720×480 for creating an NTSC DVD, or 720×586 for a PAL DVD.

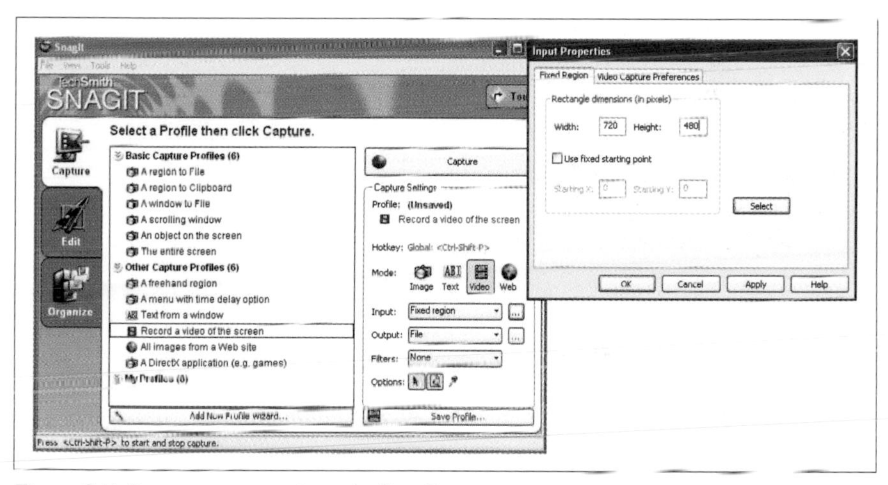

Figure 8-7. Screen capture settings in SnagIt

From the Input menu, make sure that the Include Cursor option is *not* selected. That way, the cursor won't be recorded to our video. Once you've configured your settings, click the Capture button.

SnagIt will highlight the area of the screen that will be captured with a red box. Move the box over the presentation program window and left-click to select the area to grab. They won't match up, but you can resize the PowerPoint window by simply dragging the corners until the presentation just fits inside the dotted line. Figure 8-8 shows SnagIt about to capture the presentation.

Starting the video capture. To start capturing the video, click the Start button. Press the left arrow on the keyboard to step through the slides in time to the audio you produced earlier. Once the presentation is done, press Ctrl-Shift-P to stop the recording. You can now go back and watch the captured

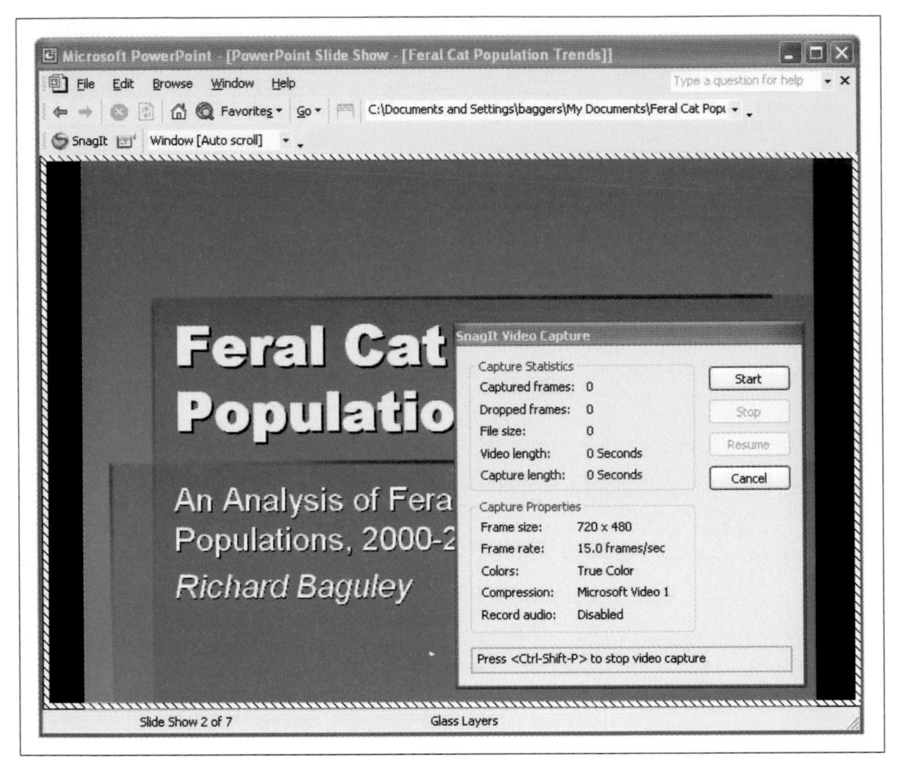

Figure 8-8. Capturing the presentation in SnagIt

video in SnagIt. If you're happy with it, click Finish and save the file. If not, press Cancel and try again.

To make the capture process as smooth as possible, close all other applications except PowerPoint and SnagIt, because the process of displaying a PowerPoint slide while simultaneously capturing the screen takes a lot of memory. It's also a good idea to keep the PowerPoint presentation as simple as possible: complex animations and transitions often don't capture very well.

Finalizing the DVD

Once your capture is complete, go into your video-editing or DVD-authoring application and import both the video and audio files, matching them up so that they start simultaneously. If the application you use doesn't support having separate audio tracks, look for an option to add background music to your video; you can import the audio as a background track. Now you're ready to write the resulting video out to DVD—ready to send out to friends, colleagues, and admirers the world over.

If you have a really complex presentation you want to put onto DVD, TechSmith offers another application, called Camtasia Studio (*http://www.techsmith.com*; $299), which automates the whole process. It's expensive, so it's intended only for those who need to convert many presentations.

Many applications also allow you to save the resulting video in a format that you can stick on a web page. Some of these applications, such as Windows Media, Real, or QuickTime, can save the video as a movie file. You can then transcode the movie file **[Hack #29]** and distribute it as a streaming movie **[Hack #83]**, a download, via email, or even to a cell phone **[Hack #89]**.

Hacking the Hack

If you really want to get fancy, you could go a step further and add the video that you took of the presenter using a picture-in-picture effect. This is the sort of thing that you see on the news, where a small video plays behind the anchors shoulder. And if you want to get even fancier, you could switch between the two, so the viewer can see both your beautiful face and the presentation in all of their glory.

—*Richard Baguley*

Watch TV on a Palm Pilot

Transfer prerecorded television shows, or any video, to a Palm Pilot.

ReplayTV (*http://www.replaytv.com*; $99–$799) is my PVR of choice for a number of reasons, not the least of which is that by using some free software it is possible to download recorded television shows onto my Mac. Once I get those shows onto my Mac, it is a reasonably straightforward process to get them onto my Treo 600 (*http://www.handspring.com*; $299–$479). Watching prerecorded shows can make the long commutes to and from work pass by a lot more pleasantly. It's also a cool way to show off my Treo!

Downloading Video from ReplayTV

The ReplayTV service distributes software that is able to upload pictures to a ReplayTV PVR (*http://www.digitalnetworksna.com/dvr/photosoftware.asp*; free), but a couple of third-party developers have created some free applications that can not only upload pictures, but also download captured video from a ReplayTV PVR:

DVArchive
: Available for Windows, Mac, and Linux at *http://www.dvarchive.org*.

ReplayTV Client
: Available for Windows, Mac, and Linux at *http://www.flyingbuttmonkeys.com/replay/*.

In fact, if you're running Mac OS X, FlyingButtMonkeys actually distributes a Cocoa-based application, called mReplay (*http://www.bentpixel.com/fbm/*; free), that makes downloading/uploading ReplayTV content a breeze. One cool feature this software possesses is the ability upload pictures from a Mac to a ReplayTV PVR for display (ReplayTV does not support Mac OS X with its software, so this is a nifty way for Mac users to access the full functionality of the PVRs). Figure 8-9 shows the mReplay application in use.

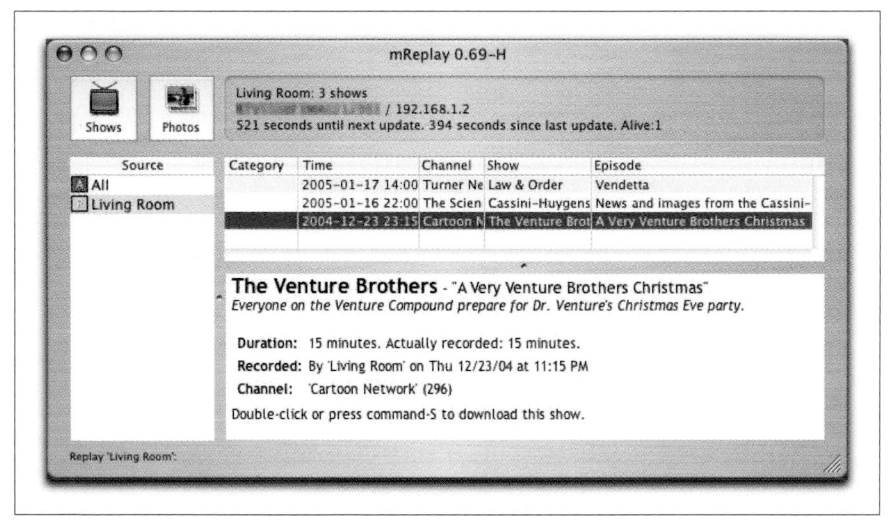

Figure 8-9. mReplay at work in the living room

When you launch the mReplay application, it looks for any ReplayTV devices on the same subnet. It can also connect to any ReplayTV device on the Internet, so long as the IP address assigned to the ReplayTV PVR is known. Downloading video is as simple as double-clicking on a show title.

The files are downloaded as MPEG-2 video. QuickTime requires that an MPEG-2 component (*http://www.apple.com/quicktime/products/mpeg2playback/*; $19.99) be installed before it will play movies in this format.

Creating Video to Play on a Palm Pilot

The next problem is getting video onto a Palm OS device, such as the Treo 600. A number of applications can play video on a Palm 5 PDA, but my favorite is made by Kinoma (*http://www.kinoma.com*). Kinoma's solution consists of a pair of applications that get video from a machine running Mac OS X or Windows XP and play it on a Palm OS–powered device.

Kinoma Player (free) and Kinoma Player EX ($19.99) play video on PDAs running Palm OS 5.0 and greater. Kinoma Player plays video stored in Palm database (*.pdb*) files. Kinoma Player EX is capable of playing media in a number of formats, including MPEG-4, in addition to *.pdb* files.

Kinoma Producer ($30) converts audio and video files into a format for playback compatible with Kinoma Player and Kinoma Player EX. Producer has options to set the number of video frames per second, the video bit rate, and sound quality. You can set it to conform to a number of different Palm screen resolutions. Figure 8-10 shows Kinoma Producer about to create a movie for the Treo 600.

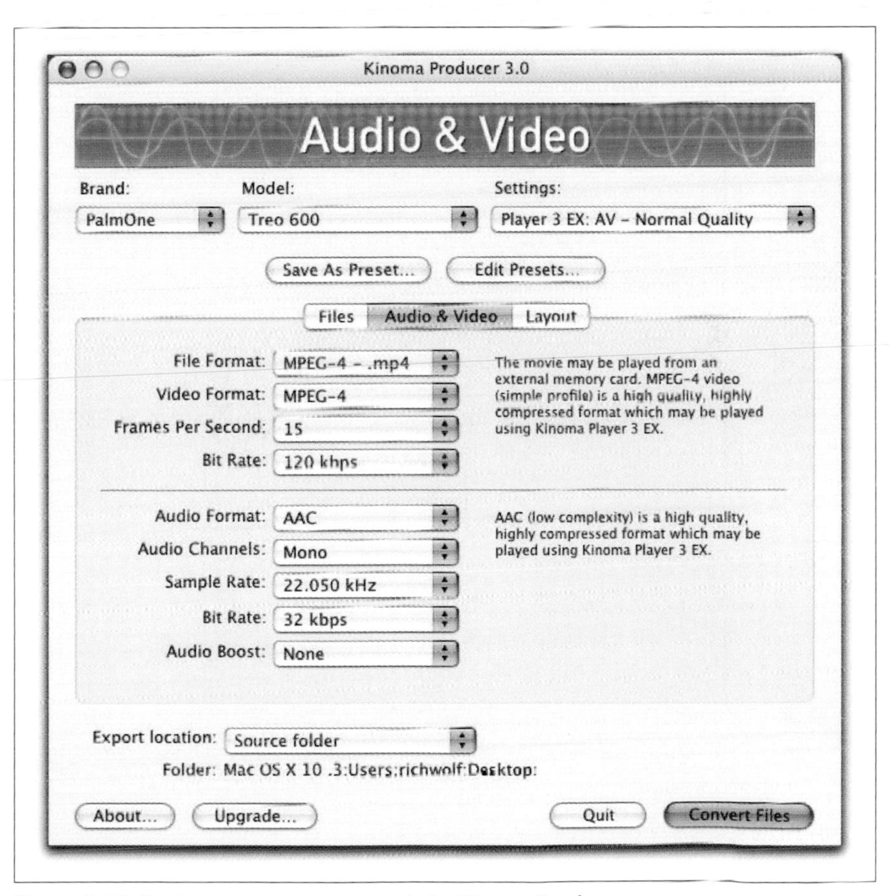

Figure 8-10. Setting up to convert a movie in Kinoma Producer

In addition, Kinoma Producer has built-in presets for a number of Palm devices. This makes configuration for a specific Palm device simple. Once you've selected your particular Palm device, click the Convert Files button. Producer will then begin your conversion, as shown in Figure 8-11.

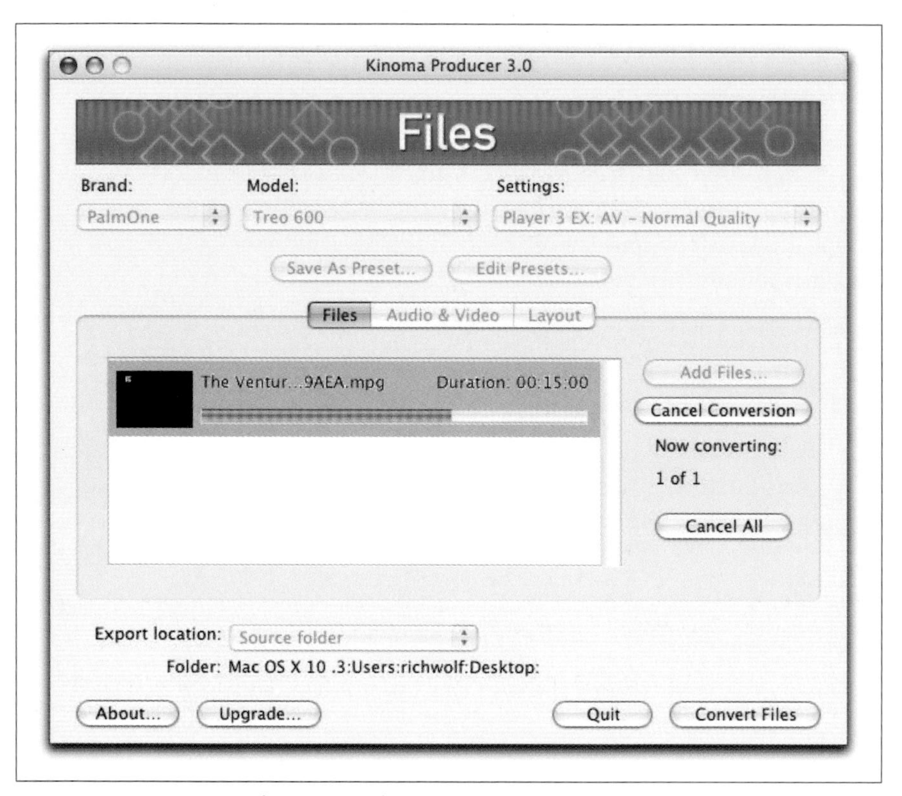

Figure 8-11. Kinoma in the process of converting a movie

Even under the Treo's 120×160 screen resolution, video can take up a lot of memory. Getting about an hour's with of video onto my PDA uses the better part of a 128MB memory card. You can get your video onto your Palm by copying the movie file Producer created to your Palm's memory card, or using the HotSync option to Install Handheld Files…

Once you've copied your movie to your Palm, you can play it using the Kinoma Player. Figure 8-12 shows the final movie being played on a Treo 600.

If you plan to play video that lasts more than a few minutes, you'll probably have to purchase the biggest compact flash card you can afford. I have found that a 512MB card is not too expensive and will play almost anything shorter than *The Lord of the Rings*: *The Return of the King*.

Figure 8-12. The final movie, being played on a Treo 600

As you can probably gather, Kinoma's solution can get your home videos onto your Palm device just as easily as it can get television shows and movies.

—*Richard Wolf*

HACK #96 Back Up Your Computer to a DV Tape

Digital videotape records digital information to create audio and video. But the data doesn't have to be audio and video.

When transferring video from a digital videotape to a computer, you should plan on using about 13GB of hard drive storage per hour of footage. Considering the footage is coming off a digital source, you might have wondered if you can write computer data to the tape. After all, a 60-minute MiniDV tape can be purchased for less than $10.

DV Backup Software

Around 2001, Tim Hewett decided to find out if he could accomplish what many of us wondered. He began to experiment with writing data from his computer to a digital videotape. The result of his experiment is the Mac OS X application DV Backup.

If you are using Windows or Linux, you might have some luck using dvbackup (*http://dvbackup.sourceforge.net/*; free, open source). You will, however, need to be familiar with using the command line. It also helps if you are familiar with ANCI C, a programming language.

DV Backup can write data to DV, DVCAM, DVCPRO, and Digital8 tapes using a FireWire connection. A typical, 60-minute MiniDV tape can store slightly more than 10GB of data. Because the application can write to DVCAM and DCVPRO, and these formats offer larger capacity tapes up to 184 minutes, it is possible to use DV Backup to store over 50GB of data.

There are two editions of DV Backup available: Standard and Lite. Both versions are full-featured with backup, verify, and restore functionality. The Standard version has additional features—most notably, those that allow you to backup to a hard disk, span a large backup across multiple tapes, create sets, and create scheduled backups.

Backing Up Your Files

Using the application for the first time is easy, because many of the more powerful features are tucked out of the way. When you launch the program, it searches for a connected digital video camera or deck. Once it finds one, it performs a brief search to see if there is a *Table of Contents* on the tape.

If you are using a digital video camera, make sure your camera is in Player mode.

Formatting your tape. If you are using a new tape, DV Backup will ask about the length of the tape, the name you would like to use for the tape, and whether you would like to use a Normal or Strict recording format. For the length of tape, you should input the length as reported by the manufacturer for SP mode, as shown in Figure 8-13.

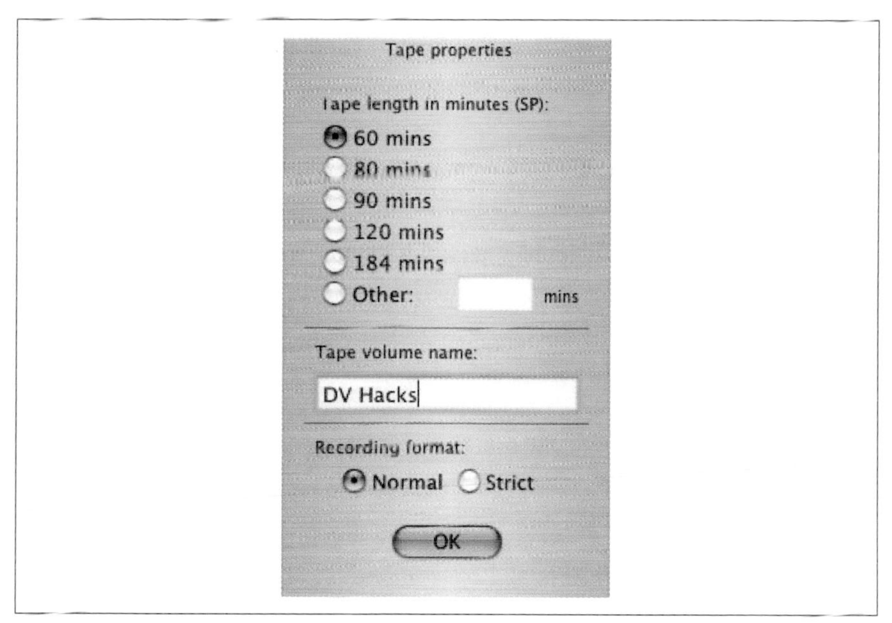

Figure 8-13. Formatting options

The Strict format will reduce the amount of storage by about 15% and is recommended for use only if the Normal format fails to work properly.

Selecting your files. Every good backup program should allow you to choose the files you would like to backup. DV Backup does just this. After you have formatted your tape, simply click the Backup... button and then click the Add... button in the resulting window, shown in Figure 8-14.

Note there is a set of options available to you in a drawer attached to the Select Files for Backup window. The options include:

Error Protection Level
: From 1:1 to none; 1:1 is full duplication and the best level of protection

Backup Comments
: Any comments you would like to note about the backup; these will appear in the Table of Contents

Compressed
: Will apply gzip compression to your backup; takes time but allows you to save more data

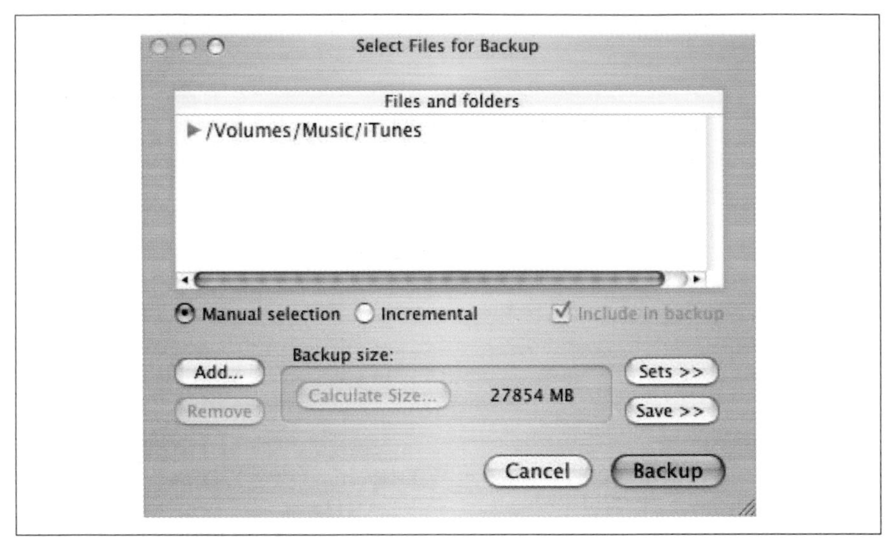

Figure 8-14. Selecting the files to back up

Follow symbolic links
: Will back up the objects/files referred to by symbolic links, instead of backing up the symbolic links themselves

SP/LP
: A radio button used to indicate which mode the camera has been set

Auto-verify
: Indicates whether DV Backup should verify the backup upon completion

After selecting your files and options, click on Backup button and the program will begin its process.

Restoring Your Files

Should you need to recover your files, the process is even easier than backing up:

1. Launch DV Backup.
2. Insert the tape that contains the files you would like to restore.
3. Select the backup you would like to restore.
4. Click the Restore... button.
5. Choose where you would like to restore the files. Optionally, select the specific files you would like to restore.
6. Click the Restore button.

Upon completion, your restored data can be found in a directory labeled with the date and time of recovery. If you already have a DV camera, using DV Backup can prove itself very useful and affordable. Additionally, you can use it to back up your project (excluding the raw footage), including any special media like sound effects and still photos.

Play "Movies" on an iPod photo

I picked up an iPod photo the other day, and most observers have asked me, "Can that thing play movies?" The short answer is no—only music and photos—but the longer answer is "sorta, but not really anything worth calling a movie."

In this hack, I'll show you how to "play" a movie on an iPod photo by exporting frames of a movie, importing them to the iPod and playing them manually by click-wheeling through thousands of images, or even viewing 3D views of objects. Pointless? Yes. Will a lot of people try this and put movie trailers and other short films on their iPod photo? You betcha and, as always, I'm here to help. Figure 8-15 shows a movie trailer along with the same trailer playing on an iPod.

Figure 8-15. The same movie trailer on two screens

Getting Started

All you'll need for this hack is QuickTime Pro, iTunes, and an iPod photo; the same instructions work on both the Mac and PC. I'll show both the PC and Mac steps in the examples.

Getting a Movie

Just about any QuickTime movie will do; for our example, we're going to use trailer for *Star Wars: Episode III*. We found one via Waxy.org (*http://www.waxy.org*). You can right-click (PC) or Control-click (Mac) to download the movie to your local system. Figure 8-16 shows the *Episode III* trailer as a QuickTime movie.

Figure 8-16. The movie trailer as a QuickTime movie

Once downloaded, open the movie in QuickTime. You might have other applications that can export sound and frames; if so, feel free to use those. Figure 8-17 shows the QuickTime file, as saved to the hard drive.

Exporting the Sound

First, we're going to export the sound. Once it's exported, we'll add it to the iPod photo so that we can listen to the soundtrack as we play the movie. Figure 8-18 shows the Sound Settings dialog box when exporting on Windows.

On a PC, with the movie open, choose File → Export. Choose Movie to WAV, click Options, and choose 44kHz, 16 bit, and Stereo.

On a Mac, with the movie open, click File → Export. Choose Movie to AIFF, click Options, and choose the same settings of 44kHz, 16 bit, and Stereo.

Import the sound into iTunes. You can export them to the iPod now, or later; it doesn't really matter.

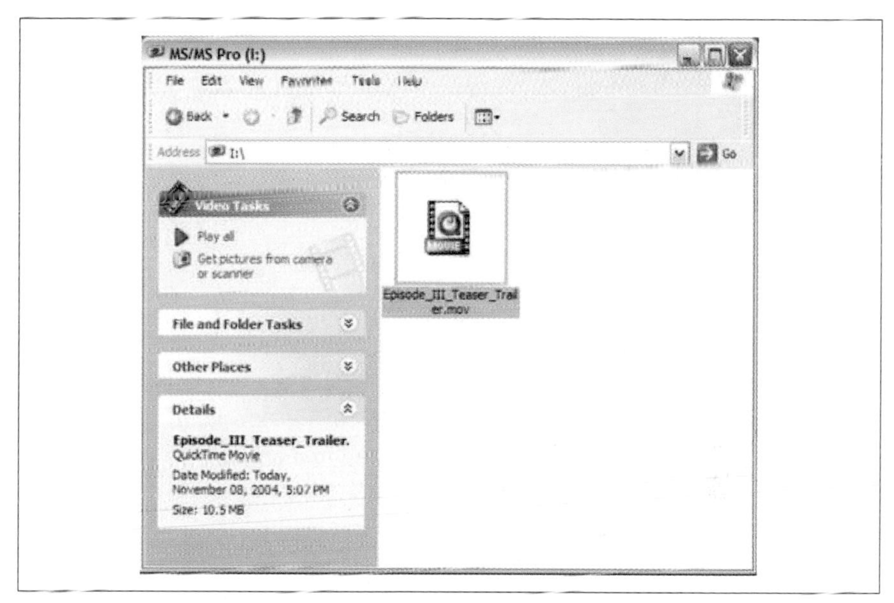

Figure 8-17. The Episode III trailer, saved to disk

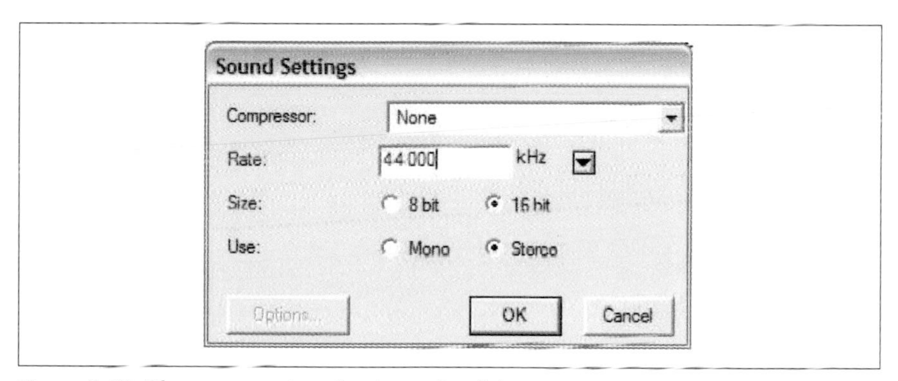

Figure 8-18. The export settings for the audio of the movie

If you don't want to export the sound and try to sync it as you play, that's fine too. H. M. Warner said it best: "Who the hell wants to hear actors talk?"

Exporting the Movie to an Image Sequence

Now that the sound is out of the movie, it's time to export each frame. The frame rate of the movie we downloaded was 15 frames per second, so we'll export the same in QuickTime. Figure 8-19 shows the Image Export settings.

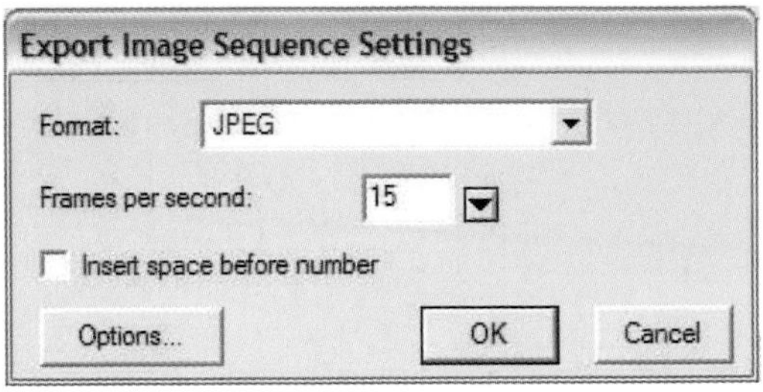

Export Image Sequence Settings

Format: JPEG

Frames per second: 15

Insert space before number

Options... OK Cancel

Figure 8-19. Using the JPEG format at 15 frames per second

On a PC, with the movie open, choose File → Export. Choose JPEG, type 15 for Frames per second, and click Options. I selected Medium for the Quality setting, but feel free to experiment. Figure 8-20 shows the JPEG compression options.

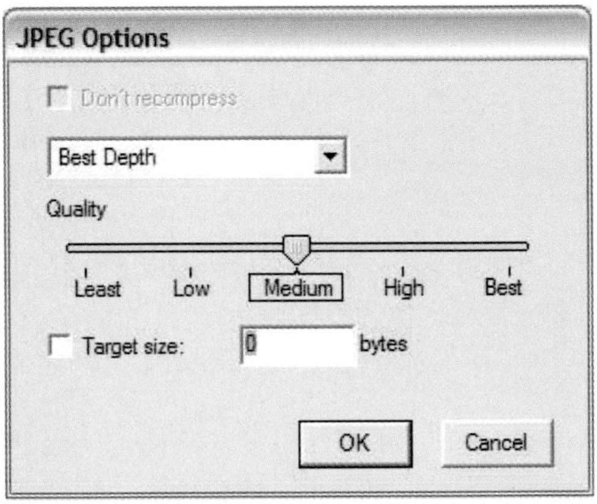

JPEG Options

Don't recompress

Best Depth

Quality

Least Low Medium High Best

Target size: 0 bytes

OK Cancel

Figure 8-20. Using Best Depth and Medium Quality for JPEG Options

On a Mac, with the movie open, choose File → Export. Choose Export Image Sequence and also select the same options as we did in the PC example. Figure 8-21 shows the export dialog box for a Mac.

I exported these to a new folder to keep them all tidy and one spot.

When you're finished, the 1 minute, 47 second movie exports to 1,616 frames (107 frames × 15.1 frames per second = 1,616). This might take awhile, so set it to export and get comfy. On a Mac, QuickTime gave us 1,627 frames, as shown in Figure 8-22.

On a PC, QuickTime gave us 1,616 frames for the same trailer—not sure why, but there it is, shown in Figure 8-23.

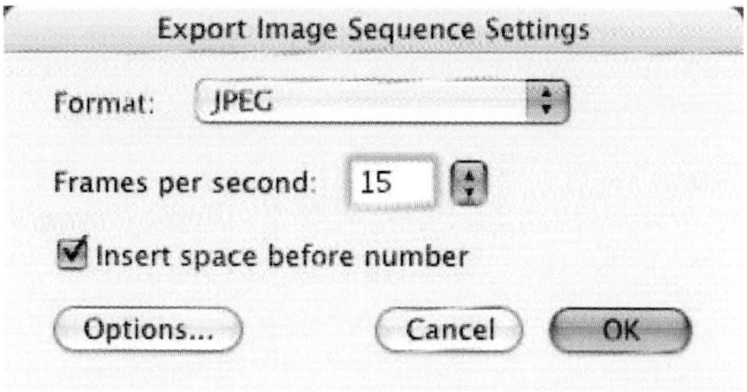

Export Image Sequence Settings

Format: JPEG

Frames per second: 15

Insert space before number

Options... Cancel OK

Figure 8-21. Mac-based image export settings

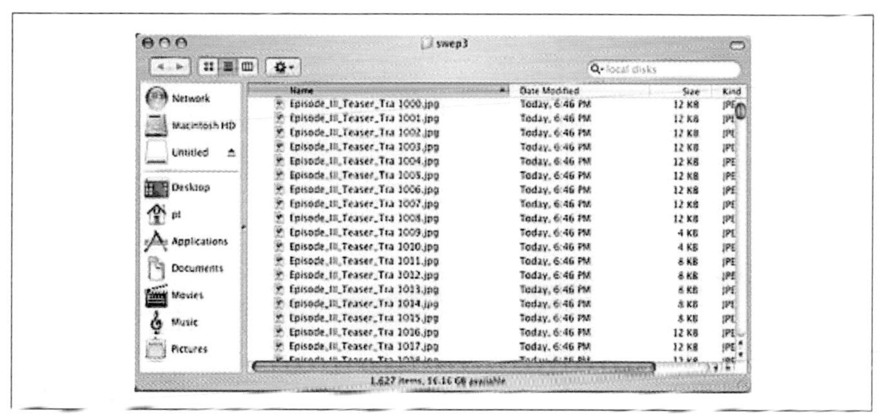

Figure 8-22. Yup, 1,627 frames of Star Wars goodness

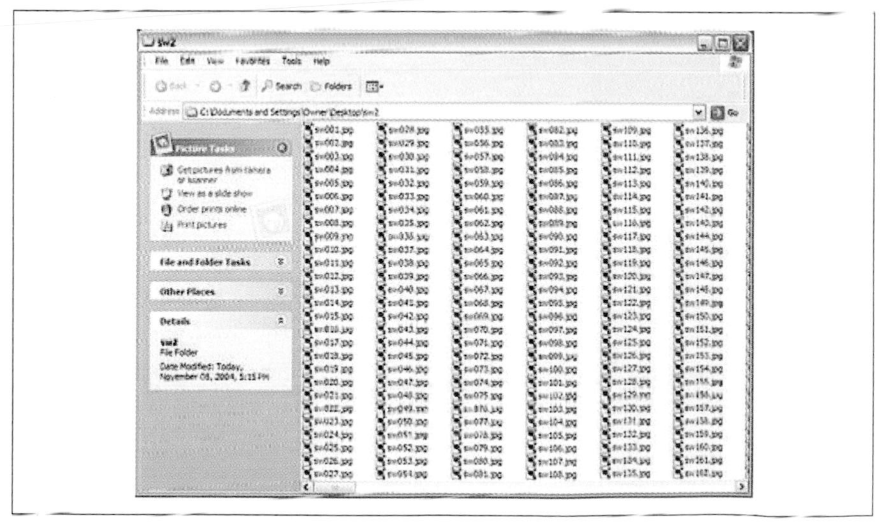

Figure 8-23. Hmm, only 1,616 frames on a PC

Transferring the Photo Sequence to the iPod photo

Now that we have 1,616 photos, we'll use the new features in iTunes to sync the photos over the iPod photo.

On a PC, choose Edit → Preferences → iPod, choose Photos and click Choose Folder. Select the folder you saved all the images to from QuickTime. Figure 8-24 shows the iTunes options on Windows.

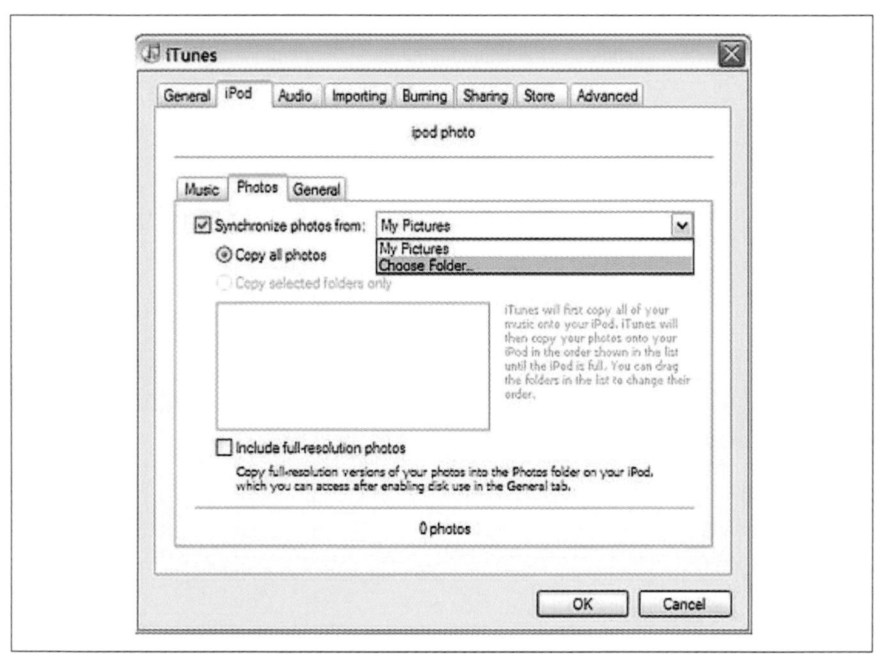

Figure 8-24. Windows options for iPod sync using photos

On a Mac, as with Windows, choose iTunes → Preferences → iPod. Figure 8-25 shows the iTunes options on Mac.

Once selected, you'll see the total number of photos that will be imported. iTunes will now convert (optimize) the images before they're sent over to the iPod photo. Once completed, they'll then be sent over, again, this might take awhile. Figure 8-26 shows the iPod syncing with the photos.

Playing the "Movie" on the iPod photo

After the photos are synced over to the iPod photo, disconnect the iPod from the dock. To listen to the sound while you play the movie, choose Music → Songs → Episode_III_Teaser_Trailer and press Play. To start the movie, choose Menu → Menu → Photos → Photo Library. The iPod photo

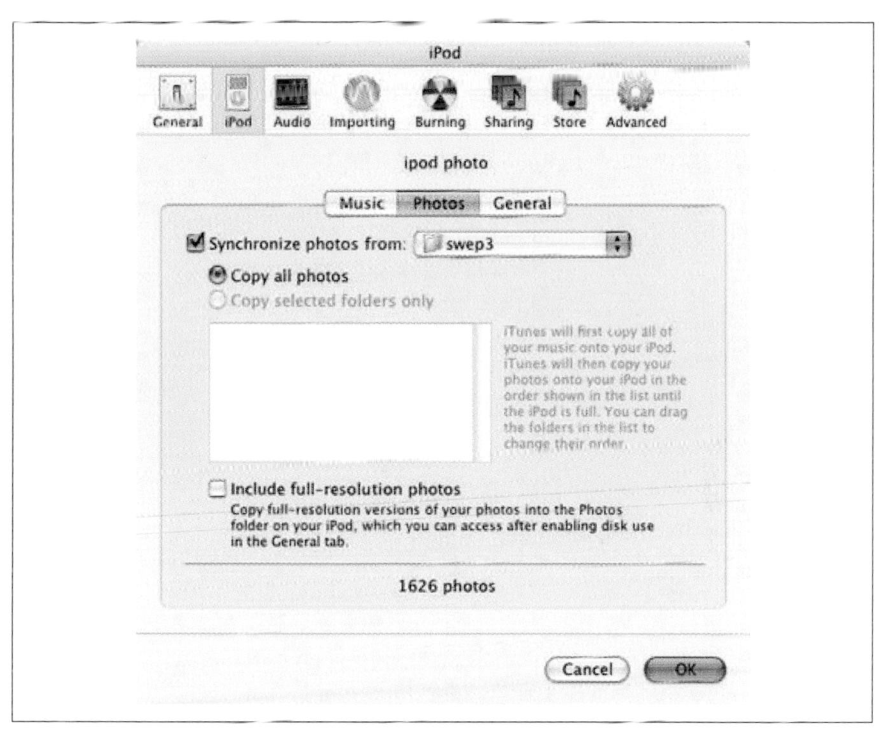

Figure 8-25. Mac options for iPod sync using photos

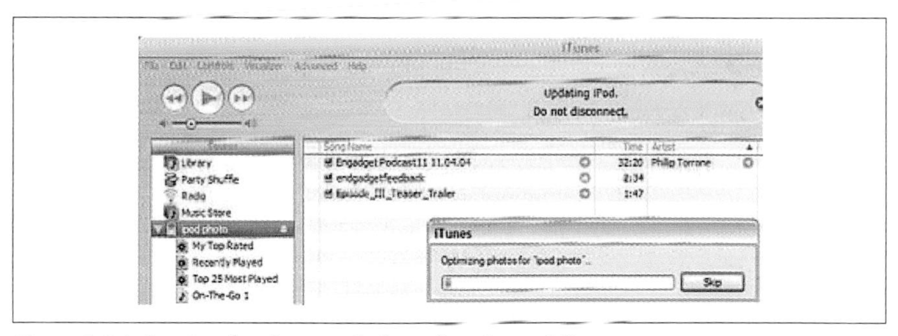

Figure 8-26. Syncing the photos, which might take a while

will spin up and display the thumbnails. Depending on where the soundtrack is, or where you want to start playing, click-wheel to a photo and press the center button. As it plays, quickly spin your finger around the wheel to get a "movie-like" playback.

With practice, you can play clips of movies and, for the most part, it looks as you'd expect: kinda crummy. So, what does it look like to play "movies" on the iPod photo? Figure 8-27 shows the trailer being played on an iPod.

Figure 8-27. Playing the trailer on the iPod photo

It seems on both a Mac and a PC, the photos (frames) do not import in order. It's likely a weird modification date or something else. I went in manually and removed some photos, resynced and that works (sometimes).

Hacking the Hack

A simple but very cool version of this is to create virtual reality (VR) objects or 3D views of an object and import them to the iPod. On example is an iPod on an iPod that the folks over at Griffin made and sent along to me. To make your own, just take a few photos of an object at different rotations around the object and import them in the same way we did with the movie frames. Figure 8-28 shows a VR object being rotated.

I didn't have time to document the best use of this simple trick, but I'll be working on it shortly. That use, of course, is to export *The Wizard of Oz* and play it along with the *Dark Side of the Moon*. If you don't know what we're talking about, follow the yellow brick road at *http://www.everwonder.com/david/wizardofoz/*.

Okay, Really Playing Movies...

Now that being said, the screens on the iPod are quite nice, and if real movie playback were possible, it wouldn't be that bad at all. We're hoping that Apple considers it, but if they don't, there are tons of alternatives. And, of course, if you'd really like to play movies on a portable device, you can hold

Figure 8-28. Figure 8-28. 3-D in print (it's there . . . really)

off on that $500 iPod photo purchase and pick up an Archos, Creative Zen, or about dozen other choices. But hey, there's nothing like doing an old fashion trick on a new shiny iPod. I like the Creative Zen Portable Media Center, for what it's worth.

—*Phillip Torrone*

Inventory Your Home

Your home is filled with your possessions, and most of them are valuable or priceless. Using your video camera, you can inventory your home quickly.

Your digital video camera doesn't have to be reserved for capturing family moments, such as your child's first steps, or for creating your own movies. It can also be used as a practical device for keeping people honest [Hack #93] or even recording an inventory of your home. Recording an inventory of your home is particularly useful before packing to move to a new home, or if you ever need to make an insurance claim.

This hack was inspired by Derrick Story's "Create a Home Inventory" in *Digital Photography Hacks* (O'Reilly).

Unless you're just starting out in life, you've probably already gathered a bunch of possessions. And more than likely, you couldn't remember every single item. Luckily, armed with a video camera and your voice, you can document everything for future reference.

To get started on creating a home inventory, use a new videotape and write the month and year on the label. Then, select a room to begin with and record its contents. If the item you are focusing on has a make, model, and/or serial number, try to zoom in on it. Figure 8-29 shows a television's make, model, and serial number, as recorded on a MiniDV tape.

Figure 8-29. About as detailed an inventory as you can get

Videotape records audio, in addition to video, so you can verbally annotate each item. So, if you happen to know when and where you purchased and item, as well as how much you paid, you can just speak that information. For example, while focusing on your television, you can simply speak "Uh, let's see, we bought the TV at Best Buy last Christmas…umm, it was on sale for $349."

Proceed from room to room, recording every possession you own. Also be sure to capture those priceless possessions, such as family photos, because those tapes provide a low-level of backup. Even though an image on a videotape could never replace the photos, children's artwork, and countless other items we all hold so dearly, being able to *see* those items (should they ever be lost) can provide some level of satisfaction.

Depending on the size of your home, or your organizational obsessions, you might want to record each room on a different tape. This can prove useful, because you can easily reinventory a room by simply recording onto the appropriate tape, as opposed to having to rerecord your entire home. Additionally, you won't have to concern yourself with which tape has the most recent version of a particular room.

When you have completed your home inventory, store the tape(s) in a safe deposit box. If you're really motivated, you can also transfer the tape to DVD **[Hack #79]** or put it on a web site **[Hack #83]** in order to have back-up copies.

To keep your inventory up-to-date, you should record your home at least once a year—such as every January, right after the holidays.

HACK #99 Capture Life's Little Moments with Camera-Phone Video

No, you're not going to record camera-phone video of your son's entire graduation ceremony. But you might catch him receiving his diploma.

Some camera phones can record video as well as still images. However, various hardware factors often limit the video segments to just a few seconds. Here are some of the limitations of recording video with camera phones:

- Not all video camera phones can record sound with the video.
- The visible frame size is small.
- The video clips are considerably more grainy and pixilated than video clips taken with still digital cameras that have video-recording capability.
- The 3GPP (3rd Generation Partnership Project) file-storage format can be viewed but not readily edited by common video-editing applications. You can find more information about 3GPP at *http://www.3gpp.org*.
- The maximum recording time for a video segment is usually measured in seconds. For example, the Nokia 3650 used for the hacks in this chapter can record a maximum of 95KB (about 15 seconds) of video with sound. My experience is that 10 seconds is the maximum clip duration.

A lot can happen in 10 seconds. The world's fastest human can run 100 meters, with a fraction of second to spare. Your child can scamper through a good portion of your home or yard. And, don't forget that you can edit multiple 10-second clips together **[Hack #39]** to create a several minutes of memorable video.

You can view 3GPP video natively using a number of desktop applications. Apple QuickTime (*http://www.apple.com/quicktime/*), Nokia Multimedia Player (*http://www.nokia.com*), and Real Video (*http://www.real.com*) are all capable of playing 3GPP video files.

Although viewing 3GPP files on a desktop is not a problem, you have to do a little preparation before you edit and splice the video files. Many video editors don't work with 3GPP files. However, you can work directly in QuickTime or convert the 3GPP video files to a more familiar format, such as Audio Video Interleaved (AVI). Let's cover the conversion process first and then touch on QuickTime.

First, you can use the free 3gpToRawAvi application for Microsoft Windows to convert 3GPP camera-phone video files to the Raw AVI format. You can find 3gpToRawAvi here:

http://www.allaboutsymbian.com/downloads/3gpToRawAvi.zip

The files should be unzipped and installed in a dedicated folder. The filenames for the folder and any of its parent folders cannot contain any spaces. 3gpToRawAvi does not have any installation routine. You use the application by launching *Convert3gp.vbs* (a VBScript file) from its installation folder or, more conveniently, by placing a shortcut in the Windows Start menu. After you've launched the application, you'll be greeted with the interface shown in Figure 8-30.

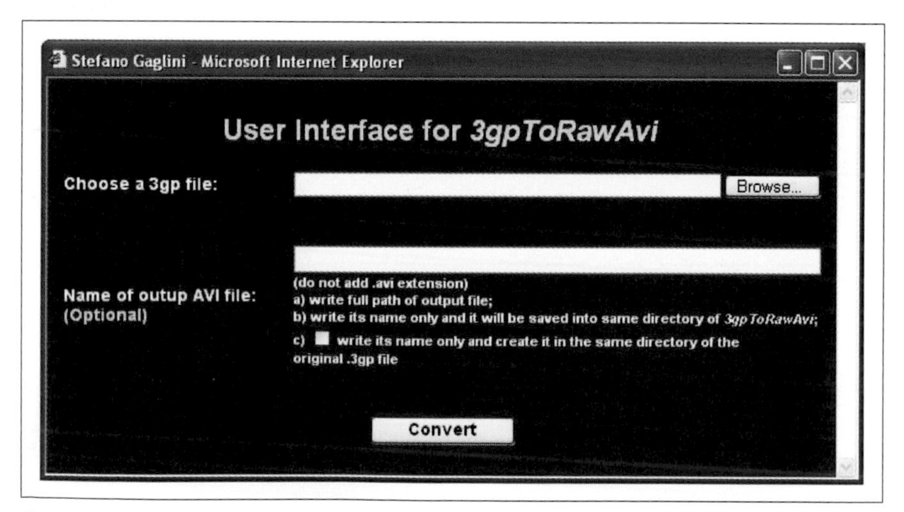

Figure 8-30. The 3gpToRawAvi user interface

Import the converted AVI video files into Windows Movie Maker (bundled with Windows XP) to turn several 10-second video segments into a single video, as shown in Figure 8-31.

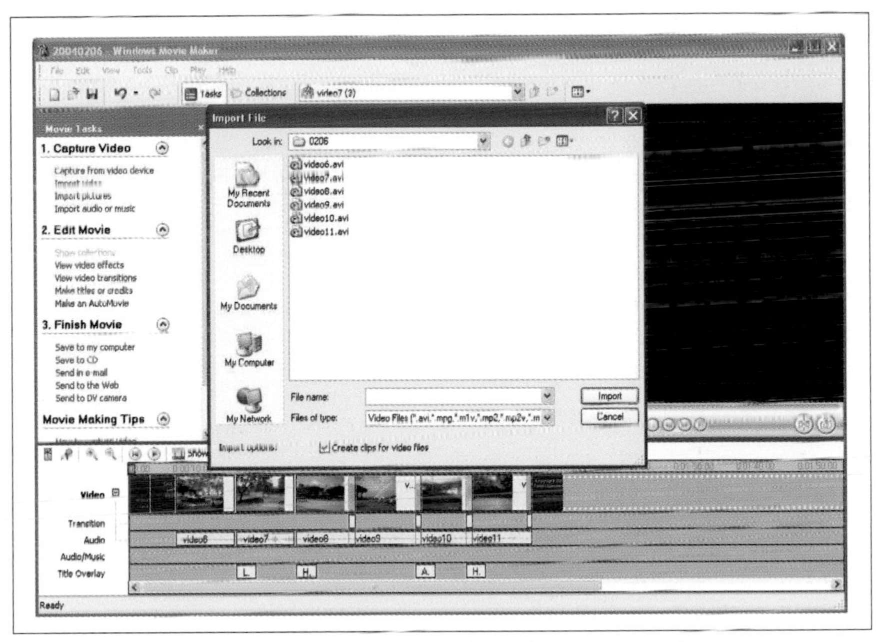

Figure 8-31. Editing AVI video files in Microsoft Windows Movie Maker

The resulting Windows Media Video (WMV) file can be played on platforms supported by Windows Media Player: Windows 98SE/ME/2000/XP, Macintosh OS 9 and OS X, and Pocket PC. Of course, you could produce MPEG video files for use with other video editors.

As mentioned earlier, you can also edit your video files in QuickTime Pro **[Hack #39]** and export your completed movie as a QuickTime *.mov* file that can be played on any Windows or Macintosh computer. Most editing systems can import the video files, so you can add titles, sound, and visual effects to them, just like any other video. As you can see, if you have the right tools in place, you can do much with those 10-second video clips from your phone.

It's also possible to send video files directly from your phone to some moblog sites **[Hack #81]**. This lets you create a series of camera-phone videos (with sound) that can be viewed by friends, family, and the general public on a public web site.

—*Todd Ogasawara*

Weekend Project: Create a Documentary

Using free footage and still photos from the Internet Archive, and a little creativity, you can make an historical documentary worthy of PBS.

You have a camera, you have a system to edit on, and you have the time and the motivation to create a masterpiece. You do not, unfortunately, have a plot, a script, actors, or financing. However, an historical documentary will allow you to flex your creative muscles while using someone else's footage and photos. No actors need apply.

Choose a Subject and Research It

It might seem like a simple and obvious step, but it is possible to do it wrong. It's important that you choose a subject that will engage you. If you don't care about your subject, it will show through in the finished product, assuming you can bring yourself to finish the project at all. It's also important to pick a subject about which you will be able to find ample material. Many have wondered why The History Channel airs so many documentaries about World War II. It's because footage and photos of World War II are abundant, whereas if you want shots of Genghis Khan, you'll have to put men in funny hats on horses and get them yourself.

The Internet Archive (*http://www.archive.org*) is a nonprofit digital library of video, audio, and text. They have a huge array of media, and it is quite possible to create movie without leaving your desk **[Hack #35]**. Browse through the archive, and look for something that you find interesting. Download any footage you think might be useful. If you don't think you've got enough moving pictures, don't worry. We'll pick up some still pictures shortly.

You should research your potential subjects as you go. Sometimes even a perfunctory web search will reveal that a theme you thought was interesting really isn't, and something you thought might be boring fascinates you. Google (*http://www.google.com*) is your friend, in more ways than one. As you dig deeper into your potential subjects, use Google's Image Search (*http://images.google.com*) to find photos that you can use to complement the video you've already gathered. Photos are invaluable to documentaries; still cameras can often go places where video cameras can't. They're good for filling in blanks.

When you've got your photos and footage gathered, make sure you also know enough about your topic to present it in an interesting way. You might want to create a storyboard **[Hack #6]** to help you visualize how your project is going to look. You might also want to log your footage, too **[Hack #5]**. After all, you didn't shoot this footage, so you'll need to familiarize yourself with it.

Add Motion to Your Pictures

You might notice that you've got more photos, and less video, that you thought. Many a documentary filmmaker has faced that dilemma, and the solution they found was *motion control photography*, which is also known as the "Ken Burns Effect."

Ken Burns is an Academy Award–winning documentary filmmaker. He is perhaps best known for his PBS miniseries *The Civil War* and *Jazz*.

This effect was originally created by panning and zooming on actual still photos with a motion picture camera, giving them a sense of motion. The effect is more easily created these days with digital photos and software. If you have the money and the time to master it, the professional choice is Adobe's After Effects (*http://www.adobe.com*; $699). However, if you're using a Mac, it's built right into Apple's iMovie (free with a new Macintosh, available as part of iLife suite; $79), with no plug-ins required.

Import your pictures into iPhoto. First things first: take the still photos you plan to use, and import them into iPhoto, by choosing File →Import..., and browsing to where your pictures are. Once they've been added to your photo library, create an album for your project and drag the photos into that album. Remember, keeping yourself organized is a good thing. You should be able to go right to your project's pictures without having to sort through vacation pictures.

Apply the effect in iMovie. Now open iMovie, and if you haven't already done so, create an iMovie project for your documentary. Then, import your video by choosing File → Import... and browsing to your video collection. Once your video is finished importing, click on the Photos button. Figure 8-32 shows iMovie in photo editing mode.

The drop-down menu allows you to choose albums from your iPhoto library, showing you the pictures from the chosen album in the box. When you click on a picture, it immediately begins previewing the effect we're after in the box in the top-right corner. To customize the effect, first click on the Start radio button. Now use the top slider to adjust the zoom for the start of the effect, and the bottom slider for the duration of the effect. You'll notice that when you move the cursor over the preview, it becomes a hand; you can also adjust on what part of the picture the zoom or pan starts and ends. When you're satisfied with your starting point, click the End button, and set the zoom and focus on which you want to effect to end. When you're

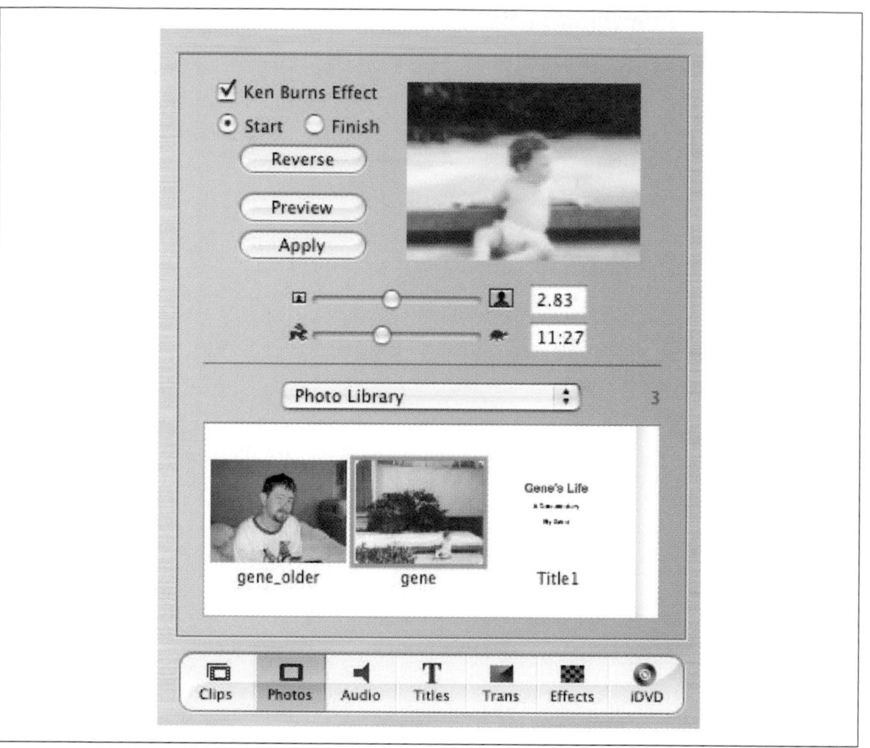

Figure 8-32. Applying the Ken Burns Effect

satisfied, click Preview to make sure you're actually satisfied. Then click Apply, and iMovie will place the photo on your timeline. Click the Clips button, and drag the newly zoomed and panned photo into your clip library. Voilà! The Ken Burns Effect.

Add Audio

There are still some things missing from your documentary: music, and narration. If you're planning on distributing this documentary of yours, your best bet is to avoid the hassles of licensing and royalties and use free music **[Hack #58]**. If you're doing this for the pleasure of it, then raid your music library and use what you like. With the music taken care of, we move on to the narration.

I can assure you, from personal experience, that you're going to feel silly when you record narration for your documentary. That doesn't matter: documentaries need narration. It's a fact of life, especially when using still photos. Get your timeline cut together, write what needs to be said, and practice a few times to get the timing right. You can capture your audio with iMovie and your Camera.

Add Your Titles and Put it to Bed

After your narration and music are done, you just need to add some titles. There's a lot of ways to go about this, from creating fancy slates in Photoshop and importing the images, to using cloth and sharpies [Hack #69]. Since we've already established that a) you have no financing, and b) you're using a Mac, we'll try a low-cost alternative.

Open TextEdit, which is in the *Applications* folder, and create your first title slate. Make it as pretty as you like. When you're done, open Grab, installed *Applications/Utilities*. Choose Capture → Selection, and select the section of your TextEdit document with your title on it. Leave yourself some room around the edges. Save this image to your Desktop, and then import it into iPhoto like you did with your other pictures. Apply a slow vertical pan with the Ken Burns Effect, and you've got professional rolling titles, as shown in Figure 8-33.

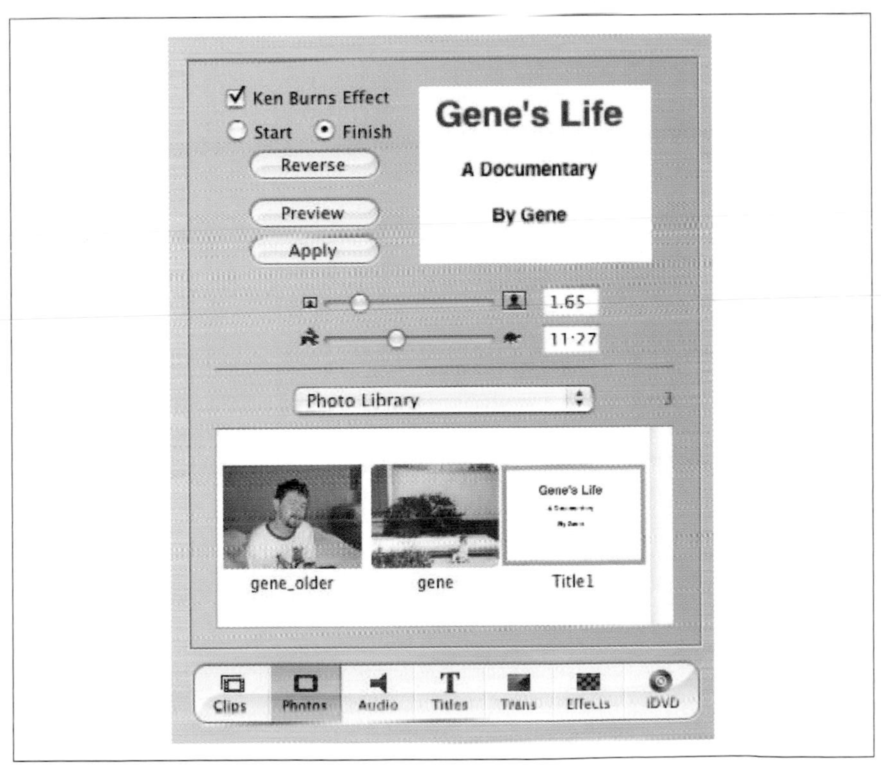

Figure 8-33. The titles go rolling by . . .

Congratulations; your documentary is finished! Output it, make some copies, and you're ready for distribution. Good luck on the festival circuit!

—*Gene Sullivan*

Index

Numbers

A

We'd like to hear your suggestions for improving our indexes. Send email to *index@oreilly.com*.

D

E

F

G

H

I

J

K

L

M

N

O

P

Q

R

S

T

U

V

W

X

Colophon

Our look is the result of reader comments, our own experimentation, and feedback from distribution channels. Distinctive covers complement our distinctive approach to technical topics, breathing personality and life into potentially dry subjects.

The object on the cover of *Digital Video Hacks* is a film clapboard, also known as a film slate. This small board is usually black and white, and contains sections for displaying information such as the title of the film, the director, the editor, scene/take numbers, and time information. At the top of this board sits the clapper, a hinged section that is clapped together at the start of each take to provide a sound cue that can be used during the editing process.

Philip Dangler was the production editor and proofreader for *Digital Video Hacks*. Nancy Reinhardt was the copyeditor. Marlowe Shaeffer and Darren Kelly provided quality control. Julie Hawks wrote the index.

Mike Kohnke designed the cover of this book, based on a series design by Edie Freedman. The cover image is from Corbis; the layout was produced with Adobe InDesign CS using Adobe's Helvetica Neue and ITC Garamond fonts.

David Futato designed the interior layout. This book was converted by Joe Wizda to FrameMaker 5.5.6 with a format conversion tool created by Erik Ray, Jason McIntosh, Neil Walls, and Mike Sierra that uses Perl and XML technologies. The text font is Linotype Birka; the heading font is Adobe Helvetica Neue Condensed; and the code font is LucasFont's TheSans Mono Condensed. The illustrations that appear in the book were produced by Robert Romano and Jessamyn Read using Macromedia FreeHand MX and Adobe Photoshop CS. This colophon was written by Philip Dangler.

Keep in touch with O'Reilly

1. Download examples from our books

To find example files for a book, go to:

www.oreilly.com/catalog

select the book, and follow the "Examples" link.

2. Register your O'Reilly books

Register your book at *register.oreilly.com*

Why register your books? Once you've registered your O'Reilly books you can:

- Win O'Reilly books, T-shirts or discount coupons in our monthly drawing.
- Get special offers available only to registered O'Reilly customers.
- Get catalogs announcing new books (US and UK only).
- Get email notification of new editions of the O'Reilly books you own

3. Join our email lists

Sign up to get topic-specific email announcements of new books and conferences, special offers, and O'Reilly Network technology newsletters at:

elists.oreilly.com

It's easy to customize your free elists subscription so you'll get exactly the O'Reilly news you want.

4. Get the latest news, tips, and tools

http://www.oreilly.com

- "Top 100 Sites on the Web"—PC Magazine
- CIO Magazine's Web Business 50 Awards

Our web site contains a library of comprehensive product information (including book excerpts and tables of contents), downloadable software, background articles, interviews with technology leaders, links to relevant sites, book cover art, and more.

5. Work for O'Reilly

Check out our web site for current employment opportunities:

jobs.oreilly.com

6. Contact us

O'Reilly & Associates
1005 Gravenstein Hwy North
Sebastopol, CA 95472 USA

TEL: 707-827-7000 or 800-998-9938
(6am to 5pm PST)

FAX: 707-829-0104

order@oreilly.com
For answers to problems regarding your order or our products.
To place a book order online, visit:

www.oreilly.com/order_new

catalog@oreilly.com
To request a copy of our latest catalog.

booktech@oreilly.com
For book content technical questions or corrections.

corporate@oreilly.com
For educational, library, government, and corporate sales.

proposals@oreilly.com
To submit new book proposals to our editors and product managers.

international@oreilly.com
For information about our international distributors or translation queries. For a list of our distributors outside of North America check out:

international.oreilly.com/distributors.html

adoption@oreilly.com
For information about academic use of O'Reilly books, visit:

academic.oreilly.com